Emilie du Châtelet between Leibniz and Newton

ARCHIVES INTERNATIONALES D'HISTOIRE DES IDÉES

INTERNATIONAL ARCHIVES OF THE HISTORY OF IDEAS

205

EMILIE DU CHÂTELET BETWEEN LEIBNIZ AND NEWTON

Edited by

Ruth Hagengruber

Board of Directors:

Founding Editors:
Sarah Hutton
Director:
M.J.B. Allen (Los Angeles); J.-R. Armogathe, Paris; S. Clucas, London;
G. Giglioni, London; P. Harrison, Oxford; J. Henry, Edinburgh;
M. Mulsow, Erfurt; G. Paganini, Vercelli; J. Popkin, Lexington;
J. Robertson, Cambridge; G.A.J. Rogers, Keele; J.F. Sebastian, Bilbao;
A. Sutcliffe, Oxford; A. Thomson, Paris; Th. Verbeek, Utrecht

For further volumes:
http://www.springer.com/series/5640

Ruth Hagengruber
Editor

Emilie du Châtelet between Leibniz and Newton

Editor
Ruth Hagengruber
Universität Paderborn
Fak. Kulturwissenschaften
Abt. Philosophie
Warburger Str. 100
Paderborn 33098
Germany
ruth.hagengruber@uni-paderborn.de

ISSN 0066-6610
ISBN 978-94-007-2074-9   e-ISBN 978-94-007-2093-0
DOI 10.1007/978-94-007-2093-0
Springer Dordrecht Heidelberg London New York

Library of Congress Control Number: 2011939188

© Springer Science+Business Media B.V. 2012
No part of this work may be reproduced, stored in a retrieval system, or transmitted in any form or by any means, electronic, mechanical, photocopying, microfilming, recording or otherwise, without written permission from the Publisher, with the exception of any material supplied specifically for the purpose of being entered and executed on a computer system, for exclusive use by the purchaser of the work.

Printed on acid-free paper

Springer is part of Springer Science+Business Media (www.springer.com)

Acknowledgements

Together with the physicist and Leibniz researcher, Hartmut Hecht, we organised the conference on Emilie du Châtelet in honour of her 300th birthday. All contributors were conference participants. Special thanks to Hartmut Hecht, who has supported my research on Emilie du Châtelet for many years. Brunhilde Wehinger made the conference at the Research Center for European Enlightenment in the autumn of 2006 in Potsdam possible. My thanks to her. I thank the contributors, who expand the scientific horizons of the Emilie-Du-Châtelet-research with their own studies. Thanks also to Ana Rodrigues, who has been helping to establish the archives on Emilie du Châtelet since 2006 and who compiled the bibliography published here. Further, I would like to thank Dr. Alexander Mangold and Kirsten Bolte for their untiring zeal in the translations of the manuscripts. Thank you to Anita Fei van der Linden from the Springer publishing house for holding on to this project so tightly for so long, in spite of the many delays. Above all, I thank my friend Otti Stein at Château Merl for her indefatigable support of this subject.

Paderborn / Château Merl

Editor's Introduction

What is knowledge and how do we achieve confident knowledge? Questions like these are as hotly disputed today as they were two centuries ago. This book makes a deliberate effort to answer these questions by going back approximately 270 years in time and by relating them to a debate between classical advocates of Leibnizianism and Newtonianism. The catalyzing function of Emilie du Châtelet's research and the influence of her work on colleagues and contemporaries in this discussion still is widely unknown today. This book intends to contribute to a more complete understanding of her work and, thus, to what could widely be called a more accurate history of science and philosophy.

The *Institutions physiques*, written by Emilie du Châtelet in the mid 18th century, leads the reader into a lively discussion on Leibniz und Newton, in which their ideas, as well as those of Descartes, John Locke and many others, are attacked or praised. The most outstanding philosophers and mathematicians of her age, Voltaire, Diderot, Maupertuis, Euler, the father and son Johann Bernoulli, were among her friends. Others received deep insights thanks to her analyses, as the writings of Julien Offray de la Mettrie and Immanuel Kant demonstrate.

The goal of this volume is to make the intellectual strength of Emilie du Châtelet apparent and to facilitate a better understanding of the singularity of her work. Emilie du Châtelet is to be established as a contributor to the Leibniz-Newton debate in the first half of the 18th century. The ingenuity of her work and the distinctiveness of her thought clearly come to light when it is analysed within the context of her time and in the discussion with prominent fellow-thinkers.

This book contributes to completion of the analysis of 18th century Newtonianism and Leibnizianism, which has not been written up to now. Where the ideas of Emilie du Châtelet can repeatedly be found in the center of a pulsing discussion, new insights about persons and their connections with each other come to light. Her contribution to the differentiation of physical and philosophical questions does not only concern the discussion surrounding Newton und Leibniz, although this discussion provides a significant contribution to the clarification and concretion of concepts still focused on today and is considered fundamental in the areas of philosophy and science. The differentiation between "science" and "natural philosophy" and the understanding of philosophy as its own discipline alongside physics has been accepted since the 18th century. The meaning of these concepts however, has continually undergone changes since then. In the middle of the 18th century, philosophers grappled with the meaning of philosophy, then opposing the metaphysical approach against the scientific one. The materialism and sensualism of the French Enlightenment is a battle against the old metaphysicians. In this discussion, Emilie du Châtelet appears as a driving force arguing against this fashionable attitude, holding on and rediscovering the fundamental role of metaphysics. The idea of metaphysics as a basis for scientific research had its beginnings in the works of Emilie du Châtelet, in this approach she probably influenced the philosopher Kant. In contrast to the anti-metaphysical tendency of her philosophical entourage, she defended her perception that there was and had to be something that formed the contextual basis for science, being a prerequisite and the reason for its validity. Emilie du Châtelet was eager to perform the methodic approach of a modern "Newtonian" scientist. But additionally, she strived for the foundations of science, which she certainly understood as experiential, "phenomenological" and observational, but for which she was in search of a basis for validity which could not be found in experience and observation alone. Even in her early years in Cirey she addressed the ideas of the philosopher John Locke critically. In her writings from that time she dismisses Locke's concept of experience-based ideas and pleads a principle that gives order to the variety of experiential contents and defends them as "innate". A main thesis is built on Du Châtelet's intellectual independence of Voltaire's thought. It procures with her early criticism against Locke, which consequently shows that the Marquise's scientific universalism was already outlined before she came to know Leibnizian ideas. With this background, she remained skeptical towards a phenomenalism as represented by the Newtonians. Emilie du Châtelet criticized the ban of the Newtonian scholars of the hypothesis, stating that Newton, Kepler and many others used them to constitute their insights. To the importance and scientific function of hypotheses she dedicated a complete chapter in her *magnum opus*. Emilie du Châtelet researched a new metaphysical system and at the same time criticized the old one of the scholastics, taking distance repeatedly from the metaphysics of "empty words", although without giving up the idea of an assumptive philosophical foundation, which she postulates as a methodical necessity for all experience-based insight. Her attempt to reconcile an empirical and a rationalistic approach shows her to be a genuine member of the "Bernoulli" circle, to which her friend and longtime scientific colleague, Pierre Louis Moreau de Maupertuis, had introduced her. However Maupertuis' and Du Châtelet's interpretation of Leibniz started to drift apart, which can be observed in Du Châtelet's letters to Maupertuis. While Maupertuis ignored her discussion on solid bodies and the problems of conservation within the correspondence, he took up her ideas in his thoughts concerning his *principle of least action*. The discussions about the value of metaphysics for knowledge were intensively and touched the most rigorous and materialist philosophers of the time, as we see in the writings of La Mettrie.

Nonetheless Du Châtelet's approach to metaphysics shows certain similarities and interesting relations to La Mettrie.

In his contribution "In the Spirit of Leibniz – Two Approaches from 1742" Hartmut Hecht delineates the positions of Maupertuis and Du Châtelet who distance themselves from each other within this changing scientific context. Hecht illustrates how "Maupertuis turns out to be a master of scientific imagination"; but he knew that his eloquent scientific explorations could not convince Emilie du Châtelet, as Maupertuis' *esprit systématique* must have seemed "closer to *esprit* than it was *systématique*" as Hartmut Hecht ironically comments. "Mme Du Châtelet's teachings on nature could be seen as a statement in favour of a systematic philosophy, in spite of and as a consequence of the breathtaking scientific developments of her time". While Maupertuis was primarily inspired by the Leibnizian "possible world" problem, which he transformed into a kind of scientific methodology, Du Châtelet examined the principles of contradiction and sufficient reason as fundamental for a systematic reconstruction of a metaphysical base for science. Their different approaches to finding a connection from Leibiniz to early 18th century problems paved the way for unique and distinctive metaphysical solutions by Du Châtelet and Maupertuis as well. These solutions have to be understood as the first steps to redefine the relationship between metaphysics and science under the conditions of Newtonian physics, 15 years before the publication of Kant's *Allgemeiner Naturgeschichte und Theorie des Himmels*. The Leibniz-Clarke correspondence in Sarah Hutton's contribution "Between Newton and Leibniz: Emilie du Châtelet and Samuel Clarke", shows surprising concurrence between Emilie du Châtelet's work and that of Samuel Clarke. Historical dates and facts of interest are made known, as seen in the information about the connection between Voltaire und Clarke. In her *Institutions* Du Châtelet took Leibniz's side against Clarke, but she did not reject Newtonian physics. In her essay "Between Newton and Leibniz: Emilie du Châtelet and Samuel Clarke Sarah Hutton examines Du Châtelet's position in the wider context of the fluid state of scientific theory in the early eighteenth century. This is reflected in Du Châtelet's revisions to her *Institutions* as well as in another work of Clarke" with which she was well-acquainted: his translation of Rohault's *Physics*. This is a hybrid text by virtue of the fact that Clarke added a Newtonian gloss to a textbook of Cartesian physics. This hybridity is paralleled in Du Châtelet's response to Newton and Leibniz in her *Institutions*, which was also conceived as a textbook. Sarah Hutton suggests that Madame Du Châtelet's views on Newtonianism in her *Institutions* is coloured by three things – the hypothetical character of Newton's as yet unproven theories, theological concerns influenced by her then mentor Maupertuis, and her concern with the underlying problems of physics, which both Newton and Leibniz sought to address. She refutes the theory that Du Châtelet had ever given up Newton, but was "highly conscious of a number of unresolved issues in Newtonianism, and the discussions surrounding them". Reconciling Leibniz and Newton meant, in Sarah Hutton's words, "reading Newton through Leibnizian spectacles", as she aptly notes. In this content Du Châtelet's *Essay on Optics* has assumed a more meaningful position, which it also received within the framework of the conference, because Fritz Nagel from the Bernoulli-Edition in Basel discovered Du Châtelet's complete *Essay*, which had so far only been known as a fragment of the documents from St. Petersburg Library. Forty-three original letters written by Emilie du Châtelet to Johann II Bernoully exist today. Nagel's discovery substantiates the relationship of trust that Emilie du Châtelet shared with the Bernoullis, especially with the son Johann II., as Nagel expounds in his contribution "'Sancti Bernoulli orate pro nobis'. Emilie du Châtelet's Rediscovered *Essai sur l'otique* and Her Relation to the Mathematicians from Basel". Johann I Bernoulli, a famous mathematician in his time and also a friend of Leibniz, was Maupertuis' and Leonhard Euler's teacher. Euler had already studied at the University of Basel at the age of 13 and was tutored by Father Johann I. Bernoulli. He received his master's degree at the young age of 16 and replaced Johann's son, Daniel Bernoulli, at the Academy of Sciences in Saint Petersburg. Dieter Suisky describes the expanded context of this scientific atmosphere with the analysis of the Eulerian texts in comparison to Du Châtelet's: "all authors treated simultaneously the legacy of Descartes, Newton and Leibniz and in all cases, the authors developed their own point of view". Euler's *Mechanica sive motus scientia analytice exposita* was published in 1736 and most probably read by Du Châtelet. Euler und Du Châtelet, both almost the same age, "wrote their first comprehensive treatises on the basic principle of mechanics almost simultaneously during the very short period of the fourth decade of the 18th century". Like Du Châtelet, Euler claimed a common measurement for *dead* and *living forces*, building a new framework for the interpretation of the forces. Du Châtelet translated Newton into Leibniz and Leibniz into Newton, "partially modifying the terminology" but also discovering the "hidden common basis of both theories", writes Suisky in his contribution "Leonhard Euler and Emilie du Châtelet. On the Post-Newtonian Development of Mechanics". The core of the *vis viva* controversy is analyzed by Andrea Reichenberger, writing on "Leibniz's Quantity of Force: a 'Heresy'? Emilie du Châtelet's *Institutions* in the Context of the *Vis Viva* Controversy". Her portrayal of Du Châtelet's synthesis of the *vis viva* is a contribution to the many analyses in this field which have thus far left out the importance of Du Châtelet's contribution. In "From Translation to Philosophical Discourse – Emilie du Châtelet's Commentaries on Newton and Leibniz" Ursula Winter delineates a broad perspective on Du Châtelet's commitment to aspects of the philosophy of Leibniz and Newton from 1738 onwards, including its consequences in the following periods. Winter shows how Du Châtelet's work was recognized and appreciated by later philosophers, especially Kant. She is convinced that Du Châtelet's correspondence proves that she supported Leibniz's philosophy up to the end of her life – including his controversial theory on monads. Also she indicates analogies in the *Institutions* to Newton's *Principia*, "Apart from the first three chapters, the 38 versions of the *Institutions* follows the structure of Newton's *Principia*", Winter states.

The history of the publication of the *Institutions de physique*, which was later renamed to *Institutions physiques* by Emilie du Châtelet herself, constitutes its own research area because of the various drafts of the manuscripts and printed editions. An all-embracing discussion of this is not the purpose of this volume. The contributors use quotations from different editions of this text to support their arguments.

Most of the citations are taken from or translated into English, some remained French, due to retain the subtlety or ambiguity of the expression.

I am happy and grateful to the researchers, who subjected themselves to this undertaking with such joy, endurance, and effort, in order to highlight Emilie du Châtelet as an important figure in the Leibniz-Newton debate, and who through their knowledge have contributed to the effort to position Emilie du Châtelet within the philosophical context of her time.

Contents

Emilie du Châtelet Between Leibniz and Newton:
The Transformation of Metaphysics 1
Ruth Hagengruber

In the Spirit of Leibniz – Two Approaches from 1742 61
Hartmut Hecht

Between Newton and Leibniz: Emilie du Châtelet
and Samuel Clarke 77
Sarah Hutton

"Sancti Bernoulli orate pro nobis". Emilie du Châtelet's
Rediscovered *Essai sur l'optique* and Her Relation to the
Mathematicians from Basel.. 97
Fritz Nagel

Leonhard Euler and Emilie du Châtelet.
On the Post-Newtonian Development of Mechanics 113
Dieter Suisky

Leibniz's Quantity of Force: A 'Heresy'? Emilie du Châtelet's
Institutions in the Context of the *Vis Viva* Controversy 157
Andrea Reichenberger

From Translation to Philosophical Discourse –
Emilie du Châtelet's Commentaries on Newton and Leibniz 173
Ursula Winter

Emilie du Châtelet, a Bibliography 207
Ana Rodrigues

About the Authors... 247

Index 251

List of Abbreviations

[LetChBI] + [LetChBII]	Du Châtelet, E. 1958. *Les Lettres de la marquise du Châtelet: publiées par Theodore Besterman.* 2 vols. Genève: Institut et Musée Voltaire.
[LetSav1738]	Du Châtelet, E. 1738. Lettre sur les Eléments de la Philosophie Newton. Paris. *Journal des sçavans*: 534–41.
[DissCh1739]	Du Châtelet, E. 1739. Dissertation sur la nature et la propagation du feu. In *Recueil des pièces qui ont remporté le prix de l'Académie royale des Sciences en 1738*, ed. Académie royale des Sciences, 85–168. Paris: Imprimerie royale.
[Inst1740]	Du Châtelet, E. 1740. *Institutions de Physique.* Paris: Prault.
[MaiCh1741]	Du Châtelet, E. 1741. *Réponse de madame la marquise Du Chastellet à la Lettre que M. de Mairan, secrétaire perpétuel de l'Académie royale des Sciences, lui a écrite le 18 février 1741 sur la question des forces vives.* Bruxelles: Foppens.
[MaiCh1741D]	Du Châtelet, E. 1741. *Zwo Schriften, welche von der Frau Marquise von Chatelet, gebohrner Baronessinn von Breteuil, und dem Herrn von Mairan, beständigem Sekretär bei der französischen Akademie der Wissenschaften, das Maaß der lebendigen Kräfte betreffend, gewechselt worden: aus dem Französischen übersetzt von Louise Adelgunde Victoria Gottsched, geb. Kulmus.* Leipzig: Bernh. Breitkopf.
[Inst1741Am]	Du Châtelet, E. 1741. *Institutions de physique.* Amsterdam: Pierre Mortier.

[Inst1741Lo]	Du Châtelet, E. 1741. *Institutions de physique*. London: Paul Vaillant.
[Inst1742Am]	Du Châtelet, E. 1742. *Institutions physiques de madame la marquise du Chastellet adressés à M. son fils: Nouvelle édition, corrigée et augmentée considérablement par l'auteur.* Amsterdam: Aux dépens de la Compagnie.
[Inst1742Rep1988]	Du Châtelet, E. 1988. Institutions physiques: Nouvelle édition. In *Gesammelte Werke*, ed. Jean Ecole 28, Abt. 3: Materialien und Dokumente. Hildesheim/Zürich/New York: Olms.
[Inst1743Ve]	Du Châtelet, E. 1743. *Istituzioni di Fisica di Madama la Marchesa du Chastelet indiritte a suo figliuolo. Traduzione dal linguaggio francese nel toscano, accresciuta con la Dissertazione sopra le forze motrizi di M. de Mairan.* Venedig: Presso Giambatista Pascali.
[Naturlehre1743]	Du Châtelet, E. 1743. *Der Frau Marquisinn von Chastellet Naturlehre an ihren Sohn. Erster Theil nach der zweyten Französischen Ausgabe übersetzet von Wolfgang Balthasar Adolf von Steinwehr Prof. Publ. Ord. auf der Universitet zu Frankfurt an der Oder, derselben Bibliothecario, und der Königl. Preußischen Societet der Wissenschaften Mitgliede.* Halle/Leipzig: Rengerische Buchhandlung.
[DissCh1744]	Du Châtelet, E. 1744. *Dissertation sur la nature et la propagation du feu.* Paris: Prault Fils.
[PrincChat1756]	Du Châtelet, E., and I. Newton. 1756. *Principes mathématiques de la philosophie naturelle: par feue Madame la marquise du Chastellet.* 2 vols. Paris: Desaint & Saillant.
[PrincChat1759]	Du Châtelet, E., and I. Newton. 1759. *Principes mathématiques de la philosophie naturelle de Newton: Par feue madame la marquise du Chastellet.* 2 vols. Paris: Desaint et Saillant.
[PrincChat1966]	Du Châtelet, E., and I. Newton. 1966. *Principes mathématiques de la philosophie naturelle de Newton: par feue madame la marquise du Chastellet.* Paris: Albert Blanchard.
[PrincChat1990]	Du Châtelet, E., and I. Newton. 1990. *Principes mathématiques de la philosophie naturelle.* Sceaux: Jacques Gabay.
[PrincChat2005]	Du Châtelet, E., and I. Newton. 2005. *Principia. Principes mathématiques de la philosophie naturelle. Traduit de l'anglais par la marquise du Châtelet. Préface de Voltaire.* Paris: Dunod.

Emilie du Châtelet Between Leibniz and Newton: The Transformation of Metaphysics

Ruth Hagengruber

> *Je suis persuadée que la physique ne peut se passer de la métaphysique, sur laquelle elle est fondée*
>
> Emilie du Châtelet

Who Was Emilie du Châtelet?

In the year 1743, Emilie du Châtelet was invited to contribute a scientific portrait of herself to the monumental book project of the most important living authors of her time.[1] Within the society of the learned, Du Châtelet was held in high regard. She corresponded with Wolff, Euler, Maupertuis, Clairaut, Jurin, Jaquier, and Musschenbroek about physics and natural philosophy, and with her philosophy she had an enormous effect on the thinking of her time.[2] Diderot wrote that there had been two wonderful moments in his life, one of them being due to Du Châtelet and her reply to his *Lettre sur les aveugles*.[3] La Mettrie was attached to her, and Kant wrote his first reflections on natural philosophy on the occasion of the publication of

---

[1] The reference goes to Brucker. 1745. *Bilder-Sal heutigen Tages lebender und durch Gelahrtheit berühmter Schrifftsteller*. See also LetChBII 302: June 3 rd, 1743. "Je suis très flattée de la compagnie dans laquelle on veut me mettre…"

[2] LetChBII 322: May 30th, 1744. To learn more about Du Châtelet as a busy correspondent, see: Bessire. 2008. Mme Du Châtelet epistolière. 25–35.

[3] Diderot. 1955. *Correspondance*. 155.

R. Hagengruber (✉)
Universität Paderborn, Fak. Kulturwissenschaften, Abt. Philosophie Warburger Str. 100, Paderborn 33098, Germany
e-mail: ruth.hagengruber@uni-paderborn.de

her pamphlet on the *living forces*.[4] Ampère spoke of her as a "Genius in Geometry" as she was said to have been able to simply "multiply nine figures by nine others in her head".[5] And while some of her interpreters at that time attributed her outstanding achievement more to her rank than to her philosophy, as is also the case today, her contemporary, a young poet named Helvétius, admired the "sublime Emilie" for her dedication to science and her disregard for social standing, being an example that rebutted the prejudices of her time.[6] Du Châtelet's achievements set the prejudices of her time swaying, but they also go far beyond the normally accepted historical scope of women today. Without a doubt, Du Châtelet may be considered an excellent representative of the Age of European Enlightenment. The most famous writers and scientists of the time were among her friends. Her companion, Voltaire, as well as Maupertuis and many other distinguished philosophers, literates, and scientists were included in her inner circle. She knew her value und she expected to be judged accordingly: "I am my own person and only responsible to myself for everything I am, what I say, and what I do. There may be metaphysicians and philosophers whose knowledge is greater than mine. I haven't met them yet. But even they are only weak human beings with faults, and when I count my gifts, I think I may say that I am inferior to none".[7]

At the beginning of a study on Du Châtelet, the scholar is challenged to disprove many of the prejudices which existed in her time, as well as those which exist today. It is not possible to gain an understanding of her by idolizing her or by following the modern assumption about the segregation of women. Neither perspective is adequate at giving a clear picture of Du Châtelet. She did not perceive herself to be a singular phenomenon, but to be part of a tradition of women philosophers who have dedicated their lives to philosophy and science. Du Châtelet compared herself to Elisabeth of Bohemia and Christine of Sweden, who had also devoted themselves to philosophy.[8] In her *Institutions* she refers to the Hanoverian princesses, and the portrait in the *Bilder-Sal* places her within the tradition of Diotima and Aspasia, of Theano and again with Elisabeth and Christina.[9]

---

[4] The relation between Du Châtelet and La Mettrie is analysed in: Jauch. 1998. *Jenseits der Maschine*. 314–335. See Kant. 1902. Gedanken von der wahren Schätzung der lebendigen Kräfte.

[5] Mozans. 1913. *Women in Science*. 151.

[6] Learn from Helvétius' poem: "Viens servir ton pays, viens, sublime Emilie/Enseigner aux Français l'art de vivre avec eux*/Qu'ils te doivent encor le grand art d'être heureux/Viens, dis leur que tu sus, dès la plus tendre enfance/Au faste de ton rang préférer la science". See Mauzi. 1961 *Discours sur le Bonheur*. IX-CXXVII. See also Whitehead. 2006. An Analysis of the 'Discours sur le bonheur'. 255–277. Hagengruber. 2010. Das Glück der Vernunft – Emilie Du Châtelet's Reflexionen über die Moral. 119 f.

[7] Cited from a letter to the crown prince Frederick of Prussia, in: Edwards. 1989. *Die göttliche Geliebte Voltaires*. 7.

[8] LetChBI 69: July 18th, 1736.

[9] Du Châtelet's portrait is cited according to Iverson. 2006. Du Châtelet's portrait in the 'BilderSal (…) berühmter Schriftsteller'. 53. See also note 1.

Voltaire praised her as the *Minerve* of the French Enlightenment, comparing her wisdom to that of the goddess who belonged to an era in which the intellectual capacity of women was mirrored in this name.[10]

Much has been written about Emilie du Châtelet within the last few decades, including much about her reversal from Newtonianism to Leibnizianism and possibly back. I will introduce Du Châtelet as an independent philosopher, and try to reconstruct her philosophical approach by examining her writings and the reflections of her colleagues' answers and reactions to her. From this perspective, her relationships with and the differences between her and her closest friends and colleagues, Voltaire and Maupertuis, can be viewed in a different light. I will argue that when Du Châtelet entered the philosophers' circle she was already quite self-determined in her approach to science and philosophy. This becomes obvious in her early writings, with a specific focus on her Locke critique. From this point of view, her critical view on Voltaire und Maupertuis can be commented on. At the same time, we find the foundations for her own direction here, which she develops within the Leibniz-Newton controversy. Du Châtelet rebels against the blind devotion to Newton and calls for his classification within the history of theories. From the Leibniz-Newton debate she follows that it is not vital to adhere to any one's theory, but to find criteria that can be used to test the theories and to prove their claims to truth. Her critical reflections on the scientific and philosophical contents led her to her own conclusions about the relationship between science and philosophy and, finally, to a more differentiated position regarding the function of physics, mathematics and metaphysics. She develops this standpoint through her critical reading of the empiricists and phenomenalists, for example Locke and Newton, and the rationalists and scholastics, turning against the blanket degradation they had been subject to. In this way, Du Châtelet transformed the philosophical and metaphysical understanding of her era.

In the following, single steps on the way to a new positioning of the sciences and philosophy shall be traced. In the first part, the objections she raised with regard to the philosophies of the empiricists and rationalists will be explained, with special attention to the examples of John Locke, Descartes, and the philosophy of the then so called scholastics. The tenet of the hypothesis, which she presents in the preamble and in the fourth chapter of her *Institutions* is to be seen as the result of this critical reflexion. Her methodical contribution is to establish a new metaphysics, which satisfies the demands of rationality as well as the standards of the experientially dependent contents. In the second part, the way in which Du Châtelet defends her position against opposition will be shown. This can be seen in her correspondence with the mathematician Maupertuis and later in the references to her earlier works by the physician and philosopher La Mettrie. The comparison makes clear how

---

[10] The renaissance of gods and goddesses in the times of Enlightenment and the hidden tradition of noble women, identifying themselves with goddesses of the Roman and Greek period, Hagengruber. 2010. Von Diana zu Minerva. Philosophierende Aristokratinnen des 17. und 18. Jahrhunderts und ihre Netzwerke. 11–32.

deeply Du Châtelet's thought is rooted in the discussions that were current during her lifetime.

Du Châtelet was a respected intellectual factor in her time. In spite of this, the reception of her writings was neglected after her death and they almost sank into oblivion. The fragility of her scientific and personal achievements was impressively and symbolically demonstrated shortly after her death.[11] In 1749, the year of Du Châtelet's death, Rousseau won the prize given by the Academy of Dijon with his strict "no" to the combination of the progress of morals with the progress of science and arts. This was, in effect, also a "no" towards women's education. Rousseau was not the only one who emphatically wrote that "research on abstract and speculative truths, on principles and on axioms of science, all of which tend to a generalization of concepts, is not for women".[12] This critique on the culture of the Enlightenment and its impact on women, regarding it as a trend "against nature", has influenced our perspectives for more than two centuries and is still present today. The truth, however, is that there have always been women who influenced culture and the sciences in an important way. Du Châtelet is one of them. Emilie du Châtelet was part of the scientific community of her period; her work crosses borders by means of her knowledge and independent wisdom as well as her liberal views on sex, religion, and dynastic traditions. Du Châtelet challenged the morals of the distinguished society and the scientific convictions of the society of the learned. She acted as a philosopher, a physicist and mathematician, as a member of the noble class and mainly as herself.

Striving for a Metaphysics of Science

Becoming a Philosopher

Du Châtelet was certainly well educated.[13] She read Latin and Philosophy, Horaz, Virgil, Lucretius, and was familiar with the writings of Cicero.[14] As an outstanding adept of the classics, she had also read Plato and Homer, and at the age of 17 was

---

[11] Hagengruber. 2002. Gegen Rousseau - für die Physik: Gabrielle Emilie Du Châtelet (1706–1749). 27–30.

[12] Rousseau. 1978. *Emile oder über die Erziehung*. 447.

[13] See Judith Zinsser's introduction in: Du Châtelet. 2009. *Selected Philosophical and Scientific Writings*. 1–25.

[14] Voltaire. 1878. *Œuvres complètes*. Vol. XXIII. 7: "Son père, baron de Breteuil, lui avait fait apprendre le latin, qu'elle possédait comme mme Dacier; elle savait par cœur les plus beaux morceaux d'Horace, de Virgile, et de Lucrèce; tous les ouvrages philosophiques de Cicéron lui étaient familiers. Son gout dominant était pour les mathématiques et pour la métaphysique." Information on Du Châtelet's youth and education can be found in Zinsser. 2007. *Emilie Du Châtelet. Daring Genius of the Enlightenment*. 13–62.

already familiar with the essays of John Locke.[15] Her studies were dedicated to morals, natural philosophy and metaphysics. When Voltaire's *Letters on England* were to be prohibited in France, Du Châtelet's family was there to offer protection to the author at Cirey. In 1735, Du Châtelet decided to dedicate her life completely to science and followed Voltaire to Château Cirey. After her arrival, she and Voltaire established quite a singular academy there, an academy for science and arts. The latest and best books, the newest and best scientific instruments were acquired, experiments were conducted and a range of research was undertaken.[16] At night, Du Châtelet rehearsed the leading parts in Voltaire's plays, which were presented in the small theater. She skillfully sang the most avant-garde operas of her time and presented herself as an indefatigable scientist and artist.[17] While it is said that Voltaire initially needed Maupertuis' advice for his *Letter* on Newton's law of attraction, because of this new partnership he was able to write the *Eléments de la philosophie de Newton* with Du Châtelet's assistance.[18] She became Voltaire's companion and the mathematical advisor of his *Eléments*, published in 1738. Addressing the value of her contributions, Voltaire confirmed this in an early letter to Frederic in January 1737, "… Madame Du Châtelet had her part in the work. Minerva dictated and I wrote".[19] Du Châtelet acted as the wise goddess Minerva for Voltaire, bringing Newton's inspirations to him. She directed Newton's spiritual light towards Voltaire as pictured on the frontispiece of the *Eléments*, followed by Voltaire's dedication to her in the first edition.[20]

Several texts originated in the cooperation between Voltaire and Du Châtelet, and it is fascinating to discover more clearly where their convictions and ideas matched and what had been written in competition.[21] An examination of the Bible,

---

[15] Brown and Kölving. 2008. À la recherche des livres d'Emilie Du Châtelet. 111–121 and Kölving and Brown. 2008. Emilie Du Châtelet, lectrice d'une Apologie d'Homère. 135–167 and Vaillot. 1978. *Madame du Châtelet*. 59.

[16] Gandt. 2001. *Cirey dans la vie intellectuelle: la réception de Newton en France*. See also Zinsser. 2007. *Emilie Du Châtelet: Daring Genius of the Enlightenment*. 105–152.

[17] Her capacities as a musician are presented in Adelson. 2008. La belle Issé: Mme Du Châtelet musicienne. 127–134.

[18] Du Châtelet to Maupertuis, LetChBI 12: April 28th, 1734: "Voltaire, à qui j'ai dit que je vous écrivais, me prie de vous dire mille choses pour lui. Il est inquiet et avec raison du sort de ses lettres, il est bien flatté de ce que ses ennemis croyent que vous avez eu part à celle de monsieur Neuton, et si les lettres de cachet ne s'en mèlaient pas, je crois que votre approbation lui tiendrait lieu de tout le reste."

[19] "J'avois esquissé les principes assez faciles de la philosophie de Neuton et Madame du Chastelet avoit sa part a l'ouvrage. Minerve dictoit et J'écrois". Letter to Frederick II., January 1st, 1737 in Voltaire. 1968–1977. *Complete works*. 88, 196. See also Voltaire's letter to Cideville December 23 rd, 1737, Voltaire. 1968–1977. *Complete works*. 88, 429 and Voltaire's letter to Berger in July 1736: "J'étudie la philosophie de Newton sous les yeux d'Emilie", Voltaire. 1968–1977. *Complete works*. 88, 16.

[20] Voltaire. 1738. *Eléments de la philosophie de Newton*. Avant-propos 9–13.

[21] See Brown and Kölving 2003. Qui est l'auteur du Traité de métaphysique? 85–94.

observations on grammar, annotations to Descartes, and texts on poetry and politics reflect the variety of topics Du Châtelet was occupied with. Her writings extended from texts on happiness to discussions about natural philosophy.[22] Voltaire's legacy in the St. Petersburg National library comprises hundreds of manuscript pages belonging to Du Châtelet, and in addition to this particular literary inheritance, previously unknown works of Du Châtelet are still being discovered today.[23]

She started her career as a writer by commenting and translating Bernard de Mandevilles' scandalous *Fable of The Bees*.[24] Several texts were then published by Du Châtelet herself. In 1738 she had already applied for the approbation to her *Institutions de physique,* first published in 1740 and titled *Institutions physiques* in the 1742 Amsterdam edition. In 1739, her *Dissertation sur la nature et la propagation du feu* was printed together with the other prize winners.[25] In 1741 Du Châtelet was acclaimed for her *Réponse* to Mairan, *About the Living Forces,* which was promptly translated into German.[26] In 1743, following the augmented second edition from 1742, the *Institutions physiques* was translated into German and Italian.[27] In 1746 she earned the honour of becoming a member of the *Accademia delle Scienze di Bologna*. An extensive article about her subsequently found its way into the *Journal Universel* in August 1746. She was widely cited in several other works.[28] La Mettrie referred to her and dedicated his writings to her in 1747. Her commentary and translation of Newton's *Principia* was finished in 1749, the year of her death but published only years later. Beginning in 1754, several articles from the *Encyclopédie* were directly taken from the *Institutions*.[29] In September 1749, shortly after her death, Voltaire wrote to their friend, the comte d'Argental: "Je viens de relire des matériaux immenses de métaphysique que madame du Châtelet avait assemblés avec une patience et une sagacité qui m'effraient…C'était le génie de Leibniz avec de la sensibilité".[30]

---

[22] Schwarzbach. 2008. Mme Du Châtelet et la Bible. 197–213; Schwarzbach. 2006. Mme Du Châtelet's *Examen de la Bible* and Voltaire's *La Bible enfin expliquée*. 142–164. Douay-Soublin. 2008. Nouvel examen de la *Grammaire raisonée* de Mme Du Châtelet. 173–196. Kölving. 2008. Émilie Du Châtelet devant l'histoire. 1-13. Waithe. ²1991. *A History of Women Philosophers*. Vol. III. 127–147.

[23] Because of the increasing interest in Du Châtelet's philosophy, many documents are expected to be coming up in the future, as Fritz Nagel's discovery of Du Châtelet's complete *Optics* shows. The discovery of the *Essai sur l'optique*, which we were happy to hear of for the first time at our conference in Potsdam in September 2006, was discovered by Nagel in the Bernoulli archive.

[24] Du Châtelet. 1947. Mme. du Châtelet's Translation of the *Fable of the Bees*. 131–187.

[25] DissCh1739 and DissCh1744.

[26] MaiCh1741; MaiCh1741D.

[27] Naturlehre1743; Inst1743Ve.

[28] All bibliographical details on Du Châtelet's writings can be found in Kölving. 2008. Bibliographie chronologique. 341–385.

[29] Some of these articles have been more or less rewritten, others kept in their original form. Maglo. 2008. Mme Du Châtelet, l'Encyclopédie et la philosophie des sciences. 255–266.

[30] LetChBI: Notes préliminaires. 17.

In the Center of the Debate

Du Châtelet had hinted about the close collaboration between herself and Voltaire. Her anonymous review, *Lettre sur les éléments de la philosophie de Newton* in the *Journal de sçavants* from 1738, rallied a great deal of support among the general public.[31] In the introduction to her *Institutions,* however, she marks the constraints of Voltaire's *Elémènts* from 1738, saying that she would not confine herself to such narrow boundaries as he, but would embrace a vaster terrain.[32] Du Châtelet, whose contribution to the *Eléments* was popular among the *philosophes*, strove to draw a clear demarcation between Voltaire's thoughts and her own. On several occasions she transgresses the confinements of her collaborator's thinking. This is obvious in her redraft of Voltaire's dictum on metaphysics.[33] If something is available to our senses and thus to reason, it can be known. If not, it cannot be known, says Voltaire. Du Châtelet judges the realm of metaphysics in dependency to our mental endeavour. "Metaphysics contains two types of things: the first, that which all people who make good use of their minds can know, and the second, which is the most extensive, that which they will never know".[34] Du Châtelet's definition of knowledge refers to a rational endeavor, which results from "the good use of mind"! The meaning of this remark becomes clear in the light of the critique she exercises in her *Institutions* and which does not singularly or essentially target Voltaire. One reason for their intellectual differences is to be found in their contrasting views on the philosophy of John Locke.

---

[31] See also her remark in LetChBI 129: June 21st, 1738: "Les deux derniers chapitres de la philosophie de Neuton ne sont pas de mr. de V. Ainsi vous ne devez rien lui attribuer de ce qu'on y dit sur l'anneau de Saturne. Son dessein n'etait pas d'en parler du tout".

[32] Du Châtelet. 2009. *Selected Philosophical and Scientific Writings*. 119. Inst1742Am Avant-propos VI. "You can draw much instruction on this subject from the Elémens de la philosophie de Newton, which appeared last year. And I would omit what I have to tell you about that – Newton's system – if the illustrious author had embraced a vaster terrain; but he confined himself within such narrow boundaries that he made it impossible for me to dispense with my own exposition of this matter".

[33] While Voltaire says: "[t]o my mind, all of metaphysics contains two things, the first, that which sensitive men know, the second, that which they can never know", he acknowledges a sensualist positivism, which he also expressed in his *Letters on England.* Voltaire's remark, dating from 1737, citation slightly changed taken from Waithe. ²1991. *A History of Women Philosophers*. Vol. III. 145. For Du Châtelet. 2009. *Selected Philosophical and Scientific Writings*. 124, note 46: "This sentence reflects an interchange late in the 1730s between Voltaire and Frederick of Prussia, in which Voltaire made this distinction. See D1376, Voltaire to Frederick of Prussia (15 October 1737), *Oeuvres complètes*, v. 88, 381."

[34] Du Châtelet. 2009. *Selected Philosophical and Scientific Writings*. 124. Inst1740/Inst1742Am Avant-propos XII.

Critisizing Locke

Voltaire penned his *Letters on England* – later known as *Philosophical Letters* – in England. These *Letters* were to be prohibited in France, and Voltaire was threatened yet again with imprisonment in the Bastille. In his book, English liberalism was favoured over French nationalism and philosophy, also presenting an appealing understanding of Locke's work. When Du Châtelet decided to protect Voltaire, it can be taken for granted that she was familiar with John Locke's *Essays Concerning Human Understanding,* which Voltaire had praised in his book, and had read them in the French translation.[35] For Du Châtelet and Voltaire, John Locke's writings did not remain without effect. Writing from Cirey where Voltaire was hiding, Du Châtelet confirms that the philosophy of John Locke was still at stake: "Je partage mon temps entre des maçons et m. Lock".[36] During this period of time or shortly after, Du Châtelet started to compose her commentary on the philosophy of Mandeville's *Fable of The Bees*.[37] In this commentary, Du Châtelet takes the opportunity to oppose some of Locke's basic reflections. Her criticism is important to understand her developing concept of philosophy and her critiques against Voltaire. It is also a decisive critique of the philosophical empiricism of John Locke and the materialistic inspiration later deduced from it by La Mettrie. Du Châtelet rejects John Locke's adversion of innate ideas and a priori principles. She also reverses Locke's negation of the principle of contradiction, which would constitute the basis of her methodic reflections in the *Institutions.*

Her commentary on Mandeville furthermore makes obvious how familiar she is with John Locke's musings about innate ideas and how authoritatively she disapproves it, saying that Locke had deceived himself when he "destroyed" the philosophical assumption of *innate ideas*.[38] She confirms her belief in ideas which are indigenous and a priori to all people. In order to limit the relativistic guideline of cultures and circumstances and of relativistic and positivistic ethics, there is a need

---

[35] Vaillot. 1978. *Madame du Châtelet.* 59. Zinsser. 2007. *Emilie Du Châtelet. Daring Genius of the Enlightenment.* 60, referring to Pierre Coste's French translation of John Locke's *Essay on Human Understanding* that was published in 1729. See also Voltaire's Letter to Formont June 1734: "she understands Locke better than I", citation taken from. Zinsser. 2007. *Emilie Du Châtelet. Daring Genius of the Enlightenment.* 78.

[36] LetChBI 24: October 23 rd, 1734.

[37] Although she opposes several of Mandeville's ideas in her commentary, Du Châtelet judges the *Fable of the Bees* to be "the best book of ethics ever wrtitten". Du Châtelet. 2009. *Selected Philosophical and Scientific Writings.* 50. For the original French text see: Du Châtelet. 1947. Mme. du Châtelet's Translation of the *Fable of the Bees.* 131–187.

[38] Du Châtelet. 1947. Mme. du Châtelet's Translation of the *Fable of the Bees.* 145. "Mais tous les homes s'accordent a observer les loix etablies chez eux, et a regarder les actions comme bonnes ou mauvaises selon leur relation ou leur opposition a ces loix (…) et je crois que le sage Lock a esté trop loin, quand apres avoir *detruit les idées innées* (RH), il a avancé qu'il ny avoit point d'idées de morale universelle". "Le besoin de la societé éxige cette loy comme son fondement, et les besoins qui sont differents dans les differents pays, se reunissent tous dans cette maxime."

for universal principles. There has to be something serving as the vanishing projection point, which must be valid for everyone in the world: "Le besoin de la societé éxige cette loy comme son fondement", she says, at this point defending the Golden Rule as being more than a natural principle, in contrast to John Locke's assumption.[39] It is necessary to presuppose a *universal law*, because there has to be something which is valid under all circumstances, for all cultures and religions, and which is valid for women and men.

The genesis of her writings makes it clear that her critique on Locke originates in her Mandeville commentary. She confronts us with her resolute statement in favour of universal principles which precondition human knowledge and action, and maintains that this kind of law is innate. With this statement she provokes another controversy with her companion, Voltaire. In *Letter XIII* Voltaire had cited Locke, trivializing the fact that knowledge and ideas were present at birth, praising Locke for having "destroyed" innate ideas, relegating them because if they would exist also "children and idiots" would know. Voltaire defends Locke's radical approach to the origins of knowledge, arguing that knowledge must be gained by "history and experience". To base knowledge on innate principles makes no sense at all, says Voltaire, paraphrasing Locke.[40] Du Châtelet claims the necessity of a universal presupposition, because if there is no such beginning, all our knowledge is relative: The foundations of knowledge would then have to be proven again and again and would never be secure in any respect. Some years later in her *Institutions,* she answers this argument in more detail. There she states that the possibility that something can be known must be based on a metaphysical principle which is beyond experience and the natural development of reasoning. Otherwise there can be no conclusion, no order and no continuity, no before and no after. Without such a principle, there would be no "chaîne qui y conduit", no chain of knowledge, no "truth", no "numbers", and "two and two could then make as well 4 as 6".[41]

Above and beyond her critique on the negation of metaphysical principles, her commentary on Mandeville and her critique of Locke is accompanied by some reflections about her female experience. Du Châtelet widens her philosophical insight to rebuke yet another philosophical statement of John Locke, this time repulsing his negation of the value of the principle of contradiction, which she claims to be necessary to evaluate both experienced and factual knowledge. Writing about her personal experience, she states: "I feel the full weight of prejudice that excludes us (women) so universally from the sciences, this being one of the contradictions of this world which has always astonished me, as there are

---

[39] The Golden Rule is not a "natural rule" but a kind of universal law: "Il y a une loy universelle pour tous les homes que dieu eu a luy mesme grave dans leur Coeur. Cette loy est, *ne fais pas a autruy ce que tu ne voudrois pas qui te fust fait*." Du Châtelet. 1947. Mme. du Châtelet's Translation of the *Fable of the Bees*. 145.

[40] "Mr. Locke, after having *destroyed* (RH) innate ideas …" Voltaire. 1894. *Letters on England*. Letter XIII. Compare Locke. 1975. *An Essay Concerning Human Understanding* I. 2, 5 and 8.

[41] Inst1740/Inst1742Am Chap. I §§ 1, 4.

great countries whose laws allow us to decide their destiny, but none where we are brought up to think".[42] It is not easy to understand why Du Châtelet here combines the law of contradiction with the exclusion of women. Her argument becomes more evident from the methodic point of view. It is a phenomenon we can observe that there have been no women of importance in the history of culture and human reasoning. Du Châtelet sums up: "... for so many centuries, not one good tragedy, one good poem, one esteemed history, one beautiful painting, one good book of physics, has come from the hand of women". Does this phenomenon justify further exclusion, she asks?[43] Harshly describing the absence of women in her society's culture, she conversely does not accept the apparent phenomenon as an argument. Explanations for phenomena can be astonishingly varied. Had she not considered herself to be a non-thinking being, not educated, "taking extreme care of my teeth, of my hair and neglecting my mind and my understanding". The mind has to be educated, it is like a muscle which has to be trained: Like "The fakirs of the East Indies lose the use of the muscles in their arms ... Thus do we lose our own ideas when we neglect to cultivate them. It is a fire that dies if one does not continually give it the wood needed to maintain it". This statement makes clear that she did never agree to Locke and Voltaire's refutation of innate ideas. To presuppose innate principles does not free humans to educate the mind. The fact that women were not educated does not prove that they are not able to become. Du Châtelet questions the facts and observations, motivated by her experience as a woman. Through this critical distance, she realizes the disharmony of facts and possibilities, "this being one of the contradictions of this world, which has always astonished me, as there are great countries whose laws allow us to decide their destiny, but none where we are brought up to think".[44] Although her claim to the principle of contradiction is supported by the framework of Leibnizian metaphysics in her *Institutions*, we learn from this commentary that she was aware of this basic philosophical principle much earlier and had a critical position towards those who denied it. In § 4 of her first chapter of the *Institutions*, after having explained the necessity of the principles of knowledge, she comments on the principle of contradiction as being the first axiom on which all further truths are founded. "On appelle *Contradiction*, ce qui affirme & nie la même chose en même tems".[45]

The analysis of the early commentary on Mandeville shows us that Du Châtelet had received John Locke's philosophy critically at quite an early stage. This also suggests that she was never completely in accordance with Voltaire's philosophy from the beginning of their friendship. The *Institutions*, therefore, were written

---

[42] Du Châtelet. 2009. *Selected Philosophical and Scientific Writings*. 48.

[43] Du Châtelet. 2009. *Selected Philosophical and Scientific Writings*. 48.

[44] Du Châtelet. 2009. *Selected Philosophical and Scientific Writings*. 44.

[45] "This axiom is the foundation of all certainty in human knowledge. For, if one once granted that something may exist and not exist at the same time, there would no longer be any truth". Du Châtelet. 2009. *Selected Philosophical and Scientific Writings*. 126. Inst1740/Inst1742Am Chap. I § 4.

within the context of an ongoing debate, not as the start of this debate. This impression is confirmed by another example taken from the *Institutions* in which she clarifies her position and amplifies her criticism towards phenomenalism.

In her third chapter, which precedes the one on the "Hypotheses", she writes about "Essences, Attributes and Modes". Why this ontological clarification is necessary is also made evident by a critique on Locke. His philosophy, as she writes there, is muddled. "Locke is confused", she says. Locke has no idea how essences and modes fit together; he proposes substances to be what the senses and the imagination ordinarily tell us. The basis of the confusion about these principles of knowledge is to be found in the way Locke constitutes substances out of perceived qualities. But, on the contrary, the senses and the imagination present a wrong picture of things and do not suffice as a validation for ascertainable knowledge. Locke interprets a substance as a kind of backing and as a supporter or container of attributes, having no idea that the essential, necessary, and unchanging attributes are vitally connected to the substance, whereas the changing modes are not necessarily part of the substance and are understood mostly through perception. To know the difference between necessary and unnecessary attributes is elementary for Du Châtelet's metaphysics. Either something is *A* or *not A*; if it is *A*, it is essentially *A* and not *not A*. Du Châtelet refuses to follow Locke's "confused ideas" on substance. As Locke construes substances from confused ideas about individuals, he confuses the essential attributes of the substances, thus holding it possible that matter could have mental qualities.[46] It is in this early version of her *Institutions* that Du Châtelet identifies a philosophical problem which would become famous some years later. With her remark against the idea of a *thinking matter* she again rebuts a thesis Voltaire also set forth in his *Letters*.[47] Du Châtelet's argument is found in her 1740 version of the *Institutions*. La Mettrie started to write his provocative and materialistic interpretation of this in about 1748. It is remarkable, however, as will be seen in the last part of this essay, that he does it with strong reference to Du Châtelet.

---

[46] "Mr. Locke lui-même s'arrête à la notion imaginaire de la Substance, telle que les sens & l'imagination la donnent au vulgaire, il dit: *que la Substance n'est autre chose qu'un sujet que nous ne connoissons pas, & que nous supposons être le soutien des qualités dont nous découvrons l'éxistence, et que nous ne croyons pas pouvoir subsister, sine re substante, sans quelque chose qui les soutienne, et que nous donnons à ce soutien le nom de Substance qui, rendu nettement en François veut dire, ce qui est dessous, ou ce qui soutient*. On voit aisément que cette notion de la Substance est entierement confuse, comme M. Locke l'avoue lui-même & c". Inst1740/42Am Chap. III § 51. See Locke 1975. *An Essay Concerning Human Understanding*. II, 23, 4 and 5: "The idea then we have, to which we give the general name substance, being nothing but the supposed, but unknown, support of those qualities we find existing, which we imagine cannot subsist sine re substante, without something to support them, we call that support substantia". "It is plain then that the idea of corporeal substance in matter is as remote from our conceptions and apprehensions as that of spiritual substance, or spirit; and therefore from our not having any notion of the substance of spirit, we can no more conclude its non-existence than we can, for the same reason, deny the existence of body etc.".

[47] Voltaire1894. *Letters on England*. Letter XVIII.

Du Châtelet dedicated her attention to this problem within the context of her criticism of Locke, reflecting the issue of necessary attributes. Locke's idea of the possibility of *thinking matter* is, as she says, an idea as abstruse as many of the monsters within philosophy, having been dreamt up by the imagination and having blocked the growth of science and human reason. Locke's concept of substance is more fantastic than real; he is much more of a story-teller than the old metaphysicians. It is indispensable to conclude that thought can never be an attribute of matter, she says: "il est indispensable de conclure qu'il est impossible que la pensée puisse être un attribut de la matiere". Even God cannot change the essences of substances according to His will, not even God can relegate essential attributes to another substance. If the necessary attributes of matter are extension and impenetrability, it is senseless to assert that God would subtract or add an essential attribute at will. Those who want to make us believe that thinking can be an attribute of matter according to God's Will at least admit that it is not essential. Secondly, they ruin the rationality of natural philosophy, as then all rational determinations could be confused with the Will of God.[48] Thus Locke's remark, that "[w]e can attribute their connection to nothing else but the arbitrary determination of that all-wise Agent who has made them..." also must have caught Du Châtelet's eye. Locke concedes in his *Essay* that he has no answer to some of the essential issues of understanding, such as how one could demonstrate the connection between ideas and our everyday reality. It is at this point that Du Châtelet makes obvious the weakness of Locke's experience-guided epistemology, which cannot provide a foundation for a science that would take his words seriously.[49] Referring to God as the source of reasonability does not yield valid scientific results, Du Châtelet repeats indefatigably in her writings. Science that calls on God to provide the explanations it requires effectively loses its academic standing. In the context of this discussion we can approach Du Châtelet's developing scientific methodology. In her writings on this topic Du Châtelet affirms her arguments in favour of the necessity of a priori and universal principles. Moreover, she offers a strong plea in favour of science and against the voluntaristic convictions of Locke, Clarke and other Newtonian scholars such as Maupertuis and, to some extent, Voltaire, who more than once asserted that God had initiated and created these principles, guaranteed them, and was able to change them according to His Will. Du Châtelet rebuts key issues of empiricist epistemology and repulses the voluntaristic approach.

Arguments taken from her Mandeville commentary are continued and broadened in the *Institutions*. Her conviction about the essentiality of rational and a priori

---

[48] Inst1740/Inst1742Am Chap. III § 47. "Mais ils disent, que Dieu a peut être donné à la matière l'attribut de la pensée, quoiqu'elle ne l'ait point par son essence, et qu'ainsi, comme on ne sait point ce qu'il a plu a Dieu de faire, on ne peut savoir non plus si ce qui pense en nous est matière ou non."

[49] Locke. 1975. *An Essay Concerning Human Understanding* IV. 3, 28: "we can have no distinct knowledge of such operations beyond our experience; and can reason not otherwise about them than as effects produced by the appointment of an infinitely wise agent, which perfectly surpass our comprehensions".

principles was obviously never at stake. Innate ideas such as the epistemological value of contradictions are obviously prerequisites for her scientific claim. Furthermore, these principles which Du Châtelet articulates in her critique of the philosophy of John Locke are repeated in her *Institutions*, and are presented in conformity and in analogy with some aspects of Leibnizian metaphysics. This exemplifies the continuity of her thinking as seen from the perspective of the genealogy of her writings. Her argumentation is in concordance with some of Leibniz's principles and shows a certain correspondence to the rationally and hypothetically inspired metaphysics of Leibniz and Wolff. In the context of Du Châtelet's criticism of Locke's philosophy it is also interesting to note Leibniz's rebuttal of John Locke's *Essays*. Leibniz had designed his *Nouveaux Essais sur l'entendement humain*, as a chapter-by-chapter refutation of Locke's system. Written mostly between 1700 and 1704, Leibniz's arguments were rapidly diffused throughout the world via countless pieces of correspondence.[50] Leibniz presents his critique by means of a dialogue between Theophilius and the Lockean Philalethus. It is an interesting fact that this dispute is inspired by an Anti-Leibnizian article taken from the world's most famous dictionary of the time, from Bayle's *Dictionnaire historique*.[51] This article headed an attack on Leibniz's theory of pre-established harmony, which could easily have been known to Du Châtelet. If this were the case, it can be noted that Du Châtelet even then was not completely convinced by the Anti-Leibnizian arguments.

Scholastic Voidness

Emilie du Châtelet criticizes Locke's philosophy in an elementary respect. She adheres to universal laws, a priori principles, and to the law of contradiction as a primary principle. With her criticism she positions herself within the tradition of the rational metaphysicians. To understand her approach, it is indispensable to delimitate her position towards this tradition of metaphysics. She does not contest all of Locke's insights, but rather holds fast to the necessity of the verification of knowledge through experience, and she uses the criteria of experience in order to emerge as a critic of the "old" metaphysicians. Voltaire's statements in his *Letters on England* can again be taken as a starting point in understanding Du Châtelet's critical remarks and her delimitation of a "scholastic" kind of metaphysics. Concurrently, the aspects of Voltaire's and the Newtonians' criticism of the old metaphysicians which she did not fully share, also become manifest; this makes explicit why she grouped Voltaire, Maupertuis und La Mettrie as the "bad" metaphysicians in the end.

---

[50] The work itself remained unpublished until a German edition in 1765; the first complete English edition was not published until 1895.

[51] "Cet auteur ... croit que la matière pourrait penser, qu'il n'y a point d'idées innées, que notre esprit est tabula rasa, et que nous ne pensons pas toujours." Leibniz takes the citation from the article "Rorarius" in Bayle. ²1702. *Dictionnaire historique et critique*.

In his *Letter XIV,* Voltaire compares the philosophy of Descartes to that of Newton, thereby creating not only the vision of two contrary systems, but also marking these two different intellectual traditions, that of the Continentalists and that of the Islanders, as the realm of "scholastic impertinence" versus the realm of "reason".[52] Descartes became a synonym for misled scholasticism, whose philosophy was fanciful, scarcely proven, and lacking in the observation of reality, while the English style "was quite different". Voltaire's comparison of Descartes with the scholastics is continued in part in Du Châtelet's criticism. It is telling that although she postulates the necessity of principles at the beginning of her *Institutions*, she clearly distinguishes her metaphysical claims from the understanding of the scholastics and Descartes: "principles came into disregard", as they have been "abused" by the scholastics and Descartes. It was the fault of the scholastics to have chosen "unintelligible words", which keep "man away from truth".[53] Although the scholastics admitted the necessity of reasoning, they did not endeavor to prove their ideas through experience, but only reasoned using empty words, "mots vuides de sens".[54] She admits that Descartes was aware of these scholastic deficiencies and that he tried to solve the problem, establishing "clear ideas". In this Descartes failed in a different way, she says. He tended towards a sort of "internal sense" as a category of clarity, and towards evidence that should serve as a basis for reasoning, but he failed to prove his thoughts by demonstration. Despite all his exertions, Descartes could not confer the scholastic thought with any conclusiveness.[55] Du Châtelet criticizes Descartes' unproved metaphysics, in which he identified matter by extension, without "prov(ing) the possibility of this idea". He postulated a material plenum, explaining the phenomena in the world as the result of circular movement and matter, and he remained true to his mechanistic idea of the universe, ignoring inertia and a *living force*. She criticizes the ill-conceived ideas of Cartesian intuitionism, the empty use of terminology and the Cartesian scholasticism of reasoning which is deprived of experimental proof. Descartes was a dreamer; in this point she agrees with

---

[52] "A Frenchman who arrives in London, will find philosophy, like everything else, very much changed there. He had left the world a plenum, and he now finds it a vacuum. At Paris the universe is seen composed of vortices of subtile matter; but nothing like it is seen in London". "The progress of Sir Isaac Newton was quite different. ... It was his peculiar felicity, not only to be born in a country of liberty, but in an age when all scholastic impertinences were banished from the world. Reason alone was cultivated". Voltaire. 1894. *Letters on England.*
[53] "On a beaucoup abuse du mot de Principe, les Scholastiques qui ne démontroient rien donnoient pour principes des mots inintelligibles." Inst1740/Inst1742Am Chap. I § 2.
[54] Inst1740/Inst1742Am Chap. I § 10.
[55] Inst1740/Inst1742Am Chap. I § 2. "Descartes qui sentit combien cette manière de raisonner éloignoit les hommes du vrai, commença par établir qu'on ne doit raisonner que sur des idées claires; mais il poussa trop loin ce principe: car il admit que l'on pouvoit s'en rapporter à un certain sentiment vif & interne de clarté & d'évidence pour fonder nos raisonnemens. Ce fut en suivant ce principe que ce Philosophie se trompa sur l'essence du Corps ... sans se mettre en peine de prouver la possibilité de cette idée."

Voltaire. The Book of Nature, which should have been filled with truth, was "filled with fables and reveries".[56]

Voltaire derided Descartes' "fancy" in his *Letter XIV*.[57] Du Châtelet concurred with this criticism in several points. He had also made fun of Descartes because he so completely ascribed the mathematical laws to God; he was mistaken "to declare that two and two make four for no other reason but because God would have it so".[58] Voltaire demonstrated just how unfounded Descartes' metaphysics were. God was required to guarantee for the validity of mathematics and also for the human soul. But exactly at this point, Du Châtelet shifts the argument in the direction of the Newtonians. The Newtonians themselves are becoming metaphysicians, she says, and they also forget to question the reasons behind phenomena and, in doing so, make God responsible for gravity. In Du Châtelet's opinion, these Newtonian scholars who are not interested in finding out why the laws of gravity are valid, belong to the "bad" metaphysicians just as much as Descartes does. They are all dreamers: those who believe that God created numbers, as well as those who believe that God is the reason for attraction and those who believe God could have made matter a thinking thing, only by transferring an attribute![59] "Why do the planets move from east to west, and not the other direction?" There is no reason for it, it is the way it is, as God wants it to be, Voltaire says in his *Eléments*. Du Châtelet repeatedly states that referring to God in scientific explanations, whether for attraction, for the communication of essences, or for the validity of numbers does not yield any valid scientific results.[60] God can only be the final reason, the reason behind everything that exists, but cannot be used as a scientific argument. Science which calls on God to provide the explanations it requires, effectively loses its academic standing. Nothing which necessitates citing God as the only possible reason for something, can be proven as true.[61] This is the starting point of Du Châtelet's scientific approach towards a metaphysics of experiential reason.

---

[56] "Descartes ... gave the whole learned world a taste for hypotheses; and it was not long before one fell into a taste for fictions. Thus, the books of philosophy, which should have been collections of truths, were filled with fables and reveries." Inst1740/Inst1742Am Chap. IV § 55. Du Châtelet. 2009. *Selected Philosophical and Scientific Writings.* 147.

[57] "Descartes was mistaken in the nature of the soul, in the proofs of the existence of a God, in matter, in the laws of motion, and in the nature of light. He admitted innate ideas, he invented new elements, he created a world; he made man according to his own fancy; and it is justly said, that the man of Descartes is, in fact, that of Descartes only, very different from the real one." Voltaire. 1894. *Letters on England*.

[58] Voltaire. 1894. *Letters on England*. Letter XIII.

[59] See Inst1740/Inst1742Am Chap. III § 51: "M. Locke lui-même s'arrête à la notion imaginaire", etc. as well as § 50: "l'immuabilité des essences, bannit tout d'un coup de la Philosophie .. tous les monstres sortis de l'imagination des hommes … comme je l'ai dit ci-dessus (§ 47) l'idée du célébre Locke sur la posiblité de la matière pensante."

[60] Inst1740/Inst1742Am Chap. III §§ 18–31 De l'Existence de Dieu.

[61] Inst1740/Inst1742Am Chap. III § 49. "On doit donc dire que l'actualité des choses dépend de la vonlonté de Dieu, car ayant donné l'éxistence à ce Monde …, mais non pas dans a volonté qui ne peut se déterminer que consequemment à ce que son entendement se représente. Ainsi, on ne doit rien admettre comme vrai en Philosophie, quand on ne peut donner d'autre raison de la possibilité que la volonté de Dieu, car cette volonté ne fait point comprendre comment une chose est possible."

Hypotheses

Although the mechanistic Cartesian explanation of vortices had been blacklisted in Rome and Paris in 1663, at the time of Voltaire and Du Châtelet it surprisingly became acknowledged as a valid (French) alternative to Newton's system of attraction. However, Newton's thoughts had long taken both Europe and the French Academy of the Sciences by storm. Some French mathematicians of the time, however, remained loyal to Descartes because Newton's forces of attraction seemed to possess as little explanative value to them as the Cartesian vortices. As Voltaire expressed, none of the systems was completely convincing.[62] While the "old metaphysicians" turned against Newton's laws of gravity, which they regarded as obscure because they could not be explained by mechanical means, the Newtonians were able to increasingly devote themselves to the beauty of their mathematical formulas. For the first time, one single law, the law of universal attraction, was used to explain both terrestrial and celestial physics. Newton's natural philosophy seemed to be the "masterpiece", to which nothing could be "truly compared", as Voltaire says in his *Letter*. The beauty of the mathematical laws and the fact that they so wonderfully correspond with observable occurances seemed to prove the validity of Newton's law of attraction. The search for further proofs had to remain speculative. This, at least, was Newton's viewpoint. He had published his *Philosophiae naturalis principia mathematica* in London in 1687. However, the reasonability of the laws of attraction accentuated the overall problem of his argument. Newton finally decided to avoid the problem, denying any further speculation about it in his Scholium Generale for the second edition of the *Principia*, with his famous "hypotheses non fingo," which interdicts further reasoning about the primary cause of gravitation.[63]

Du Châtelet was impressed by Newton's findings, but she was also eager to find a convincing reason for the validity of this law, besides the fact that its theses could be perfectly represented in mathematical laws. She held strongly to her conviction that these findings had to be connected to the truths already known, as well as to those that preceded them and that the validity of the Newtonian laws also depends on hypotheses. She demonstrates that Newton's system was built upon hypotheses,

---

[62] "According to your Cartesians, everything is done by means of an impulse that is practically incomprehensible, according to Mr. Newton it is by a kind of attraction, the reason for which is no better known". Voltaire. 1894. *Letters on England*. XIV.

[63] "I have not as yet been able to discover the reason for these properties of gravity from phenomena, and I do not feign hypotheses. For whatever is not deduced from the phenomena must be called a hypothesis; and hypotheses, whether metaphysical or physical, or based on occult qualities, or mechanical, have no place in experimental philosophy. In this philosophy particular propositions are inferred from the phenomena, and afterwards rendered general by induction." This remark was appended to the second edition of the Newton's *Principia*. Newton. 1999. *Philosophiae Naturalis Principia Mathematica*. 943.

believing it to be so convincing because of that very reason.[64] Newton used hypotheses up to the point where these hypothetical connections appeared to be plausible and probable: "jusqu'où la connexion & la vraisemblance".[65] The Cartesian scholars' misuse of hypotheses is as bad as the negation of hypotheses by the Newtonian scholars. "M. Newton, and above all his disciples, have fallen into the opposite excess: disgusted with suppositions and errors that they found filled books of philosophy, they rose up against hypotheses".[66] Newton is not willing to address anything beyond that which can be scientifically or mathematically proven. The basis of these calculations and their relation to reality remains open; Newtonian scholars, did not want to go beyond his mathematical description. "One of the mistakes of some philosophers of our time is to want to banish hypotheses out of physics".[67] However, without hypotheses, she says, science is not possible; to ban them is "to delay the progress of the sciences".[68] "Descartes and Newton divide the scientific world, but one has to acknowledge both; since both are also wrong; the Cartesians … and the Newtonians … !"[69] Numbers and mathematical laws are applied to reality by means of hypotheses and without hypotheses, "the application of the geometric principles of mechanics to physical effects … remains imperfect."

Science has to be built on hypotheses, although the "unintelligible jargon" of the scholastics made hypotheses the "poison of philosophy". Thus hypotheses became empty words. It is true that this 'babble' from the fields of natural sciences and philosophy must be avoided, and she agrees with the Newtonians who call them the 'poison of philosophy' if they are taken too literally. To differentiate between "good" and "bad" hypotheses she then explains how and why wrong hypotheses arise and how they differ from the good ones. It is necessary to understand this difference,

---

[64] "Therefore, the good hypotheses will always be the work of the greatest men. Copernicus, Kepler, Huygens, Descartes, Leibniz, M. Newton himself, have all imagined useful hypotheses to explain complicated and difficult phenomena". Du Châtelet. 2009. *Selected Philosophical and Scientific Writings.* 155. Inst1740/Inst1742Am Chap. IV § 71.

[65] Inst1740/Inst1742Am Avant-propos VI. Du Châtelet. 2009. *Selected Philosophical and Scientific Writings.* 119. Du Châtelet uses here the word "vraisemblance" which occurs in classical literature demanding that actions and events in a play should be believable; much more it seems that Du Châtelet's reference is to be understood in a philosophical interpretation. See Giambattisto Vico, that not truth but only the *verosimile*, is accessible to human knowledge. Otto. 1989. *Giambattisto Vico. Grundzüge seiner Philosophie.*

[66] Inst1740/Inst1742Am Chap. IV § 55. Du Châtelet. 2009. *Selected Philosophical and Scientific Writings.* 147.

[67] Inst1740/Inst1742Am Avant-propos VIII. Du Châtelet. 2009. *Selected Philosophical and Scientific Writings.* 121.

[68] Inst1740/Inst1742Am Avant-propos VII. Du Châtelet. 2009. *Selected Philosophical and Scientific Writings.* 120.

[69] Inst1740/Inst1742Am Avant-propos VII. Du Châtelet. 2009. *Selected Philosophical and Scientific Writings.* 120.

as many great thinkers fell into the mistake of starting with "an ingenious and bold hypothesis, which has some initial probability (vraisemblance)" and finally failed:

> "Most great men who have made systems provide us with examples of this failing. These are great ships carried by the currents; they make the most beautiful manoeuvres in the world, but the currents carries them away".[70]

How to Make Good Hypotheses

"Il est vrai que ... il reste encore bien des obscurités dans la Métaphysique". Proving knowledge is difficult and metaphysics is still obscure, says Du Châtelet. However, there is "a possible compass which is guiding hypotheses".[71] The main problem of hypotheses-based scientific reasoning is the necessity to agree that our knowledge of the world can never be complete, "it seems that there are some unknowns for which no equation could ever be found". But it is also true that the metaphysics has not yet been developed to such a high standard as mathematics or geometry: "il nous manque un calcul pour la Métaphysique".

> "We lack a system of calculation for metaphysics similar to that which has been found for mathematics, by means of which, with the aid of certain givens, one arrives at knowledge of unknowns".[72]

There is no archimedic point of departure for knowledge and phenomena do not lead to the discernment of necessary attributes. The modes of being are confused with essential attributes, she states. As a critique of Descartes, she cites his claim that substances could be determined only by reason. But she refuses knowledge could be gained only by experience. As her very special critique of these two metaphysical directions in Chapter 4 of her *Institutions,* she presents a theory of how knowledge may be acquired and justified, built on hypotheses: "Philosophers frame hypotheses to explain the phenomena, the cause of which cannot be discovered either by experiment or by demonstration".[73] The empirical approach failed, since every explanation about what has been observed is necessarily a hypothesis itself. This must not result in scepticism, as "Without doubt there are rules to follow and pitfalls to avoid in hypotheses". The first rule is that hypotheses cannot be

---

[70] Wrong reasoning with hypotheses, is, that "an ingenious and bold hypothesis, which has some initial probability (vraisemblance)," leads human pride to believe it, the mind applauds itself for having found these subtle principles and next uses all its sagacity to defend them. Inst1742 Avant-propos VIII. Du Châtelet. 2009. *Selected Philosophical and Scientific Writings.* 121.

[71] Inst1740/Inst1742Am Avant-propos XII. Du Châtelet. 2009. *Selected Philosophical and Scientific Writings.* 123.

[72] Inst1740/Inst1742Am Avant-propos XII. Du Châtelet. 2009. *Selected Philosophical and Scientific Writings.* 123.

[73] Inst1740/Inst1742Am Chap. IV § 56. Du Châtelet. 2009. *Selected Philosophical and Scientific Writings.* 148.

"in contradiction with the principle of sufficient reason, nor with any principles that are the foundations of our knowledge". The second rule is to have certain knowledge of the facts within reach, and to know all circumstances attendant upon the phenomena to be explained; those who do not follow these precautionary rules run the risk of seeing their hypotheses overthrown by new facts they had previously neglected.[74]

Hypotheses, Du Châtelet says, are not necessarily true and therefore require set methods to be used correctly. She describes how to make good use of them and how to correct them. Only "when one can hope to know the greatest number of circumstances attendant upon a phenomenon, then one can seek the reason for it by means of hypotheses, taking the risk of having to correct it or having to be corrected". Indeed, it was a very modern insight, when she concretized that "one experiment is not enough for a hypothesis to be accepted, but a single one suffices to reject it when it is contrary to it". A theory is strongly connected to its applicability, she often confirms. To underline her assumption, she supports her theses by some historic examples of relevance. For example, the hypothesis that the diameter of the sun, thought to be equal at all times of the year, was proven false by an experiment that showed that this indeed changes. And "from this observation, one can therefore conclude with certainty that the hypothesis, of which equality is a consequence is false".[75] Here the third rule comes into play, which states that even one rejection of an hypothesis may falsify the hypothesis. However, it is important to be observant. One rejection alone does not necessarily rebut the whole and complete hypothesis, but could affect only part of it. A hypothesis can be true in a certain part but false in another "then the part that is found to be in contradiction with experiment must be corrected".[76] This process of argumentation, refutation and verification is a necessary process to obtain as many of the relevant determinants of the facts in question as possible:

> "Thus, in making a hypothesis one must deduce all the consequences that can legitimately be deduced, and next compare them with experiment; for should all these consequences be confirmed by experiments, the probability would be greatest".[77]

---

[74] Inst1740/Inst1742Am Chap. IV § 61. Du Châtelet. 2009. *Selected Philosophical and Scientific Writings*. 151.

[75] Inst1740/Inst1742Am Chap. IV § 61. Du Châtelet. 2009. *Selected Philosophical and Scientific Writings*. 152.

[76] Inst1740/Inst1742Am Chap. IV § 65. Du Châtelet. 2009. *Selected Philosophical and Scientific Writings*. 152. She explains this with the example of Descartes' theory of vortex of fluid matter, by which Descartes was falsified because it could not be explained that bodies fall towards the centre of the Earth, as Huygens demonstrated, but one cannot conclude from this that the vortex of fluid matter does not cause the phenomenon of the fall of bodies at all!

[77] Inst1740/Inst1742Am Chap. IV § 66. Du Châtelet. 2009. *Selected Philosophical and Scientific Writings*. 153.

Consequences and the Case of Probability

The next step Du Châtelet takes is to refine the methodical and heuristic idea of hypotheses as related to probabilities. When she uses the concept of the *vraisemblance* in her *Avant-propos*, it is to evaluate hypotheses. Her concept of the verosimilie seems to support a certain realistic approach in her reasoning. Hypotheses are built on and must be in accordance with observations. There are some observations that are more reliable than others; it is this conformance to our observations and experiences that makes hypotheses reliable as "phenomena are very far from their causes".[78]

The relationship of phenomena to their causes and principles rests on "probabilities". As Newton's theory satisfied some of those relationships, it could be accepted and was successful. Astronomy was mostly built upon hypotheses, as "to calculate the path of the celestial bodies, astronomers had waited until the true theory of the planets had been found, there would be no astronomy now".[79] So we have a science of hypotheses, not a science of final truths![80] Hypotheses, then, are only probable propositions that have a greater or lesser degree of certainty, depending on whether they satisfy a more or less great number of circumstances attendant upon the phenomenon that one wants to explain by their means. And, as a very great degree of probability gains our assent, and has on us almost the same effect as certainty, hypotheses finally become truths when their probability increases to such a point that one can actually present them as a "certainty."[81]

Phenomena provide an arbitrary set of circumstances which do not necessarily constitute a scientific outcome. These phenomena are a necessary means to obtain knowledge and to prove assumptions within hypotheses which are related to each other by reason. It is the realm of probability that makes ideas truth like. Science is a process! It is the process of putting hypotheses together, connecting them with the goal to constitute truth like knowledge. Du Châtelet's idea of a historic and growing body of knowledge refers to the hypotheses as an instrument used to change phenomena into reliable probabilities. Hypotheses must then find a place in the sciences, since they promote the discovery of truth, for when a hypothesis is

---

[78] "Les veritables causes des effets naturels & des Phénomènes que nous observons, sont souvent si éloignées des principes sur lesquels nous pouvons appuyer. qu'on est obligé de se contenter de raisons probables pur les expliquer: les probabilités ne sont donc point à rejetter dans les sciences". Inst1742Am Chap. IV § 53. Du Châtelet. 2009. *Selected Philosophical and Scientific Writings*. 147.

[79] Inst1740/Inst1742Am Chap. IV § 57. Du Châtelet. 2009. *Selected Philosophical and Scientific Writings*. 149.

[80] Inst1740/Inst1742Am Chap. IV § 62: "degré de probabilité" and § 67 "degré de certitude"; Dieter Suisky thus stated her theory close to a theory of probability, comparing her convictions to the theory developed by Leonard Euler. Suisky. Forthcoming. Emilie Du Châtelet und Leonhard Euler über die Rolle von Hypothesen.

[81] Inst1740/Inst1742Am Chap. IV § 57. Du Châtelet. 2009. *Selected Philosophical and Scientific Writings*. 154.

once posed, experiments are often done to ascertain if it is a good one, experiments which would never have been thought of without it. Experiments confirm hypotheses about phenomena and all the consequences drawn from them, and if these agree with the observations, the probability increases. Thus we come closer to truth, since a highly probable truth is "almost equivalent to a demonstration."[82]

Science as a Process

"Thus, it is evident that it is to hypotheses first made and then corrected that we are indebted for the beautiful and sublime knowledge of which astronomy and its subsidiary sciences are filled at present. It is impossible to see how men could have arrived there by other means."[83] Hypotheses are the basis for preliminary scientific explanations, the scaffolding needed for any form of construction! And, as is generally the case with scaffoldings for buildings, they also supersede their initial use and lose overall significance once the building is finished. Hypotheses are of good use where reasons are still unknown to the scientist.[84]

Newton is not an incomparable thinker, outstanding and singular; he learned from Kepler. Huygens and Leibniz learned from Descartes and Galilei. Thus, it is not true that Newton's theory surpasses and is incomparable to earlier scientific insights. Much more, it is built on the wisdom of the scientific ancestors. In this context, again Du Châtelet opposes Voltaire's criticism of Descartes, claiming that also the Cartesian thoughts had in fact enlightened the universe.[85]

It would be a misunderstanding to think of Du Châtelet's theory of science as historic, as if science were a continuous historical process, but this is not true at all. Her analytic approach focuses on the necessity of the highest probability of related phenomena, which makes it obvious why Du Châtelet could admire and agree to Newton's law of attraction. We crest the summits of truth, she says, standing on the shoulders of Descartes and Galileo. Newton has profited from their work.[86] But the reason for her admiration is completely different from that of the Newtonian scholars who speak of its singularity. She cannot fall in line with Newton's entourage of popular admirers and idolatrizers as she accepts the Newtonian theses as far as they fit into the scientific context. To her, these kinds of metaphysical insights represent an important aspect of scientific research and, in turn, they also constitute an important factor in a progressive approach to a growing body of knowledge. One theory refers to another, as "each philosopher sees something and no one sees everything".

---

[82] Inst1740/Inst1742Am Chap. IV § 58. Du Châtelet. 2009. *Selected Philosophical and Scientific Writings*. 149.

[83] Inst1740/Inst1742Am Chap. IV § 58. Du Châtelet. 2009. *Selected Philosophical and Scientific Writings*. 149.

[84] Inst1740/Inst1742Am Chap. I § 10.

[85] Inst1740/Inst1742Am Avant-propos VII. See Klens. 1994. *Mathematikerinnen im 18. Jahrhundert*. 112: "Mit Descartes wendet sie sich gegen eine ausschließlich empirisch verfahrende Wissenschaft".

[86] Inst1740/Inst1742Am Avant-propos V. Du Châtelet. 2009. *Selected Philosophical and Scientific Writings*. 118.

There can be nothing in this world which is completely true, and nothing which is completely false. "Chaque Philosophe a vu quelque chose, & aucun n'a tout vu". There is not a book which is so bad that nothing can be learned from it, and no book is so good that one might not improve it. There are sound ideas in Aristotle, right beside the greatest absurdities, and reading Newton's questions at the end of the *Opticks* "I am struck with a very different astonishment".[87] One philosopher is not right in every respect, as well as it is not true that every scientist or philosopher starts from the beginning: "The greatest philosopher may well add new discoveries to those of others, but once a truth has been found, he has to follow it".

Scientists plan, build and construct upon each other and science itself is comparable to this architecture. It is a building to be constructed, puzzles to be solved, but not heavenly truths to be discovered. This process of doing science is not linear and the findings are not laid down by a God to be picked up by the discoverer. Doing science and gaining knowledge is a process and by no means a one-way exercise. It is not guaranteed to be continuously successful nor is it possible to move forward continuously. One may bring a single stone; another may build a complete wing. No one contributes in the same way, but everyone must refer to certain stable insights, to metaphysical and universal principles while constructing this structure of knowledge. There is no end to this process, although there are results. Metaphysics is the ridge of the building, the most excellent, but often unclear and confused top, not always easy to be seen or found, yet always the most important part to view.[88] Truth is an issue of metaphysics, however we may approximate and get partly knowledge. Du Châtelet imagines herself in the role of an architect, with an eye to the fact that the blueprint must be planned using a much broader concept than "only clay and brick". She defines knowledge in the scientific sense. Science is established through the work of many over time, using different theories; truth needs to be established within this net of related theories that result in their specific probability, nearly as sure as demonstrations.

An Open System

"If you have reason, do never believe someone because of his words, examine always on your own!" Du Châtelet writes, and "[d]o not subordinate to the meaning of the most, and do not idolatrise the ideas of the Greatest".[89] This particular constraint to the autonomy of reasoning allows Du Châtelet quite a critical view of the history of science and the results of the research done up to that point. The indispensability of

---

[87] Inst1740/Inst1742Am Avant-propos X. Du Châtelet. 2009. *Selected Philosophical and Scientific Writings*. 122.

[88] Inst1740/Inst1742Am Chap. I § 53. "Les veritable causes des effets naturels et des Phénomenes que nous observons, sont souvent si éloignées des principes sur lesquels nous pouvons nous apuyer, et des Expériences que nous pouvons faire, qu'on est obligé de se contenter de raisons probables pour les expliquer: les probalités ne sont donc point à rejetter dans les sciences ..., mais ... elles frayent le chemin qui mène à la verité".

[89] Inst1740/Inst1742Am Avant-propos VI. Du Châtelet. 2009. *Selected Philosophical and Scientific Writings*. 119.

experience is "le bâton que la Nature a donnée a nous autres aveugles".[90] Here she willingly concedes experience to be the staff of knowledge and the principle that bans the mere reasonings of the scholastics. Yet in her Mandeville introduction she prepared her argument out of her own experience and, in line with this she admonishes her son to trust in his own capacity.[91] Experience based on the reliability of universal principles leads to new insights. It opens the possibility of detecting unforeseen connections and also mistakes: "à en tirer de nouvelles connaissance et de nouvelles lumières."[92] Du Châtelet's concept of experience is quite different from that of John Locke's philosophy. Her metaphysically based way of doing science is tied to experience, and the experiences are tied back into metaphysics: "So making hypotheses is allowed, and it is even very useful in all cases when we cannot discover the true reason for a phenomenon and the attendant circumstances, neither a priori, by means of truths that we already know; nor a posteriori, with the help of experiments".[93] Hypotheses are the chosen instruments for approaching knowledge, as neither a priori nor a posteriori knowledge is sufficient for acquiring a more extensive understanding. Experience itself is not the archimedic starting point for gaining knowledge; it is a means to accomplish the a priori. While the old metaphysicians, as Voltaire says about Descartes, "(were) systematically wrong, and with logical coherence", Du Châtelet drafts a metaphysics which is based on a priory principles that are meant to guide experience! She rebuts the skepticism of Pyrrhoniens and the relativistic approach of Locke and Mandeville. The path she takes between empiricism and rationalism makes her a forerunner to Kant's scientific metaphysics, as had even been noted at that time.[94] Emilie du Châtelet adheres to a realistic ontology and applies a critical realism. Her idea of truth is built upon a metaphysical construction in which a priori knowledge and experience are bound together.

Numbers and Hypotheses

We noted Voltaire's derision of Descartes' mathematical idealism and we have seen that Du Châtelet thought that numbers and mathematical laws had to be understood as hypotheses. Strong metaphysical convictions, such as the principle of contradiction,

---

[90] Inst1740/Inst1742Am Avant-propos XI. Du Châtelet. 2009. *Selected Philosophical and Scientific Writings*. 122 f.

[91] Inst 1740/Inst1742Am. Avant-propos I.

[92] Inst1740/Inst1742Am Avant-propos IX. "à en tirer de nouvelles connaissance et de nouvelles lumières".

[93] Du Châtelet. 2009. *Selected Philosophical and Scientific Writings*. 151. "Il est donc permis, & il est même très utile de faire des hipotheses dans tous les cas, où nous ne pouvons point découvrir la véritable raison d'un Phénomene & des circonstances qui l'accompagnent, ni à priori, par les moyen des vérités que nous connoissons déjà; ni à posteriori, par le secours des Experiences". Inst1740/Inst1742Am Chap. IV § 60; see also § 56.

[94] Eberhard. 1789. Über den wesentlichen Unterschied der Erkenntnis durch die Sinne und durch den Verstand. 300.

are necessary to allow us to combine insights and hypotheses, to form a chain of facts, since otherwise no "truth", no "numbers" would be possible and "two and two could then make as well 4 as 6" if a priori principles did not exist. Du Châtelet's argument on numbers is based on her theory of knowledge: the truth of numbers and mathematical operations results from hypotheses: "Il en est de meme dans les nombres"! Numbers are based on hypotheses. "Division ... is founded on hypotheses only" as the whole operation is done by means of hypotheses. Descartes' belief, and before him Galilei's, that numbers are of ideal certainty, is not true. Her hypotheses-based metaphysics embraces all sciences. Besides the a priori principle and the a posterori experiences, our knowledge, even our mathematical knowledge, is built on them.[95] It is this conviction which marks the separation between her and her former friend and teacher, Maupertuis, as the allusions in the *Institutions* and her letters inform us and as will be shown in the following.

Whereas Maupertuis is convinced that mathematical laws demonstrate truth, Du Châtelet holds that numbers are hypothetical. As good hypotheses are not to be confounded with truth, this is also true for all insights which are gained through numbers and experiences.[96] This is a very important point which will have an impact when Du Châtelet argues against scientific justification by means of mathematical laws and their application to reality. Numbers work similarly to essential attributes: they can be combined but their connection is not necessary but hypothetical![97] The fact that the law of attraction can be expressed mathematically is not sufficient to prove the truth of it, as Du Châtelet will outline in her correspondence against Maupertuis. At this stage it can be seen that Du Châtelet's reflections on hypothetical numbers can be traced back to her critique of John Locke's theory of experience and history. Locke acknowledged a stronger certainty for mathematics than for ideas resulting from experience.[98] Locke's empirical approach leads to a skeptical view. It is quite probable that Du Châtelet had reflected on these problems intensively and thus developed her repudiation of his arguments. She explicitly rebuts skepticism. Her concept of what science is and how knowledge is gained is quite contrary to that of John Locke. According to Du Châtelet, it is possible to obtain truth, as strong hypotheses are "nearly as convincing as demonstrations"; experiences, however, are not the only means. For Du Châtelet there is a threefold way to truth: the a priori knowledge "par le moyen des vérités que nous connaissons déjà", the a posteriori "par le secours des Expériences" and the "good use of the mind", by hypothetical reasoning. The truth she refers to

---

[95] Inst1740/Inst1742Am Chap. IV § 60.

[96] Inst1740/Inst1742Am Chap. IV § 59. Du Châtelet. 2009. *Selected Philosophical and Scientific Writings*. 151.

[97] Inst1740/Inst1742Am Chap. IV § 46.

[98] Locke. 1975. *An Essay Concerning Human Understanding* IV. 12, 10. "This way of getting and improving our knowledge ... only by experience and history, which is all ... we ... can attain to, makes me suspect that natural philosophy is not capable of being made a science. We are able, I imagine, to reach very little general knowledge ...".

must be exhausted by a scientific process. It tests probabilities and stabilizes itself within this process. Knowledge can be true, however it is only partly true and never complete.

Advocates and Adversaries

With his *Letters on England* Voltaire gave quite an appropriate presentation about the "Islanders and Continentalists", who live, according to Voltaire, in different universes. Many witnesses can be found to testify to the apparently insurmountable gap that existed during that period.[99] It is this sort of "national prejudice" which played a certain role in the Leibniz-Newton debate of that time, and against this Du Châtelet directs her introductory remarks.[100]

Du Châtelet presents herself as an enlightened philosopher, stating clearly that in the realms of science the quest for truth does not allow for ethnocentric reasoning. Consequently, Du Châtelet argues her ideas along the lines of universal principles, regardless of their English or French origins. This is one aspect Du Châtelet is praised for in Germany when her *Institutions* are recognized. In her *Eloge,* the German author Luise Gottsched emphasizes that Du Châtelet has no nationalistic prejudices.[101] It is especially remarkable that Luise Gottsched refers to Du Châtelet as a woman, "thinking with the power of a man", escaping all prejudices and finally being more "useful to her nation than a thousand men" on her way to truth.[102] Gottsched also refers to Du Châtelet's theory of science as a process, pointing out that it was not only Newton, but also his many predecessors such as Kepler, Galilei

---

[99] Thomas Burnett, to whom Leibniz had sent his Locke-critical Nouveaux Essais confirmed this view. Burnett replied: "Me thinks that we live in harmony with men from Germany because they don't know our work and we don't read their books". Cited in Specht. 1989. *John Locke.* 20.

[100] Inst1740/Inst1742Am Avant-propos VII. Du Châtelet. 2009. *Selected Philosophical and Scientific Writings.* 119. "Guard yourself ... whichever side you take in this dispute among the philosophers, against the inevitable obstinacy to which the spirit of partisanship carries one: this frame of mind is dangerous on all occasions of life; but it is ridiculous in physics."

[101] Eloge to The Marquise du Châtelet: "Den nenn ich noch sehr klein Der bloß durchs Vaterland berühmt und groß will seyn"; see MaiCh1741D. Maglo. 2008. Mme Du Châtelet, l'Encyclopédie et la philosophie des sciences. 255–266. He underlines the important role of Maupertuis und Du Châtelet within the virulent nationalism of the Newton debate. Differently argues Iverson. 2008. Emilie Du Châtelet, Luise Gottsched et la Société des Alétophiles: une traduction allemande de l'échange au sujet des forces vives. 289: "deux petits textes qui donnent à l'ouvrage un caractère fortement nationaliste".

[102] „Frau, deren kühner Geist mit Männerstärke denkt (…)/vernimm von deutscher Hand ein wahrheitsliebend Lied,/das, so wie Du gethan, die Vorurtheile flieht …"(…) „Du, die du jetzt den Ruhm des Vaterlandes stützest/Frau, die Du ihm weit mehr als tausend Männer nützest/Erhabne Châtelet, o fahre ferner fort/Der Wahrheit nachzugehn./Sie hängt an keinem Ort". MaiCh1741D. Partly translated as "You, who now uphold the fame of your own native land/Woman! You who serve your country better than a thousand men/Sublime Du Châtelet! Oh, continue ever forward/In chasing after truth. It is confined to no one place/" In: Iverson. 2006. Du Chastelet's portrait in the 'Bilder-Sal'. 58.

and Descartes who had repeatedly contributed to the world of science. She emphasises the fact that Du Châtelet exemplifies the favourable aspects of scientific research and is by no means afraid to follow what she preaches. The fact that both Emilie du Châtelet's and Louise Gottsched's pictures are portrayed in the monumental book of the "most famous writers of the epoch", the *Bilder-Sal*, is further proof that it was especially this particular circle of German academics who supported Du Châtelet's work.

It was not at random that the German intellectuals who suffered from the Leibniz-Newton struggle applauded Du Châtelet's Newtonian physics, which were presented within a Leibnizian metaphysical framework.[103] In addition to Du Châtelet's unprejudiced handling of scientific issues, it is her metaphysical statement, her strong arguments in favour of hypotheses, that position her as a member of this "continental" tradition. Leibniz described the natural sciences as a hypothetical discipline. Du Châtelet's defense of hypothetical scientific reasoning puts her in line with Leibniz's thoughts in his *Hypotheses physica nova* from 1671. Christian Wolff, a member of the St. Petersburg Academy of Sciences and of the *Académie des Sciences* in Paris, representative of the Leibniz tradition in Germany, worked on hypotheses, reformulating Leibniz's thoughts, and like Du Châtelet, was also in line with a more empirical approach to Newton's work. Wolff acknowledged Du Châtelet's approach.[104] The philosophers' circle surrounding Wolff realised that Du Châtelet's Newtonianism opposed Voltaire's views in many ways. Wolff's letters mentioning Du Châtelet's *Institutions* dating from 1739 state his approval of her work, and his hope that Du Châtelet might well be on her way to confronting French materialism and rescuing metaphysics single-handedly.[105] His friend von Manteuffel remains somewhat sceptical, but admits that regarding her criticism relating to Voltaire's point of view, Du Châtelet not only traces Voltaire's thoughts, but also uncovers his errors and surpasses

---

[103] For more information on the Leibniz-Newton debate see Lefèvre. 2001. *Between Leibniz, Newton, and Kant. Philosophy and Science in the Eighteenth Century.*

[104] "Die sinnliche Erkenntnis der Tatsachen ist die Rechenprobe für die Begriffsentwicklung der Vernunft" ('The sensual realisation about the nature of things is the calculation for any form of rational terminolgy'). Arndt. 1983. Rationalismus und Empirismus in der Erkenntnislehre Christian Wolffs. 33; see his commentary on Wolff. *Vernünftige Gedanken von den Kräften des menschlichen Verstandes.* Chap. 5 § 5: Wolff seems to have been of the opinion that the empirical value of Leibniz' theses of pre-established harmony and the unconscious perceptions of the monads as a mirror image of the universe actually equalled zero, since it was not falsifiable in principle, which in itself poses the subtle missing part of the hypotheses.

[105] "In Frankreich reißet der Deismus, Materialismus und Scepticismus auch gewaltig und mehr ein, als fast zu glauben stehet, und es wäre gut, wenn die vortrefflich gelehrte Marquise gleichfalls das Instrument sein könnte, wodurch diesem Übel mittelst meiner Philosophie abgeholfen würde." Droysen. 1910. Die Marquise Du Châtelet, Voltaire und der Philosoph Christian Wolff. 227. The letters Droysen mentions have been lost in the war, which was pointed out by Besterman. Also note that the correspondence between Wolff and Manteuffel is full of references to Du Châtelet from the summer 1739 onwards. LetChBI 215 comments on her letter from June 1st, 1739.

Voltaire's clarity of ideas: "son ami Voltaire, qu'elle surpasse en cent piques dans la justesse et la netteté des idées".[106]

It was also Wolff's closer circle who finally proved responsible for the German translations of Du Châtelet's work.[107] Wolff still thought highly of her; his praise becomes even broader when he speaks of the second edition of the *Institutions*, which he thought to be even clearer and more coherent than the first one.[108] As Droysen has shown, it was no other than Frederick the Great who had a hand in this fortunate acquaintance of Wolff and Du Châtelet.[109] The crown prince's affinity towards Wolff's work and the translation of his metaphysics, which he had sent to Cirey, acquainted Du Châtelet with Wolff, although Du Châtelet proved to be somewhat skeptical towards Wolff's work: "Je connais mr Wolff pour un grand bavard en metaphysique", she writes in September 1738, making at least partially clear that his kind of philosophy belonged to the kind of void words of the scholastics. An identification of her metaphysical position with that of Wolff, as Wolff himself articulated, was not evident to her.[110]

However, Du Châtelet's work did not generally meet with unanimous approval. Jean Deschamps disagreed with Voltaire's Newtonianism and praised Du Châtelet's approach.[111] Wolff refrained from corresponding with Du Châtelet

---

[106] Von Manteuffel in his letter from June 6th to Christian Wolff: "Elle rapporte même, mais sans nommer le masque, les propres paroles d'un endroit de la être que son ami Voltaire a publiée sur ce sujet et en montre solidement et poliment la fausseté. Enfin si tout le livre répond à ce que nous en avons lu, il rendra plus de service à la vérité et à la philosophie que tout ce qu'on eût pu faire et écrire en Allemagne, pour en étaler l'évidence et l'utilité ... ce qu'il y a de sûr, c'est qu'il saute aux yeux qu'elle a renoncé, comme elle vous l'avait mandé il y a quelque temps, à toutes les chimères de son ami Voltaire, qu'elle surpasse en cent piques dans la justesse et la netteté des idées." All citations taken from Droysen. 1910. Die Marquise du Châtelet, Voltaire und der Philosoph Christian Wolff. 230 f.

[107] LetChBII 281: September 22nd, 1741. "Je vous supplie aussi de vouloir bien y joindre un exemplaire de cette traduction allemande de ma dispute sur les *forces vives*, dont je vous ai l'obligation. Mon fils apprend l'allemand, et je la lui ferai lire pour s'y fortifier. Dès que l'édition que l'on fait en Hollande des institutions physiques sera finie j'aurai l'honneur de vous en envoyer un exemplaire pour vous, et un pour celui qui veut bien avoir la bonté de les traduire."

[108] "Ich verwundere mich über die Deutlichkeit, damit sie auch die subtilsten Sachen vorträgt. Wo sie von dem redet, was ich in meiner Metaphysik vorgetragen, ist es nicht anders, als wenn ich mich selbst in Kollegiis reden hörte…" Letters from Christian Wolff, May 7th, 1741 and June 14th, 1741. In Droysen. 1910. Die Marquise du Châtelet, Voltaire und der Philosoph Christian Wolff. 233.

[109] See Droysen. 1910. Die Marquise du Châtelet, Voltaire und der Philosoph Christian Wolff; Böttcher. 2008. La reception des *Institutions de physique* en Allemagne. 243–255.

[110] LetChBI 146: September 29th, 1738. She had read Wolff before and cites a passage from it to Maupertuis: LetChBI 140: September 3rd, 1738.

[111] Deschamps. 1742. Cours abrégé de la philosophie Wolfienne en formes de lettre. See Droysen. 1910. Die Marquise du Châtelet, Voltaire und der Philosoph Christian Wolff. Note 25. Kölving. 2008. Émilie Du Châtelet devant l'histoire. 5.

whenever he thought that she had turned away from Leibniz and towards the philosophy of Maupertuis and Clairaut.[112]

It is necessary to analyse Du Châtelet's importance for the German Enlightenment from various aspects.[113] Despite Voltaire's remark in the *Eloge historique*, Du Châtelet had not abandoned Leibniz's ideas.[114] Dieter Suisky's contribution in this volume suggests that Leibniz's influence was by no means negligible when considering her translation of Newton's *Principia*[115]. It is quite unfortunate that Du Châtelet could not afford the time to participate in the Berlin Academy's competition on Leibniz's Monads, given that, due to those circumstances, her contribution would certainly have left an inspiring impression, as she wrote to her friend Bernoulli "Je serais bien tentée de défendre les monades mais je n'ai point de temps"; "Je crois que les monades sont un des fondements de la saine métaphysique".[116] It would have been interesting to hear Du Châtelet's thoughts on the monads, even more so as La Mettrie, who otherwise remained highly skeptical and sarcastic towards Du Châtelet's efforts to promote Leibniz's ideas in France, surprisingly notes that in this particular case of the monads Du Châtelet at last refrained from following Leibniz's thoughts too closely!

Some arguments allow to assume that Voltaire and Du Châtelet were both familiar with the famous *Leibniz-Clarke correspondence,* possibly as early as the summer

---

[112] Böttcher. 2008. La reception des *Institutions de physique* en Allemagne. 248 ff.

[113] It is also noteworthy that Immanuel Kant's first publication in 1747 *Thoughts of the True Estimation of Living Forces* is dedicated to the problem of *living forces*, taking up Du Châtelet's dispute with Dortous de Mairan.

[114] "Ainsi, apres avoir eu le courage d'embellir Leibnitz, elle eut celui de l'abandonner". Voltaire. Préface historique. PrincChat1756.

[115] "Du Châtelet presented Newton's axioms completed by notions stemming from Leibniz's theory and, moreover, by those that Euler had introduced", Suisky writes. "Du Châtelet treated Newton's Laws as *Lois generals du movement*, i.e. preferentially for motion. Moreover, Du Châtelet interpreted Newton's 2nd Law in terms of Leibniz's principle of sufficient reason". Suisky refers to Inst1740 Chap. IX § 229; cited from Suisky. Leonhard Euler and Emilie du Châtelet. On the post-Newtonian development of mechanics. In this volume. Gardiner claims that a number of passages and apparently 'wrong' wordings indeed show her work to be indebted to Leibniz. "On peut citer bon nombre de passages semblables … où les traductions volontairement erronées de Mme Du Châtelet servaient à … introduire d'autres idées fondamentales de la cosmologie leibnizienne mais carrément étrangères à celle de Newton". Gardiner. 2008. Mme Du Châtelet traductrice. 170 ff. She compares the publication of the Newton commentary with the manuscript and presents amazing results: "On peut citer bon nombre de passages semblables, surtout dans les Défintions et les Scholies qui constituent la partie la plus philosophique des Principes, où les traductions volontairement erronées de Mme Du Châtelet servaient à éliminer la possibilité du vide, redéfinir les concepts de la matière et de la force, et introduire d'autres idées fondamentales de la cosmologie leibnizienne mais carrément étrangères à celle de Newton", ibid. 172, with a reference to Biarnais. 1982. *Les Principia de Newton*.

[116] LetChBII: January 8th, 1746; September 6th, 1746.

of 1734.[117] Maupertuis or some other English friend could have recommended the reading to Voltaire, to which he freely refers in his *Eléments* and in his *Metaphysics of Newton*. Voltaire remained an ardent follower of Newton's ideas and claimed Clarke's voluntaristic interpretation.[118] Maupertuis also seemed to be satisfied with the scientific value of Newton's law of attraction, since its mathematical application corresponds to its scientific outcome. According to Voltaire, God is both the source of truth and the initiator of the 'why' in the knowledge of the physical world, ignoring and partly disapproving of Du Châtelet's endeavour for a more reasonable theory of the validity of scientific findings. Thus in his *Eléments*, Voltaire seems to provoke Du Châtelet's theory of "it could have been otherwise" in a fictitious dialogue: "Why do the planets move from east to west, and not the other direction?" Instead of assuming that our knowledge of physics is dependent on the information available and the reasons we have to argue in favour of something, Voltaire repeats Clarke's arguments, saying that not everything provides a reason for its particular state in the world.[119] Voltaire openly contradicts Du Châtelet's concept in the *Institutions*, as Du Châtelet repeatedly states that referring to God as a reason for scientific explanations does not yield any valid scientific results. God can only be a final reason, a reason behind all being, but not a scientific argument. Science which calls on God to provide the explanations it requires, effectively loses its academic standing. Without leaving any doubts as to her own argumentation, she states: If the reason for scientific phenomena lies with God, then neither Newton nor the natural sciences can provide any valid predictions about the nature of the world. The week reasonability of Newton's grounding of attractional force and its reference to God as reason of conservation becomes a challenging element of her methodology.[120]

Emilie du Châtelet was part of an interesting atmosphere of mixed Leibnizianism and Newtonianism that is proved by her contact to Maupertuis and the Bernoullis.[121]

---

[117] See Voltaire's remarks in 1968–1977. *Complete works*. D 764: June 27th, 1734; D911: September 11th, 1735, D978, January 3rd, 1736, D985: January 10th, 1736; The letter Voltaire included in his tragedy, Alzire, also provides conjectures about his earlier knowledge of Leibniz's letters to Clarke; "qui sait se fortifier avec Locke, s'éclairer avec Clarke et Newton ... qui (profite) publiquement des lumières des Maupertuis, des Ráumur, des Mairan, des Dufay et des Clairaut (…)" Kölving. 2008. Émilie Du Châtelet devant l'histoire. 2. Sarah Hutton argues differently in this volume.

[118] The reception of Newton as a religiously inspired writer during this period, see Hutton/Force. 2004. *Newton and Newtonianism*. See also Newton. 1999. *Principia Mathematica*. 483: "et haec de Deo; de quo utique ex Phaenomenis disserere, ad Philosophiam Experimentalem pertinet".

[119] Voltaire. 1997. *Elemente der Philosophie Newtons. Verteidigung es Newtonianismus. Die Metaphysik des Neuton*.

[120] See Inst1742Am Chap. I § 10. Du Châtelet. 2009. *Selected Philosophical and Scientific Writings*. 131.

[121] One even could name Fontenelle, who followed the invitation to the house of Breteuil, wrote on Leibniz; Fontenelle had written a well-known appreciation of Leibniz's work in 1717, which she possibly had opportunity to become familiar with. Towards Fontenelle she is quite sceptical. She gave Algarotti the name "le sange de Fontenelle" and wondered about Algarotti's dedication of his *Neutonianismo per le Dame* to Fontenelle in 1738. See also in Zinsser/Hayes. 2006. *Emilie du Châtelet: rewriting Enlightenment philosophy and science*. 28, 30. See also: LetChBI 135: Feb 17th (or August), 1738. Interesting instructions on this topic can be gained from: Hecht. 1994. Pierre Louis Moreau de Maupertuis oder die Schwierigkeit der Unterscheidung von Newtonianern und Leibnizianern. 331–338.

Maupertuis was an influential friend of Voltaire and Du Châtelet and his acquaintance with Leibniz's work dated back to his studies with the Bernoulli family in Basel.[122] Johann I Bernoulli wrote a famous *Discours* cited by Du Châtelet, which includes a study dedicated to the *vis viva*, a term that was coined by Leibniz in an effort to challenge the Cartesian theory of the conservation of motion.[123] Some years after this, Johann I Bernoulli moved in to support Maupertuis' work on the *Discours sur les differentes figures des astres*, which entailed a juxtaposition of Cartesian and Newtonian ideas and their respective effects on the workings of our planet, an approach which would later lead to Maupertuis' famous Lapland expedition.[124] Du Châtelet herself had already read Maupertuis' work in late 1732 or in early 1733.[125] Maupertuis was evidently impressed by Du Châtelet's mathematical abilities and forwarded her several of his early manuscripts.[126] He was one of the 'modernists' who engaged in discussions with the English philosophers. In 1728, he met Samuel Clarke, the notorious defender of Newton's thoughts, during one of his trips to England. In the 1730s he had already gained quite a reputation as a member of the Paris Academy, and since his teacher was none other than Johann I. Bernoulli, friend and colleague of Leibniz, he had already been influenced by Leibniz before the publication of his own *Discours* in 1732.

Maupertuis, Voltaire and Du Châtelet shared a common respect for Newton's work, and they soon became what could be described as a group of modern 'Newtonians' who disagreed with many of the traditional metaphysical views of

---

[122] Maupertuis' teacher, Johann I. Bernoulli, won the grand prize of the Paris academy in 1730 with his *Discours sur les lois de la communication du movement*, in which he illustrates the consistencies of Kepler's third law of motion with Descartes' vortex theory. Bernoulli. 1742. *Opera omnia* III. See also: Tonelli. 1959. *La nécessité des lois de la nature au XVIIIe siècle et chez Kant en 1762*. 225–241.

[123] See the remarks of Washner/von Borzeszkowski Verteidigung des Newtonianismus. In: Voltaire. 1997. *Elemente der Philosophie Newtons*. 4. See footnote 120. For Du Châtelet's reception of *vis viva* See Iltis. 1977. Leibniz and the *vis viva* Controversy. 32. Walters. 2001. La querelle des *forces vives* et le rôle de Mme du Châtelet. 198–211. Rey. 2008. La figure du leibnizianisme dans les *Institutions de physique*. 231–243, and Reichenberger's contribution to this volume: Leibniz' quantity of force: a heresy? Emilie du Châtelet's *Institutions* in the context of *vis viva* controversy.

[124] Against Jacques Cassini's prediction of the world's shape to be prolate, Maupertuis based his reflections on Newton and by the help of his mentor Johann I. Bernoulli predicted that the earth was oblate. Louis XV. sent an expedition to Lapland which was led by Maupertuis. See Terrall. 2002. *The man who flattened the Earth – Maupertuis and the sciences in the Enlightenment*.

[125] It was probably about that time that she become a student of Maupertuis in mathematics. Later, in 1735, Alexis-Claude Clairaut (1713–65) became her mathematical instructor. See Zinsser and Courcelle. 2003. A remarkable collaboration: the marquise Du Châtelet and Alexis Clairaut. 107–120.

[126] "jay cru monsieur que pr etre digne de repondre à la lettre que vs mavés escrit il falloit vs auoir lu, iay été tres contente de vos deux manuscrits i'ay passé hier toute ma soirée a profiter de vos lecons…" see Voltaire. 1968–1977. Complete Works. D 696: Jan 1734. See also Vaillot. 1978. *Madame du Châtelet*. 81.

their contemporaries.[127] Du Châtelet's correspondence provides ample proof of this: "They don't want Newton to count in France, and they are afraid that Voltaire and Maupertuis working together may exert complete domination", she writes. "I should not be surprised if the Parliament launched a decree against them".[128] But even in this context we have the opportunity to observe Du Châtelet's intellectual distance to her companion, Voltaire. In a letter dating from 1735, she informs the Italian Newtonian, Algarotti, "J'ai une assez jolie bibliothèque. Voltaire en a une toute d'anecdotes; la mienne est toute (de) philosophie".[129] She is grateful towards Maupertuis and thanks him for the fact that his tutoring allowed her to support Voltaire's *Eléments* with some expertise: "My companion in solitude has written an Introduction to Newtons's philosophy which he dedicates to me. I have the advantage over the greatest Philosophers in having had you for my teacher."[130] The relationship between Voltaire, Maupertuis and Du Châtelet seems to be excellent at this particular point in time.

Leibniz's philosophy has always been considered to be a decisive factor in the majority of the interpretations of Du Châtelet's work. Du Châtelet's approach to Leibniz in her early years before the publication of the *Institutions* was, however, dark and doubtful, owing to the fact that Samuel König, who like Maupertuis was a student of Johann Bernoulli I. and a tutor of mathematics for Du Châtelet, famously accused her of having copied large sections of her *Institutions* from his excerpts of Leibniz's and Wolffs writings.[131] The preface of the first edition of the *Institutions* dating from 1740 admits this fact in part.[132] For the subsequent editions of the book, Du Châtelet deleted this reference. The question of how much and to what extent these accusations played a part in the overall public reception of her work at that

---

[127] Carboncini. 1984. Lumière e Aufklärung. 1297–1335.

[128] English citation taken from: Gooch. 1961. Mme du Châtelet and her Lovers I. Maupertuis. 652.

[129] LetChBI 44: October 10th, 1735.

[130] Citation taken from: Gooch. 1961. Mme du Châtelet and her Lovers I. Maupertuis. 652. Expecting Maupertuis' departure to Lapland, she writes to Algarotti: "Doubtless you know of the expedition of Maupertuis, of the beauty and precision of his operations which have exceeded his hopes. His trials are worthy of Charles XII. His reward has been persecution by the old Academy, Cassini and other Jesuits. They have persuaded silly folk that he does not know what he is talking about and half Paris, even three quarters, believes them. He has had endless difficulty about publishing the report of his journey, and I am not sure if he will succeed". Citation taken from: Gooch. 1961. Mme du Châtelet and her Lovers I. Maupertuis. Document. LetChBI 113: January 10th, 1738.

[131] Samuel König studied with Johann und Daniel Bernoulli at Basel University; in 1735 he studied with Christian Wolff in Marburg. In 1738 he met Du Châtelet in Paris and later went to Cirey on Maupertuis' recommendation in 1739. Samuel König's proof of the bees' efficiency earned him a membership in the Academy in Paris; he later went on to become a member of the Prussian Academy of the Sciences, of which Maupertuis was president at the time. Szabo Compare. 1997. Der philosophische Streit um das wahre Kraftmaß im 17. und 18. Jahrhundert.

[132] "Je vous explique dans les principales opinions de M. de Leibniz sur la métaphysique. Je les ai puisées dans les ouvrages du célèbres Wolff, dont vous m'avez entendu parler avec un de ses disciples, qui a été quelquetemps chez moi et qui m'en faisait quelquefois des extraits". Inst1740 Avant-propos XII.

time has not yet been investigated. It seems to be obvious that these kinds of accusations circulated mainly in the Berlin circle, with Maupertuis at its center. The relationship between Samuel König, who would later become his protégé, and Maupertuis has scarcely been taken into account in the interpretation of these facts. Even more, about 10 years later König also accused Maupertuis of plagiarising Leibniz. Maupertuis arranged that König came to Cirey as Du Châtelet's tutor. Du Châtelet, however, had asked to be tutored by Johann II. Bernoulli, the son of Johann I. Bernoulli.[133] In January 1739, Du Châtelet was still waiting for Johann II. From about the same period, we find a letter Bernoulli sent to his friend, Maupertuis, where he expresses surprise that Maupertuis had apparently neglected to instruct Du Châtelet carefully on the topic of the *living forces*.[134] This seems quite remarkable, given that Du Châtelet would later go on to become one of the most influential scholars on this topic and had thus far been seen as a diligent pupil of Maupertuis. The *living forces*, in fact, represent an important part of the argument she presents in her *Institutions*; they constitute a major part in her dispute with Mairan and her later reflections on it position her amongst the most influential figures of her times. But this negligent instruction is not the only fact that fosters a certain doubt towards Mauptertuis' relation to Du Châtelet.[135] In September 1739, Du Châtelet again asked Maupertuis to send Johann II. Bernoulli instead of König.[136] Bernoulli still seemed hesitant. Even though Du Châtelet eventually succeeded in securing a temporary positive reply and started organising Bernoulli's trip, the latter later cancelled at short notice.[137] Was it in fact Maupertuis who had an interest in keeping his friend, Johann II. from visiting Du Châtelet? Du Châtelet comments meaningfully to Bernoulli on this surprising news: "Je savais, monsieur, que les mêmes personnes

---

[133] "Vous m'avez donné un désir extrême de m'appliquer à la géométrie et au calcul. Si vous pouvez déterminer un de mrs Bernouilli à apporter la lumière dans mes ténèbres J'éspère qu'il sera content de la docilité, de l'application, et de la reconnaissance de son écolière". LetChBI 175: January 20th, 1739. More information on the relationships of Maupertuis, Du Châtelet and the Bernoullis is available in Nagel, see also footnote 136.

[134] "Je m'étonne, Monsieur, que depuis si longtemps que vous connoissez cette Dame philosophique, vous ne lui ayez pas donné de meilleures instructions sur cette importante matière". Letter dating from April 12th, 1739: Quoted in Iltis. 1977. Madame Du Châtelet's Metaphysics and Mechanics. 96.

[135] To Johann II Bernoulli she mentions in her wonderful style of attitude and irony: "Je vous avoue que les procédés de M. König me feraient hair les mathématiciens et les Suisses si je ne vous connaissais pas" LetChBI 203: September 15th, 1739. In March 1739, shortly after König's arrival in Cirey, Du Châtelet asks Johann II Bernoulli to enquire about a copy of the *Commercium epistolicum*, the letters which were exchanged between Johann I Bernoulli Father and Leibniz. LetChBI 203: March 30th, 1739. *Virorum celeberr. Got. Gul. Leibnitii et Johan. Bernoullii commercium philosophicum et mathematicum*, Lausannae & c. 1745. For more See Nagel. 2011. 'Sancti Bernoulli orate pro nobis'. Emilie du Châtelet's rediscovered *Essai sur l'optique* and her relation to the mathematicians from Basel. In this volume. König's and Du Châtelet's relationship is far from excellent and she repeatedly complains about the former. LetChBI 216: June 20th, 1739.

[136] LetChBI 223: Sept 28th, 1739; LetChBI 227: October 3rd, 1739: "Quant à Mr. Koenig il m'est bien difficile de vous en rien dire, sa conduite avec moi est un mélange de bien et de mal fort difficile à débrouiller".

[137] LetChBII 232: January 11th, 1740.

qui vous avaient proposé de venir chez moi vous en ont détourné".[138] These lines seem to provide evidence that she well knew who was behind the cancellation. In March of that same year, Maupertuis states in a letter to Johann II. Bernoulli how dangerous it could be to meet with her.[139] Although Bernoulli remains Du Châtelet's friend, he refrains from meeting her. The fact that Du Châtelet's correspondence with Maupertuis occurred much less frequently from then on and finally stopped completely, suggests a connection between these incidents. On August 2nd, 1740, she sent a letter to Bernoulli stating, "Je ne suis point en commerce avec mr de Maupertuis".[140] Twenty days later she resumed her correspondence with Maupertuis, only replying to a letter from him and angrily remarking that in Berlin people apparently thought König had dictated parts of the *Institutions* to her. She takes the occasion to demand him to set things straight and to state the truth: "On me mande de Berlin qu'il y passe pour constant que Koenig me l'a dicté. Je n'exige sur ce bruit si injurieux d'autre service de votre amitié que de dire la vérité"![141] Confirming her authorship would cost him nothing more than a word. After all, Maupertuis knew better than anyone else that she had drafted her text and the ideas and convictions it contained on her own, as it shows extensive consistency with the letters she had written to him. Matters of metaphysics had occupied their correspondence for years and Maupertuis could be surer than anyone else that her knowledge of Leibniz's ideas was by no means copied from elsewhere.[142] Yet Maupertuis remained silent.[143] As the above shows, there is ample reason to assume that Maupertuis was not dissatisfied with these occurrences and did not lift a finger to circumvent them. The reasons for this can be traced back to Maupertuis' intentions when he wanted König to teach Du Châtelet and detained Bernoulli.[144]

---

[138] LetChBII 233: January 20th, 1740.

[139] "Mme du Châtelet est une femme à qui il est dangereux d'avoir affaire". "Peu s'en faut qu'elle ne me menace... Je fus surpris de la voir tout d'un coup m'écrire d'un style fort fâche et fort extraordinaire ... Je ne regrette point d'être brouillé avec Mme Du Châtelet pour l'amour de vous ... Je ne regrette que de l'être"; Voltaire. 1968–1977. Complete Works. D 2190: March 27th, 1740.

[140] LetChBII 243: August 2nd, 1740 to Johann I Bernoulli: "Je ne suis point en commerce avec mr de Maupertuis. Il m'a fait un mal qu'il ne pourra jamais réparer et que je ne lui pardonnerai jamais en vous empêchant de venir ici, et j'ai été trop de ses amis pour conserver avec lui un commerce où il entrevoit beaucoup de méfiance et de froideur".

[141] LetChBII 246: August 21th, 1740; see also: LetChBII 252: October 22th, 1740.

[142] In order to see proof that this accusation is by all means false, see Gardiner. 1982. Searching for the metaphysics of science: the structure and composition of Madame Du Châtelet's *Institutions de Physique*, 1737–1740. Iltis. 1977. Madame du Châtelet's Metaphysics and Mechanics.

[143] Koenig to Maupertuis, "Le livre de mad. du Chàt. a enfin paru. Je vous avoue, Monsieur, qu'il faut avoir la rage d'écrire pour oser faire une folie de cette nature. On dit qu'on l'a déja refutée; je me réjouis de voir comment elle fera pour répondre sur des matières qu'elle n'entend point" sent from Bern, September 21st, 1740 see for more, in: Maupertuis. 1768. *Oeuvres*.

[144] Even Voltaire subsequently indicates that Maupertuis' conduct might have had a hand in the problematic nature of the situation: "C'est une chose déplorable qu'une Française, Mme Du Châtelet, ait fait servir son esprit à broder ces toiles d'araignée. Vous en êtes coupable, vous qui lui avez donné cet enthousiaste de König..." Voltaire. 1968–1977. *Complete works*. D 2526: August 10th, 1741.

This crisis in the relationship between Du Châtelet and Maupertuis from 1741 also relates to another incident. Du Châtelet had made several attempts at sending Maupertuis copies of her *Institutions,* which he claimed he never received. "J'espère que vous avez enfin les Institutions physiques", she wrote in June 1741.[145] It becomes clear that metaphysics was the cause of their differences. Her remark asking him to proofread at least those last 11 chapters, which did not touch the metaphysical issue illustrates this: "Je sens bien que vous me conseillerez de retrancher tout la métaphysique". Maupertuis did not acquiesce to her request. Her friendship with Maupertuis did not survive those troubled times. From 1741 on, Du Châtelet no longer mentioned anything personal in her very few and short letters to Maupertuis and later stopped writing to him altogether.[146] She remained in regular correspondence with Johann II Bernoulli until her death in 1749.

About 10 years after Samuel König had accused Du Châtelet of plagiarism, he provoked a new scandal, now accusing his mentor Maupertuis of having copied Leibniz's ideas in the formulation of his *Principe de la moindre action.*[147] In the following we will see that some of its ideas can be identified in Du Châtelet's correspondence. Both of these scandals can be attributed to a rather small social circle. Some remarkable events which occurred after her death in Berlin and which included Maupertuis, Voltaire and La Mettrie must be revised in this perspective.[148]

Defending Metaphysics

After Maupertuis' return from Lapland, Du Châtelet's letters to him bear witness to some extensive thoughts which focused on the topics of scientific methods and metaphysics. In these letters Du Châtelet expresses her ideas; her idea of physics and metaphysics in general, as well as her disagreement with certain proposals of Maupertuis' metaphysics. At the present, Maupertuis' and Du Châtelet's correspondence, on which I will focus in the following, is only available from Du Châtelet's side.

---

[145] LetChBII 273: June 26th, 1741.

[146] 1744 presents another letter; in 1741 Maupertuis was captured during the battle of Mollwitz, but was back in Paris in June 1741. Emilie du Châtelet wrote to him in Vienna during his imprisonment. The documents up to 1745 are missing; Maupertuis announced his marriage. Bessire. 2008. Mme Du Châtelet épistolière. 35. There you find an image capturing Du Châtelet's correspondence.

[147] In 1746 Maupertuis published his *Principe général* in the Memoires of the Academy. The mathematician Leonard Euler and King Frederick II. denounced the Leibniz letter as counterfeited, by which König wanted to prove Maupertuis's plagiarism. Szabo Compare. 1997. *Geschichte der mechanischen Prinzipien und ihrer wichtigsten Anwendungen.* 120.

[148] Voltaire subsequently exploited Maupertuis' compromised reputation by including him in his satire of *L'histoire du Docteur Akakia et du natif de Saint-Malo.* The second part of *Dr. Akakia* presents a fictitious truce between Samuel König and Maupertuis, which effectively sees them swearing to acknowledge and uphold German scientists such as Kopernikus, Kepler, Leibniz, Wolff, Haller and Gottsched.

Since the letters partially reflect Maupertuis' comments and also discuss or reply to his thinking, they show Du Châtelet's scientific development within the context of her arguments with him. The *Institutions* are often quite close to the material Du Châtelet discussed in those letters to Maupertuis, and they provide further evidence that Johann I. Bernoulli's and Mairan's memoirs meant much food for thought to Du Châtelet in 1738.

In 1738 both Maupertuis and Du Châtelet were exponents of Newtonianism and each of them refers to Leibniz's thoughts. I will reflect on their discussion of how stable scientific insights can be found, which is a major issue in Du Châtelet's letters. Reading the letters, it seems as if in the presence of Maupertuis she claims the "mastery" of true metaphysics for herself, since she is the one who is able to make judgements and who is able to decide when a metaphysical account is satisfied: "que la métaphysique serait contente"! It is she who starts discussing the reasonability of the two kinds of forces and it is quite renowned that it was she who took up again the Leibnizian idea of the *living forces*. They both embark on metaphysical reflections about these different kinds of forces. The conservation of forces and its impact on determination and free will are at stake. With the publication of the *Institutions*, their personal relationship seems to stop. Some years later in Maupertuis'*Lettres*, certain issues can be traced back to this correspondence. Maupertuis's recollection of Du Châtelet's arguments is, however, quite bipartison. Some of her reflections are taken into account; others are sharply criticized. However, we do not find one written reference to her.

At the beginning of 1738 Emilie du Châtelet started to familiarize herself with the theory of the *vis viva*.[149] After Maupertuis' return from his Lapland expedition she asks for his judgment. "As I have read now a lot about the *living forces*, I would like to know if you are in favour of Mr. Le Mairan or of Bernoulli!".[150] Mairan's arguments, which Du Châtelet alludes to here, were taken from the memoir he wrote for the academy in 1728, and it was her criticism of this particular work which found its way into chapter XXI of her *Institutions,* which would later initiate the famous pamphlet of Mairan and herself on this topic. Du Châtelet wants to know which side Maupertuis is on: whether he supports Mairan, who widely follows Newton, or whether he leans towards Bernoulli's position, which clearly argues in

---

[149] A wide range of literature on Du Châtelet's reception of *vis viva* exists. See Terrall. 2004. Vis viva revisited.

[150] "J'ai lu beaucoup de choses depuis peu sur les *forces vives*, je voudrais savoir si vous ètes pour mr Le Mairan, ou pour mr de Bernouilly" and: "Je n'ai pas l'indiscrétion de vous demander sur cela tout ce que je voudrais savoir, mais seulement lequel des deux sentiments est le vôtre". LetChBI 118: February 2nd, 1738. Her question refers to the memoirs of Johann I. Bernoulli and Dortous de Mairan. Du Châtelet refers to Bernoulli. 1727. Discours *sur les loix de la communication du mouvement*, Chaps V-IX. La mesure des *forces vives*, see Bernoulli. 1742. *Opera omnia*. III. 1–107. For more, see Heimann. 1977. Geometry and Nature: Leibniz and Johann Bernoulli's Theory of Motion. 1–26. For Mairan see: Mairan. 1741. *Dissertation sur l'estimation et la mesure des forces motrices des corps.* 1–50. See also: Inst1742Am Chap. XXI § 574. See LetChBI 124: May 5th, 1738.

favour of Leibniz and his *vis viva*. Her correspondence with Maupertuis in 1738 shows that she had already acquired the expertise necessary for such a dispute. As a closer look at the relevant chapters, especially XXI, shows, the *Institutions* drew much of their inspiration on the topic from these early letters. Du Châtelet moves on to analyse Bernoulli's and Mairan's memoirs, while at the same time going to great lengths in her readings of the *Acta Eruditorum of 1686,* where Leibniz explains the necessity of a different measurement for motion.[151] She turns again to the *Leibniz-Clarke correspondence*. Bernoulli's memoir is praised as one of the best works she had read in a long time, while at the same time, she criticizes Mairan for merely copying his arguments from Clarke.[152] In her opinion, Mairan treats Leibniz's ideas without using sound arguments, and yet with utter disapproval in all respects, be it in relation to force, the plenum or the monads. Her dictum that Leibniz may have erred in many things, but in the *living forces* he had discovered one of the greatest secrets of creation itself, is to be found in this context of her justification of Leibniz's position as opposed to Mairan's.[153]

Du Châtelet regarded herself as a critical reader of Leibniz's work and she illustrates that she by no means owed her expertise on the *vis viva* to Maupertuis. In her letter dated 10 February, 1738, we see just how delighted she is that they agree on the subject, but she makes clear that she was not waiting for and was not dependent on "Isaak Maupertuis'" answer to legitimize her own opinion. As she notes, she had already informed Mr. Pitot, whom she had the pleasure to "coincidentally" meet. The coincidence is of interest, as Mr. Pitot would become the 'Approbateur' of her *Institutions de physique* of 1738. At this point Maupertuis does not know that she is working on her book. Subsequently, Du Châtelet's letters provide vital evidence for the claim that it was she who had urged Maupertuis to re-address the matter of the *vis viva*.[154] Additionally, she did not make the mistake of misunderstanding Leibniz's ideas of the relationships between the forces. For her it was evident that "mv and

---

[151] Leibniz conceived of force as energy in motion - as kinetic energy - and measured it as mv2. Descartes and followingly Newton agreed, measured force as mv. With his view, Leibniz succeeded in characterizing the movement of mass in scalary measurements, which in turn interpreted the force of motion in bodies as partaking in the general conservation of energy.

[152] Compare LetChBI 124: May 5th, 1738.

[153] LetChBI 120: February 10th, 1738: "Mr. De Leibnitz à la vérité n'avait guère raison que sur les *forces vives*, mais enfin il les a découvertes, et c'est avoir deviné un des secrètes du créateur".

[154] This opinion is held by Klens. 1994. *Mathematikerinnen im 18. Jahrhundert*. 243. Also see her contradiction of Brunet's thesis that Maupertuis had convinced Du Châtelet of *vis viva*. Brunet. 1929. *Maupertuis*. 241. Hartmut Hecht supports that *vis viva* did not have the importance for Maupertuis that it had for Du Châtelet, referring to Beeson. 1992. *Maupertuis: an intellectual biography*. Johann II. Bernoulli's letter to Maupertuis, in which he rebuked the latter for instructing Du Châtelet insufficiently on the topic, appears meaningful in this light. See above letter dating from April 12th, 1739 from Johann Bernoulli to Maupertuis. "Je m'étonne, Monsieur, que depuis si longtemps que vous connoissez cette Dame philosophique, vous ne lui ayez pas donné de meilleures instructions sur cette importante matière", quoted in Iltis. 1977. Madame du Châtelet's Metaphysics and Mechanics. 46.

mv^2 were by no means the same".¹⁵⁵ As force makes itself visible by its ability to overcome obstacles, it can therefore not suffice to judge it only by the distance it has covered at a given point: If one were to assume distance only, this would invariably lead to confusing results, Du Châtelet argues.¹⁵⁶ Even those scientists who deny the value of the *living forces* acknowledge the existence of two different measures of force, she is convinced, wondering why they persistently argue against the importance of this kind of force, which is not the same as *dead force*. Mairan appears to be merely disagreeing to disagree: "qu'il combattait pour combattre".¹⁵⁷ It is obvious from her perspective that the acceptance of the *vis viva* does not necessarily jeopardise the validity of the Cartesian/Newtonian measurement. Here she confirms what other philosophers will repeat much later: this particular dispute originates primarily through words: "Au reste je crois comme vous, que ce n'est qu'une dispute de mots". Du Châtelet's analysis of the situation consequently goes ahead to d'Alembert's conclusion that the debate on these two measures of force is, in fact, a *dispute over words*.¹⁵⁸

Du Châtelet agrees with Bernoulli's argument that one has to distinguish between the quantity of motion and the quantity of force in order to obtain reliable results, as otherwise two different movements would sum up to something more than the sum!¹⁵⁹ Such an 'empty' assumption would not only lead to confusing results but also invert cause and effect, she concludes in chapter XXI of the *Institutions*: If one were to measure force by this kind of motion only, it could lead to an absurd conclusion, one in which even a *perpetuum mobile* appears conceivable.¹⁶⁰ To Du Châtelet, merely stating force in relation to overcoming distances in a certain time equals the description of force without clarifying its application or measuring the resistance it meets. She illustrates this with her demonstrative and pictorial methodology, linking the concept of force to that of wealth: A rich man, she says, is usually not judged

---

¹⁵⁵ Compare Klens. 1994. *Mathematikerinnen im 18. Jahrhundert.* 244 f. See also her comment 289: "Insofern ist das Urteil von Iltis 'Madame Du Châtelet did not adopt the validity of both measures of force in her mechanics' nicht zutreffend." See Leibniz. 1686. *Discours de métaphysique.* Sommaire § 18, where he points out that the distinction between force and the variables of motion is important and underlines that one has to be able to fall back onto metaphysical concepts … in order to explain the phenomena associated with bodies.

¹⁵⁶ "J'ai toujours pensé que la force d'un corps devait s'estimer par les obstacles qu'il dérangeait et non par le temps qu'il employait", LetChBI 120: February 10th, 1738.

¹⁵⁷ LetChBI 124: May 5th, 1738.

¹⁵⁸ See LetChBI 120: February 10th, 1738: "Au reste je crois comme vous, que ce n'est qu'une dispute de mots, que mr de Mairan aurait pu terminer tout d'un coup la dispute et qu'il semble qu'il ait voulu allonger". D'Alembert. 1743. *Traité de dynamique.* In Voltaire. 1997. *Elemente der Philosophie Newtons.* 45, note 120, where Wahsner/von Borzeszkowski inform that Kant did not realize this difference.

¹⁵⁹ LetChBI 120: Febr. 10th, 1738.

¹⁶⁰ Inst1742Am Chap. XXI § 570. "l'effet seroit plus grand que sa cause & le mouvement perpétuel méchanique seroit possible".

by whether he spends his money in a day, in a year, or over the course of his whole life.[161] The analogy with the problem of the forces is clear: force cannot be measured by its movement over a distance that does not describe its complete potential. What can be described is only what has "been spent" by one move.

Preferability of Laws

Du Châtelet's theory on hypotheses provided a methodic approach for judging if and why one scientific theory is preferable to another. The extent to which a theory correlates to other laws and rules in describing phenomena augments its probability. Yet Johann I. Bernoulli dedicated his memoir to the comparison of the various laws which explain physical phenomena, and Maupertuis wrote about this in his memoir *Sur les différentes lois d'attraction*, comparing the physical explanations of Newton and Descartes.[162] Today the question is whether Maupertuis really considered Newton's squared law to be the only conceivable law of attraction. When Du Châtelet read Maupertuis' memoir she inferred that Maupertuis was convinced that Newton's law was the only valid one. Rereading the memoir, she criticizes that this memoir is written "en français un peu obscure" and claims that his explanations are not convincing enough.[163] She is not convinced by his arguments, as he wishes metaphysics only to justify the laws established by God and discovered by Newton! Maupertuis is not interested in setting clear foundations; in this he fails! As Pilate asked Jesus, she would like to ask him: What is truth?[164]

The reasonability of a mathematical law is pleasing because it is equivalent to a rational assumption of how to understand the world: It is a great, a beautiful idea

---

[161] Inst1742Am Chap. XXI § 569. On Du Châtelet's convincing rhetorics, see: Guyot. 2008. La pédagogie des Institutions de physique. 267–282. "…l'étude approfondie … a mis en évidence les grandes qualités de pédagogue de la marquise, et la clarté de ses démonstrations", 280.

[162] In LetChBI 127: May 22nd, 1738 Du Châtelet speaks of the *mémoire* of 1732.

[163] LetChBI 124: May 9th, 1738. "Vous ne dites point pourquoi dans une attraction en raison directe de la simple distance, qui a d'ailleurs tant d'avantages, n'aurait pas celui de l'accord de la même loi dans les parties et dans le tout; vous ne dites point non plus pourquoi la raison inverse du carré des distance a cet avantage".

[164] LetChBI 124: May 9th, 1738. "Je crois donc, s'il m'est permis d'avoir une opinion sur cela, que la force de ces corps se consommerait réellement dans les efforts qu'ils feraient pour surmonter réciproquement leur impénétrabilité, et leur force d'inertie, et que cet effet qu'ils auraient produit l'un sur l'autre en surmontant la force que tout corps en mouvement a pour persévérer à se mouvoir, cet effet, dis-je, représente métaphysiquement la force qui la produit et ce serait bien alors que la métaphysique serait content. Je vous croyais réconcilié avec elle depuis que vous avez décidé pour la loi d'attraction en raison inverse du carré des distances en faveur d'une raison métaphysique, mais je vois bien que vous n'en voulez que lorsqu'elle justifie les lois établies par le créateur, et découvertes par Neuton. Vous ne voulez point éclairer ses profondeurs, vous avez cependant bien tort. Je … vous demanderais volontiers comme Pilate à Jesus, *quid est veritas* ?".

that would be in keeping with the idea of an 'eternal geometrician'. Would it not be that, after all, even God would have to follow rational principles? Is it not reasonable that Plato calls God the eternal Geometer, she asks in the *Institutions*?[165] However, as we have learned from her methodic approach, truth can be found by means of hypotheses, but hypotheses cannot be confused with the truth itself. Because of this, mathematical laws are only a means to indicate or to describe the truth! Du Châtelet agrees with the argument that the world was made to follow rational laws, because she loves the Platonic explanation that even God has to think rationally. However, she refutes Maupertuis' referral to a Godly Will which becomes visible through this law, and she insists on trying to find an alternative in this physical context.[166] Of course, also Du Châtelet is inclined to accept the preference of Newton's law over any other, but she is by no means willing to accept it as the only possibility. Hartmut Hecht argues that even Maupertuis did not view Newton's position as the complete solution.[167] If Maupertuis had agreed that the preference of one hypothesis over another is dependent on their respective plausibility, he could have actually mentioned this to Du Châtelet. The fact that she never received a satisfying answer to her questions can be assumed from Du Châtelet's remark in her *Institutions*, where she points out that Maupertuis had, in fact, never given any reasons for his preference of Newton's law.[168] Moreover, as far as the letters from her side testify, in this context he refuses to discuss this issue more deeply and to reevaluate her critical reflection regarding the conservation of forces.[169] Du Châtelet remains adamant that an assumed primacy of Newton's law would have to be grounded in a convincing explanation, as well as being proven plausible by being related to other accepted scientific insights.[170]

---

[165] "Ce n'etoit pas sans raison que Platon appelloit le Créateur, l'eternel Géomètre". Inst1742Am Chap. I § 13.

[166] "J'attends avec bien de l'impatience quelques éclaircissements que vous m'avez promis sur votre mémoire de 1732. J'en ai un grand besoin ..." LetChBI 127: May 22nd, 1738: the letters before, she spoke of the mémoire of 1734.

[167] "[a] more subtle analysis shows [...] that it had never been the aim of Maupertuis merely to legitimise Newtonian attraction, but also to find a common methodological strategy that could make apprehensible both Cartesian pressure-and-impact interaction and Newtonian long-range forces acting at a distance". Hecht. 2001. Leibniz' Concept of Possible Worlds. 35.

[168] "Car selon les principes de Mr. De Maupertuis même, s'il y a eu une raison de préférence pour la loi d'attraction que Dieu a employée, il y en doit avoir eu une pour l'attraction elle-même"; Inst1740/Inst1742Am Chap. XVI § 394.

[169] "Je comprends parfaitement à présent votre raison de préférence pour la loi d'attraction mais je n'entends pas trop la réponse que vous faites à l'avantage d'uniformité qu'aurait la loi de se conserver la même". LetChBI 139: September 1st, 1738.

[170] Another letter proves Maupertuis to become more reluctant to formulate his opinion on the matter at a certain period. LetChBII 273: June 26th, 1741: "Pour l'attraction vous m'avez paru a Cirey si modéré dans vos sentiments sur cela".

Free Will

It is quite consistent with Du Châtelet's methodic approach that she pushes the question of the preferability of scientific laws. She applies this methodic questioning quite radically, as can be observed by the way she not only questions compatibility and thereby the validity of the law of attraction, but also challenges the reasons and scientific compatibility of the theory of the *vis viva*. Thus, she asks Maupertuis about the conclusions to be drawn from the *living forces*. She admits to remaining skeptical because of its solely relative determinability, mainly from the point of view of the law of conservation, and its consequences with regard to free will.[171] Although an advocate of this system, she puts forth the question of why the idea of the conservation of force should be better explained by *vis viva* than by the Cartesian measurement of force.[172] If we presuppose the conservation of force in the universe, how is this idea to be brought into line with the freedom of all creatures? This metaphysical problem is disputable.[173] The concept of *living forces* is not convincing in all its essential aspects. The idea that a certain amount of force exists in the universe and transforms itself into various actions is intriguing, however difficult to follow in its consequences. "How can motion be initiated if the quantity of movement remains an absolute term?"

> "The only thing that puzzles me at present is free will, for in the end I believe myself free and I do not know if this quantity of force, which is always the same in the universe, does not limit this. Initiating motion, is that not to produce in nature a force that did not previously exist? Now, if we do not have the power to set something in motion, we are not free. I beg you to enlighten me on this point".[174]

Maupertuis' approach and his reference to the validity of physical laws remains incompatible with her own understanding of the human free will. "If I am in fact free, then I must be allowed to initiate movement", she argues. And if actual evidence for my freedom existed, then it would be absolutely necessary "that my own will would

---

[171] Compare Inst1742Am Chap. XXI § 570.

[172] Du Châtelet asking Maupertuis whether he could explain "que la même quantité des forces s'en conservera", LetChBI 120: February 10th, 1738.

[173] "Je vous avoue qu'il me reste une grande peine d'esprit sur ce que vous me dites, que si l'on prend pour forces les *forces vives* la même quantité s'en conservera toujours dans l'univers. Cela serait plus digne de l'eternel géomètre, je l'avoue, mais comment cette façon d'estimer la force des corps empêcherait-elle que le mouvement ne se perdît par les frottements, que les créatures libres, ne le commentassent, que le mouvement produit par deux mouvements différents ne soit plus grand quand ces 2 mouvements conspireront ensemble que lorsqu'ils seront dans les lignes perpendiculaires l'un à l'autre &cc." LetChBI 120: Febr 10th, 1738.

[174] Du Châtelet. 2009. *Selected Philosophical and Scientific Writings*. 109; LetChBI 122: April 30th, 1738. "La seul chose qui m'embarasse à présent, c'est la liberté, car enfin je me crois libre et je ne sais si cette quantité de forces toujours la même dans l'univers ne détruit point la liberté. Commencer le mouvement n'est-ce pas produire dans la nature une force qui n'existait pas? Or si nous n'avons pas le pouvoir de commencer le mouvement nous ne sommes point libres. Je vous supplie de m'éclairer dur cet article".

be capable of producing forces whose *quomodo* I cannot discern!"[175] How could this go together with the idea of the conservation of forces and the conviction of Godly Determination?

Freedom is the capacity to initiate movement. It is commonly known that any starting point of motion provides ample evidence that nature has yielded a force which, up to that point, had not yet existed. If we cannot be considered to be in a position to initiate movement and action, does this in turn not also mean that we are not free? Does then the assumption of a self-conserving amount of force not prove exactly this? Du Châtelet indicates this as a serious problem which disturbs the seductive idea of the conservation of forces. She also takes up Mairan's proposal to handle this problem he mentions in his memoir. Should we then, as he does, support the argument that possibly God has indeed created two separate laws of motion, one for the force in inanimate bodies and another one which understands animated entities to be endowed with the ability to act and with intelligence or life? With astounding wit, she points out just how improbable such a one-sided understanding of force and its overall relation to motion would be.[176]

Unfortunately it is at this point that Du Châtelet's and Maupertuis' correspondence comes to a halt. As Du Châtelet's subsequent letters show, he does not attempt to provide an answer to her questions: "I would be inclined to believe, monsieur, that my last letter bored you so much, or that you found it so ridiculous, that you judge it not worthy of a response".[177]

As her correspondence shows, Du Châtelet had a very clear idea about the scientific requirements a truly metaphysical method would entail. The capacity to add new insights to humanity's knowledge, to find and to construct scientific laws, to explain and to do research on concurring insights is one of the attributes of human beings. If nature's laws and the understanding of them were laid out independently of human will, even in the form of wonderful mathematical laws, science in such a world would assume an utterly absurd position.

Elastic and Solid Bodies

In order to come to a closer and deeper understanding of the problems that arise with reference to the question of the conservation of force, Du Châtelet asks, "as Newton had already done before her", what would happen to two solid bodies that

---

[175] Du Châtelet refers herewith to the philosophical dispute on arguments; the question is not "that" something is, but "how" it is explained. LetChBI 126: May 20th, 1738.

[176] LetChBI 124: May 9th, 1738.

[177] Du Châtelet. 2009. *Selected Philosophical and Scientific Writings*. 109. LetChBI 122: April 30th, 1738.

collided in empty space?[178] Maupertuis following Leibniz saying that there cannot be inelastic bodies she replies she invented one. "Votre idée que dieu n'a pas fait ... de corps sans ressort, m'en a fait naître une". She is convinced that indivisibility is a necessary attribute to be applied to *first bodies*: "... je crois cette indivisibilité à être des premiers corps de la matière d'une nécessité indispensable en physique".[179] Du Châtelet's letters show that this particular question of the complete elasticity of matter was of great importance to her in reaching a conclusion about the problem of the conservation of force. How could force be conserved if nothing in the environment or in parts of it exists that could theoretically absorb the impact? Regarding this topic, Maupertuis clearly maintains Leibniz's idea of the elasticity of matter to sustain the question of conservation by "metaphysical reasons". But what does this mean? According to Du Châtelet, the rules of metaphysics would only be satisfied if the effect of this collision was that the forces consumed each other while overcoming each other's inertia.[180] As it is certainly difficult to prove that solid bodies exist in the first place, and even to her this seems highly improbable, it is Maupertuis' idea to take the effect "metaphysically", which is not convincing for her.[181] Applying her hypothetical method, she challenges Maupertuis' answer to this argument. Her inquisitiveness as to the universal application of physical laws influenced Maupertuis. It is at this point that she assumed a specific core of matter which could be considered 'solid'. Years later Maupertuis would also assume this position. He would then subscribe to her view that a solid and impenetrable mass must exist and that elastic bodies are composites of this, as Du Châtelet had pointed out in her letters to him years earlier. While she prepares her *Institutions*, Maupertuis ignores her metaphysical arguments of solid bodies of matter, used to explain where force is going in order to maintain the law of conservation. To structure her *Institutions* systematically, she returns to the assumptions of the Leibnizian elastic physical world, a standpoint which is clearly at odds with the viewpoint she portrays in her

---

[178] LetChBI 124: May 9th, 1738: "Je sais qu'on ne connaît point jusqu'à présent de corps parfaitement dur, mais ce n'est pas je crois une démonstration qu'il n'y en ait point"; "Je vois (autant que je peux voir) qu'il est certain que la force ou l'effet de la force des corps est le produit de la masse par le carré de la vitesse, et que la quantité de la force d'un corps sont deux choses très différentes."

[179] LetChBI 146: September 29th, 1738.

[180] LetChBI 124: May 9th, 1738: "Je crois donc, s'il m'est permis d'avoir une opinion sur cela, que la force de ces corps se consommerait réellement dans les efforts qu'ils feraient pour surmonter réciproquement leur impénétrabilité, et leur force d'inertie, et que ce effet qu'ils auraient produit l'un sur l'autre en surmontant la force que tout corps en mouvement a pour préserver à se mouvoir, cet effet, dis-je, représente métaphysiquement la force qui la produit et ce serait bien alors que la métaphysique serait contente."

[181] LetChBI 124: May 9th, 1738: "Votre idée de prendre métaphysiquement les effets pour les forces me paraît admirable, car je ne sais si elle ne pourrait point fournir une réponse a cette objection qui m'a toujours arrêtée, et qu'à mon gré mr de Bernoulli a trop méprisée."

correspondence.[182] Her method allowed her a critical approach to elasticity and made her doubt the overall validity of Leibniz's idea of elastic bodies of matter. Her method, however, forced her to hold on to a doubtful theory only because its probability was higher within the context she took into regard.

Although Du Châtelet favoured the *vis viva* as an underlying and interesting principle of force, she remained highly critical of it. It is evident that her thoughts on movement and force are mapped out in these early remarks on the conservation of forces and motion. According to Hartmut Hecht, Maupertuis' unwillingness to examine solid bodies of matter at that time might have been due to the fact that there was no principle in sight which could be valid for all types of matter.[183] However, this is very true for Du Châtelet's argument. She agrees that it is difficult to prove that solid bodies exist in the first place, yet it is apparent that she questions this as a consequence of her methodic and heuristic approach, whereas Maupertuis remained within the context of his convictions. Thus, she risks proposing the theoretical existence of a solid body which would have to be understood as the original mass, and which she even identifies with "les premiers corps de la matière".[184] It would be very interesting to understand more of what Du Châtelet figured out on this topic. She also regrets Bernoulli's objection to the idea of solid bodies due to his general agreement with Leibniz's theory of elastic bodies. Only if one could explain that a theory of solid bodies fails with respect to all other scientific laws and insights, then and only then both metaphysics and the natural sciences would be appeased, and only then would metaphysics be satisfied. "Hypotheses, then are only probable propositions that have a greater or lesser degree of certainty, depending on whether they satisfy a more or less great number of circumstances attendant upon the phenomenon that one wants to explain by their means".[185] Maupertuis, however, follows a quite different kind of metaphysics, as becomes obvious in his *Lettres*.[186]

---

[182] Inst1742Am Chap. XXI § 588: "Mais Mr. De Leibnits par sa nouvelle estimation des forces, a accordé la raison Métaphysique trouvée par Descartes … la quantité du mouvement & la quantité de la force des corps en mouvement, & en faisant cette force proportionnelle au produit de la masse par le quarré de la vitesse, on trouve que quoique le mouvement varie à chaque instant dans l'Univers, la même quantité de force vive s'y conserve cependant toujours; car la force ne se détruit point sans un effet qui la détruise, et cet effet ne peut être que la même dégré de force communiqué à un autre corps, puisque celui qui prend, ôte toujours à celui à qui il prend, autant de force qu'il en retient pour lui; ainsi, la production du moindre dégré de force dans un corps, emporte nécessairement la perte d'un égal dégré de force dans un autre corps & réciproquement: la force ne sauroit donc périr en tout, ni en partie, qu'elle ne se retrouve dans l'effet qu'elle a produit, & l'on peut tirer de-là toutes les Loix du mouvement." Also see Klens. 1994. *Mathematikerinnen im 18. Jahrhundert*. 247 f.; also refer to Inst1742Am Chap. XXI §§ 587–590.

[183] Hartmut Hecht's view is due to the discussion on solid, elastic and soft bodies of matter at that time, discussed in Mariotte. 1673. *Traité de la percussion*, which was known to Maupertuis.

[184] LetChBI 124: May 9th, 1738.

[185] Inst1740/Inst1742Am Chap. IV § 67. LetChBI 124: May 9th, 1738.

[186] Maupertuis. 1752. *Lettres de M. de Maupertuis*.

The 'Principle of Least Action' and the Fight Over Metaphysics

Judging from the correspondence available, it seems that it was Maupertuis who eventually held on to the law of conservation in accordance with the *living forces* and its elasticity of matter. Du Châtelet's problems arising from the impact on solid bodies in this context remained unaddressed or were deliberately ignored by him. Du Châtelet's approach to assuming the existence of solid bodies and applying the consequences of such an approach to the concept of the conservation of energy was rejected by Maupertuis, according to the letters. Astonishingly, Maupertuis later explicitly states that only the primordial bodies can be thought of as completely solid, but all elastic bodies would have to be understood as composites of solid ones. This, in fact, mirrors Du Châtelet's own words. Her original question on how force can be conserved in relation to the movement of solid bodies of matter becomes a starting point for Maupertuis' thoughts on the *principle of least action*. In his *Lettre IX: Sur la nature des corps*, Maupertuis states that the first property of matter when compared to space is its impenetrability, called "solidity" by the philosophers.[187] Maupertuis adopts Du Châtelet's thoughts on the existence of solid matter, which she had explained to him in her letters of 1738. In his *Lettre IX*, Maupertuis now, too, expresses his doubts about the validity of a general law of the conservation of force that cannot be explained by *living forces* alone. There appears to be no reason for such a law, as Maupertuis states. It is only applicable in certain cases and the conservation provided by the *living forces* is also only true in certain cases![188] As Ulrike Klens and Marie-Louise Dufrenoy have pointed out, Du Châtelet's comments exercised great influence on Maupertuis' ideas on the *principle of least action*.[189] It is Klens in particular who, in this context, underlines Du Châtelet's importance for his work on a subject which would probably have remained undiscovered, if it hadn't been for Du Châtelet's inspiring discussion of mechanical problems in relation to the universal conservation of force.

---

[187] "La première propriété qui distingue le corps de l'espace, est l'impénétrabilité. ... Cette propriété est appelée par quelques Philosophes solidité, dureté, et est regardée de tous comme la propriété fondamentale de la matière....". Maupertuis. 1752. *Lettres de M. de Maupertuis*. Lettre IX. See Klens. 1994. *Mathematikerinnen im 18. Jahrhundert*. 248: It is this discussion which initiates Maupertuis' criticism in relation to the principle of the conservation of the *living forces* as a primary principle in mechanics, since he is not in a position to let go of his atomic understanding of mass and he is not convinced by Du Châtelet's suggestions.

[188] "La conservation de la quantité du mouvement n'est vraie que dans certains cas. La conservation de la force vive n'a lieu que pour certains corps. Ni l'une ni l'autre ne peut donc passer pour un principe universel, ni même pour un résultat général des loix du mouvement", Maupertuis. 1752. *Lettres de M. de Maupertuis*. Lettre IX.

[189] Klens. 1994. *Mathematikerinnen im 18. Jahrhundert*. 242, and Dufrenoy. 1963. Maupertuis et le progrès scientifique. 562. „Die Darstellung Brunets, Maupertuis habe Du Châtelet für die lebendigen Kräfte gewonnen ist daher falsch." Brunet. 1929. *Maupertuis: étude biographique*. 241.

In 1746, years after Du Châtelet and Maupertuis had exchanged their most intimate scientific correspondence, Maupertuis elaborated on the circumstances which led to his "discovery" of the *principle of least action*. In his comments he spends considerable time on the role of God in the sciences. In Maupertuis' opinion, God's existence might well be proven by the very existence of *natural laws* and he holds that *natural laws* confirm Godly Will rather than question it. He also states that any mathematical law expresses the infinity and the perpetuation of Godly Will. Mathematical truths have the advantage of being evident and proving the existence of God, he states.[190] In line with a sort of mathematical positivism, Maupertuis claims that mathematical laws provide access to divine truths and that there could consequently be no more "empty disputes" over the true nature of motion.[191] These statements are in opposition to Du Châtelet's method, as becomes clear from Du Châtelet's remark on similar operations of the mathematicians, as she explains in the *Institutions*. Maupertuis judges the advantages of uniformity and analogy to be the decisive reasons in favour of the preference for Newton's law of attraction: "ces deux avantages de l'uniformité & de l'analogie, ont paru suffisans à Mr. De Maupertuis pour déterminer le Créateur à choisir la loi d'attraction ..." pointing to Maupertuis' scathing explanation, that does not suffice to her.[192] His idea

---

[190] "Les preuves de l'existence de Dieu qu'elle fournira, auront sur toutes les autres, l'avantage de l'evidence qui caracterise les verities mathématiques". Maupertuis. 1746. Les Loix du mouvement.

[191] I call this a sort of "mathematical positivism" in accordance with d'Alemberts method, although it is obvious that a strong parallel between Maupertuis and d'Alembert cannot be drawn. However, it can be considered certain that Du Châtelet was against this sort of scientific analysis which does not regard experience and the constructability of science sufficiently; this conviction was held within the Bernoulli family against d'Alembert in any case. Hankins. 1970. *Jean D'Alembert. Science and the Enlightenment*.

[192] Inst1742Am Chap. XVI § 392; see also: "Mr. De Maupertuis est de tous les Philosophes Francais, celui qui a poussé le plus loin ses recherches sur l'attraction; il donna en 1732 à l'Académie un Mémoire, dans lequel il recherche la raison de la préférence, que le Créateur a donnée à la loi d'attraction, en raison inverse du quarré des distances ... & il trouve par son calcul, que de toutes les loix qu'il a examinées, il n'y a que celle en raison inverse du quarré des distances qui donne la même attraction, pour le tout & pour les parties qui le composent, & qui joigne à cet avantage, de lui de la diminution des effets avec l'éloignement des causes; ces deux avantages de l'uniformité & de l'analogie, ont paru suffisans à Mr. De Maupertuis pour déterminer le Créateur à choisir la loi d'attraction ..." Inst1742Am Chap. XVI § 393: "La Mémoire de Mr. De Maupertuis, dont je viens de parler, est comme tout ce que fait ce Philosophe, plein de sagacité & de finesse de calcul, il n'y donne son opinion sur la raison de préférence de la Loi d'attraction en raison inverse des quarrées des distances, sur toutes les autres loix, que comme un doute, mais ce sont assurément les doutes d'un grand-homme." Inst1742Am Chap. XVI § 394. "Si ce Philosophe avant de rechercher la raison de préférence d'une loi d'attraction sur une autre, avoit recherché la raison de l'attraction elle-même, il est vraisemblable qu'il auroit bientôt reconnu que cette attraction, telle que les Newtoniens la proposent, c'est-à-dire, entant qu'on en fait une propriété de la matière, & la cause de la plupart des Phénomènes, est inadmissible; car selon les principes de Mr. De Maupertuis même, s'il y a eu une raison de préférence pour la loi d'attraction que Dieu a employée, il y en doit avoir eu une pour l'attraction elle-même."

of uniformity negates the concept of free will as well as her conviction that science is made and constructed by human experience and based on hypotheses. Furthermore, Du Châtelet understands such arguments to be merely circular, as numbers are not considered to be essential attributes but must be applied to something. According to Du Châtelet, Maupertuis is confusing the instrument of scientific heuristics with the knowledge itself.[193] It is precisely this presumed unity of mathematics and metaphysics which is questioned by Du Châtelet yet within their correspondence and turned into a methodological distance finally. As she states in the third chapter of her *Institutions*, 'Of Essence, Attributes and Modes', mathematics can in fact never be the primary source of knowledge, since it is relying on *necessary* requirements which it cannot determine by itself. *Within* the realms of mathematics and *through* its means, it may thus well be possible to understand scientific laws, yet mathematical laws do not unveil divine truth! With this fictional dispute on mathematics and its value for metaphysical truth, the disagreement between Maupertuis and Du Châtelet went into the next round. In his *Lettres*, Maupertuis condemns those who appear to adhere to what he calls metaphysical "sects" and again it is an open rebuff of Du Châtelet's metaphysics. All metaphysical approaches are inconsistent in their demands, Maupertuis states, and that Descartes and Leibniz were, in fact, incompatible with Newton! This is clearly an affront to Emilie du Châtelet and, of course, also to his old friends, the Bernoullis. Since all of the metaphysical concepts contradict each other, none of them could be legitimate, Maupertuis says. He turns to Newton, stating that only phenomena could legitimize scientific laws. Knowledge of the phenomena replaces the invisible and he believes that mathematical consistency is sufficient to explain phenomena. Maupertuis' methodology allows and subsequently asks for or even demands unsystematic methods of perception, since their very existence is proof that they are not re-sculpted by rational or metaphysical concepts, denying all kinds of rational order, even permitting contradictions: "Je ne suivrai aucun ordre;" he writes, "je parcourrai les sujets comme ils se présenteront à mon esprit: je me permettrai peut-être jusqu'au contradictions".[194]

Years before this happened Du Châtelet assumed Maupertuis' distance towards her work, summarising: "Je sens bien que vous me conseillerez de retrancher toute la métaphysique".[195] There can be no doubt that Maupertuis argues against Du Châtelet's philosophy.

---

[193] In this context, coincidentally, the arguments of Samuel König are echoed, which he would present in his *Oratio inauguralis de optimis Wolfiana et Newtoniana Philosophandi Methodis earumque amico consensu* from 1749.

[194] Maupertuis. 1768. *Oeuvres de Maupertuis*. Vol. II. 221.

[195] LetChBII 274: August 8th, 1741.

What Does Julien Offray de La Mettrie Have to Do with Du Châtelet's Metaphysics?

It is surprising that studies on Emilie du Châtelet and her relationship with Julien Offray de La Mettrie have found up to now so less attention. Ursula Pia Jauch investigated this relationship, stating "astonishing parallels" and assuming even similarities in their character.[196] She demonstrates literary and stylistic similarities like the metaphor of "experience as the staff of knowledge".[197] La Mettrie had gotten to know "Emilie du Châtelet ... as a philosophically qualified person of rank and name whose work and discernment he had come to appreciate", judges Jauch, referring also to the fact that the edition of *Institutions physiques* from 1742 was to be found in his library. Another – however well documented – example proving coherencies to Du Châtelet's *Discours sur le Bonheur* is La Mettrie's *Discours sur le Bonheur* which indicates also parallels to Maupertuis' *Essai de philosophie morale*. All three essays on happiness have references to each other, however they differ in their conclusion.[198] Apparently, they were having or had common conversation on this topic.[199] Within the period in question, La Mettrie was mentored and protected by Maupertuis.[200] Beside the fact that the three have known each other and related

---

[196] Jauch says, both were "of courageous spirit and strikingly open for new directions of thought", strong and dominant, brave enough to say what they really thought. Jauch. 1998. *Jenseits der Maschine*. 317. No proof exists thus far that Du Châtelet und La Mettrie ever had met each other, either in Café Procope, in d'Harcourt or in the theater. But the possibility of this cannot be excluded, as both belonged to the Paris intelligentsia and had a circle of common acquaintances, including Maupertuis, Voltaire, and Diderot, as Jauch states.

[197] Inst1742Am Avant-propos IX: "L'Experience est le baton que la Nature a donné à nos aveugles". See this metaphor also in La Mettrie and Diderot according to Jauch. Jauch sees similiarities also between the *Institutions* and La Mettrie's *Ouvrage de Pénélope ou Machiavel en Médicine* from 1748. La Mettrie dedicated his volume to his son, and argued similarly, not only through his use of "à mon fils", but also in view of "empiricism, probability, skepticism, and pedagogic". Jauch. 1998. *Jenseits der Maschine*. 317, 320.

[198] A compilation of the main subjects of the three essays you find in Rodrigues. 2010. Emilie du Châtelet, Julien Offray de La Mettrie und Pierre Louis Moreau de Maupertuis. 151–161.

[199] In his critical edition John Falvey compared La Mettrie's *Discours sur le Bonheur* to Du Châtelet's writing, but also to the *Essai sur la philosophie morale* of his friend Maupertuis. Jauch. 1998. *Jenseits der Maschine*. 321. "Madame du Châtelet's discourse coincides with La Mettrie's, not only in particular ideas, but also over a large area of fundamental viewpoints". "Both hold that the natural promptings are moral; considering morality here exclusively in the context of love", Falvey is convinced. Moreover, parts of the ideas taken up by Du Châtelet and Maupertuis were already executed in La Mettrie's *Ecole de la volupté*, which was also dedicated to Emilie du Châtelet and had been written in 1746.

[200] Maupertuis' support for La Mettrie in this period. Baruzzi. 1968. *Aufklärung und Materialismus in Frankreich des 18. Jahrhunderts, La Mettrie – Helvetius - Diderot – Sade*. 22. La Mettrie had provoked a plagiarism affair and dedicated his *L'homme machine* to the author of the plagiarized text, named Haller. See introductory remarks Laska. 1988. *Der Mensch als Maschine*.

to each other in their writings on morals, there is another issue on which the interests of the three met but also converged. La Mettrie's connections to Du Châtelet including their common relationship to Maupertuis go beyond this and touch the subject unwrapped here. La Mettrie's publication in question is the second edition of the *Histoire Naturelle de l'Âme,* that appeared anonymously, a method widely used in enlightened literature. He names himself however in the *Lettre Critique de M. de La Mettrie sur l'Histoire Naturelle de l'Âme à Mme. La Marquise du Châtelet*, which is also announced on the title page of the book. The book is dedicated to Maupertuis and the *Lettre critique* follows immediately and is to be found in the first 12 pages. It does not only give proof of the reception and circulation of Emilie Du Châtelet's philosophical statements during her lifetime and within the circle of the *philosophes*, overthis it is an important reference to examine Du Châtelet's philosophical position in confronting the important discussions on metaphysical issues that are at stake especially in the realm of this three colleagues.

To grasp a bit of the intention of this letter we have to take into account that the simultaneous affair concerning *L'Homme machine* could make a dedication to Du Châtelet meaningful.[201] Does La Mettrie want to legitimize his provocative theses by alluding to the renowned Du Châtelet implying that her thoughts were similar to his? His dedication to Maupertuis leads in a different direction. Is La Mettrie arguing in favour of Maupertuis? Rarely probable but not excludable. The dedication to Maupertuis and the critical letter to Emilie du Châtelet sounds again like a dispute or an answer to a dispute, which could have taken place, factually or mentally. Is it a final exchange of blows at a time when the positions of former friends and colleagues have disunited. La Mettrie praises Maupertuis, affirms that philosophy has enlightened him concerning the nature and the attributes of the soul and only he, Maupertuis, could pass judgement about it: "je suis seulement sûr d'avoir trouvé le Philosophe le plus capable d'en juger".

Although La Mettrie in his *Lettre critique* is pronouncing the distance to Du Châtelet's position, the letter is not completely negative. On the one hand it suggests a situation of unanimity in former times insinuating a change of thought on Du Châtelet's side, on the other hand it refers clearly to a position of Du Châtelet that is incompatible with his. La Mettrie wants to make her accede to his conclusions and we get even the impression that the author defends his advancement of the issue which differs from hers. But it could also be the fact that he used this kind of a faked familiarity with a well renowned philosopher to advance his own ideas. In our context it is sufficient to locate the references to her writing, that go straight into the cernal of Du Châtelet's metaphysical reflections. This *Lettre critique* gives reference to Du Châtelet's *Institutions* and her *Dissertation on fire*. This letter shows great intimacy with Du Châtelet's critical metaphysics, and her damnation of the fable tellers of the "matière pensante". It focuses on her criticism versus Locke, her

---

[201] La Mettrie's *L'Homme machine* was published at Elie Luzac in Leyden in 1747. "L'apparition de l'Homme machine fut un événement considérable, on le lisait dans toute l'Europe" see: La Mettrie. 1921. Introduction de Maurice Solovine. 8.

remarks about ontological "confusion" and the transferability of attributes, on action, free will and sensations. And yet at the beginning it takes up her relation to Leibniz. Finally, La Mettrie defines himself by Du Châtelet's dictum of being a "bad metaphysician" as he confounds the order of being.

"I do not hesitate MADAME to submit my reflections to your judgment".[202] La Mettrie praises her great scholarship and subordinates his reflections under hers. At the beginning of this *Lettre critique* La Mettrie acknowledges her as a sage philosopher, introducing her as the one who was the reason that the metaphysics of Leibniz could again become a daily topic of conversation in the scientific world. She was clever enough to leave out the mistakes in Leibniz' philosophy held up her scepticism towards Leibniz' monads and thus make it more comprehensible.[203] She wisely knows to discern Leibniz' good from his minor ideas.[204]

Thinking Matter and être simple

La Mettrie strongly seeks to prove the accordance between his and her thoughts. Had she not confirmed within her *Dissertation on Fire*, "pour élever la matière à la faculté de sentir, que pour la faire végéter; il ne donne d'autre principe que l'Ether, ou le feu"?[205] Did she not admit sensitivity and activity to matter?[206] La Mettrie refers to Du Châtelet's *Dissertation on Fire*, the author of "the best mémoire written on this topic"! She should know how close their ideas are, as she knows everything

---

[202] "Je n'hésite pas, MADAME, de soumettre ces réfléxions à votre jugement; la justesse de votre ésprit & l'entendue de vos connoissances semblent déjà me promettre votre suffrage." La Mettrie. 1747. *Histoire naturelle de l'âme*. 2 f.

[203] "Vous connoissez partfaitement les Monades de Leibnitz; vous êtes même le premier des Philosophes François qui nous aiès dévelopé son systême avec toute la clartè … mais, MADAME, vous avez eû la sagesse de n'en parler que par raport à la Physique des Mixtes, & d'abandonner les idées de ce Philosophe sur les perceptions qu'il attribué à ses êtres Simples". La Mettrie. 1747. *Histoire naturelle de l'âme*. 3.

[204] "Que pensez-vous, MADAME, de ces admirables métamorphoses de la matière, vous, qui n'ignorez pas que Wolf même a dépouillé les Monades Leibnitiennes des perceptions qui leur avoient été prodiguées?" La Mettrie. 1747. *Histoire naturelle de l'âme*. 5.

[205] La Mettrie. 1747. *Histoire naturelle de l'âme*. 7.

[206] An interesting phrase can indeed be found in Du Châtelet's *Dissertation*: "All fire does not come from the Sun (…) Each body and each point of space has received from the Creator a portion of fire in proportion to its volume. This fire enclosed in the bosom of all bodies gives them life, animates them, impregnates them, maintains the motion among their particles, and prevents them from condensing entirely …" Du Châtelet. 2009. *Selected Philosophical and Scientific Writings*. 100. See the French text: "Tout le Feu ne vient pas du Soleil …; chaque corps & chaque point de l'espace a reçu du Créateur une portion de Feu en raison de son volume; ce Feu renfermé dans le sein de tous les corps, les vivifie, les anime, les féconde, entretient le mouvement entre leurs parties". FeuChat1744, 129.

about "the energy of fire".[207] Does La Mettrie legitimize his materialism by his reference to her thoughts? Has he drawn false conclusions? Or, should she have changed her ideas meanwhile? La Mettrie is convinced, Emilie Du Châtelet, this learned woman, could certainly not ignore the "mechanism of sensations".[208] Should this active power, this "energy" that she herself ascribes to each matter "proportionally to its volume" bring something different to matter than activity? With this argument he tries to force her to cancel her arguments she states within the *Institutions*, talking about the untransferability of attributes, which was the reason why she had denied the idea of a thinking matter! His rhetoric question "Que pensez-vous MADAME, de ces admirables metamorphoses de la matière ?" focuses on this central issue of Du Châtelet's ontology as she deduced: "From this follows that the dispute about the possibility of thinking matter is finished"![209]

La Mettrie's reference to Du Châtelet's deliberations on this topic about several years later make it obvious that her reflections gave an important starting point for the enlightened discussion. Du Châtelet's ontology combats La Mettrie's sensualist materialism. Du Châtelet might have admitted a sort of "energy" in matter, however she does not allow to take in one activity and sensations. It is exactly on the relation between the faculty of sensation which is only passive and the always active movement he would like to have with her "des Entretiens Métaphysiques sur ce sujet".[210] He is convinced that there is only a "shadow" between the two things of which "we have ideas so opposite".

What this opposition is about becomes clear in the next item La Mettrie is swapping to. In a witty remark against Descartes La Mettrie takes up the idea of the reasonable beings. He alludes to Descartes *Etres de raison* saying, that the "author of this book wants to return to the rational being, which are however only good to seed the doubt in humans!" We conclude from this that La Mettrie holds against Du Châtelet's idea of a necessary and presupposed rationality. He insists that this unity can be procured also by an *Etre sensitive* which confers its capacities out of movement. In La Mettrie's world the *Etre sensitive* is material and La Mettrie presupposes her approval for his argument. His conclusion however to endow matter not only with an intrinsic principle of motion (as she did with "energy"), but also with the capacity to feel, she disagrees. Taking this argument even farther he states that also free will could be explained as a sort of individual motion. The force could even be the force to start movement, "la puissance de se mouvoir par elle-même".

---

[207] "Vous qui connoissez si bien, MADAME, toute l'energie des propriétés du feu, sur lequel vous avez donné en mémoire mieux ecrit qu'eu n'a fait jusqu'à présent en pareil genre" La Mettrie. 1747. *Histoire naturelle de l'âme*. 8.

[208] "L'extrême variété de vos connoissances ne vous permet pas d'ignorer le Mécanisme des sensations" La Mettrie. 1747. *Histoire naturelle de l'âme*. 8.

[209] Inst1740/Inst1742Am Chap. III § 47.

[210] "Je vous avoue, MADAME, que je saisirois avec plaisir l'occasion d'avoir avec vous des Entretiens Métaphysiques sur ce sujet". La Mettrie. 1747. *Histoire naturelle de l'âme*. 6–7. La Mettrie alludes to the Leibniz reception of Fontenelle in his *Entretiens sur la pluralité des mondes*. This circle of enlightened philosophers was sceptical towards Fontenelle's Leibniz interpretation as we know also from Du Châtelet herself.

Free will, if humans really have, as he sceptically remarks – shows itself to be a force of motion according to the kind of matter something is constituted with.[211] With this rebuttal of human freedom La Mettrie attacks another principle of Du Châtelet, referring to her remarks on the necessity of human free will as it realizes itself in starting a movement and thus forging the dilemma towards the philosophy of conservation of the forces that is not compatible to this conviction. While she sees become disorder greater and greater by transferring thought to matter, he, the "anatomist of the soul" sees only "nerves, blood and spirits". In his philosophy all is "forme substantielle, matérielle", which includes also humans. Like repeating her accusations, he signs himself as a "palpable & autant plus dangereux Metaphysicien se perdre de plus en plus & accumuler erreurs".[212]

La Mettrie's reference to Du Châtelet's deliberations on this topic about several years later makes it obvious that her reflections gave an important starting point for the enlightened discussion on the issue of the *thinking matter*, which has not yet found much regard.[213] Moreover – and this seems to confirm the context of La Mettrie, he put's this in the Leibniz context of a Monad, he speaks of the Cartesian "être de raison" and suggests thus that the question of a presupposed rational unity turns out to be an important point of reference in Du Châtelet's metaphysics at this time. Furthermore, La Mettrie's completion suggests that Du Châtelet's theses of the *Institutions* and the *Dissertation on Fire* were well known and intensively discussed in their philosophical circles. His discussion even gives us a bright insight why Du Châtelet is discussing in chapter III on "Essences, Attributes and Modes", which has not much been scrutinized up to now. The materialistic context which is widely debated within this letter of La Mettrie gives us an insight how the debate of the time and between the protagonists of Enlightenment went ahead. Emilie du Châtelet negates the materialistic approach for metaphysical reasons.

Disaster in Berlin

Voltaire, König, and Maupertuis, met each other in Berlin at the court of Frederick II. after Du Châtelet's death. It happened, however, that very soon the former friends were at odds with each other and parted under great contention. The productive

---

[211] "… et la liberté (si l'homme en a) vient d'une force motrice coessentielle à la matière dont l'Ame est formée" La Mettrie. 1747. *Histoire naturelle de l'âme*. 11.

[212] La Mettrie. 1747. *Histoire naturelle de l'âme*. 9.

[213] For more information about La Mettrie, D'Holbach and Diderot on this, see Vartanian. 1960. *La Mettrie's L'Homme Machine: A Study in the Origins of an Idea*. 149–165. "L'essor du matérialism au siècle des Lumières s'explique en très grande partie par l'ascendance de la méthode psychophysiologique de matérialiser l'âme. L'événement décisif à cet égard fut la parution en 1747 de l'homme-machine, où La Mettrie a consacré, pour ainsi dire, le triomphe de la formule cartésienne, tout en lançant la version de l'âme matérielle que devaient adopter à leur tour Diderot et d'Holbach". See also Robinet. 1999. Les trois modèles cartésiens d'automates. 41–54.

round of thinkers steered towards a disastrous finale. Samuel König, greatly indebted to Maupertuis and having become a member of the Berlin Academy, accused his mentor of having copied his *principle of least action* from one of Leibniz' manuscripts.[214] The accusation of plagiarism by Samuel König against Maupertuis shook the reputation of Maupertuis as president of the Academy; Voltaire subsequently exploited Maupertuis' compromised reputation by including him in his satire of *L'histoire du Docteur Akakia et du natif de Saint-Malo*.[215] The second part of *Dr. Akakia* presents a fictitious truce between Samuel König and Maupertuis, which effectively sees them swearing to acknowledge and uphold German scientists such as Kopernikus, Kepler, Leibniz, Wolff, Haller and Gottsched. In this final dispute between the most influential scientific figures of their time, metaphysics assumed an exceptionally important role.

Du Châtelet played a considerable role in the lives of these three scholars. Her importance in this situation has not been examined up to now, but facts and data point to several conclusions. La Mettrie, now the protégé of Maupertuis had dedicated his *École de la Volutpé* to his greatly admired "chère amie" in 1747.[216] He formulated a "love letter to a fictitious lover", written in part to hurt Voltaire, in his masculine pride by the publication of this romance, interprets Jauch.[217] The relationship between Voltaire and La Mettrie was evidently strained. It has been said that after that, Voltaire completely rewrote his letters from the time in Berlin, possibly even forging them.[218] What did Maupertuis with his letters he had sent to Du Châtelet? The considerable importance of these evening dinners and the French circle is a historic fact and has been highlighted repeatedly in many contributions on the subject.[219] But it remains a fact that almost all of them neglect Du Châtelet's position in

---

[214] In 1746 Maupertuis had written in the Memoires of the Academy the *principe général* as follows: "Tritt in der Natur irgendeine Änderung ein, so ist die für diese Änderung notwendige Aktionsmenge die kleinstmögliche". For more information on this see Knobloch. 1995. Das große Spargesetz der Natur: Zur Tragikomödie zwischen. Euler, Voltaire und Maupertuis. See also Szabo Compare. 1997. *Geschichte der mechanischen Prinzipien und ihrer wichtigsten Anwendungen*. 120. Euler and King Frederick II. denounced the letter as counterfeited, that should have proven Maupertuis's plagiarism.

[215] Docteur Akakia is Voltaire's scathing reply to the *Lettres d'un academicien*, which Frederick II. had published anonymously as a defence of Maupertuis.

[216] In the later edition, he extended this explicitly as a dedication to *A Madame la Marquise de* ***. See Jauch. 1998. *Jenseits der Maschine*. 325.

[217] Jauch. 1998. *Jenseits der Maschine*. 325 and 322, see also note 14. The mean remark calling Du Châtelet a "petite Maitresse ridicule", as La Mettrie writes in his *Discours sur le bonheur*, might thus not be directed against Du Châtelet, according to Jauch. Rather it was La Mettrie's intention to hurt Voltaire by damaging the reputation of Emilie du Châtelet, his official lover.

[218] Jauch. 1998. *Jenseits der Maschine*. 321, see note no. 13.

[219] Häseler. 2005. Friedrich II. von Preußen – oder wie viel Wissenschaft verträgt höfische Kultur? 73–81.

Frederick of Prussia's closer circle. Although most scholars agree on the importance of names such as Voltaire, Maupertuis, La Mettrie and even Samuel König or Leonard Euler when it comes to the philosophers who surrounded Frederick, Du Châtelet is mentioned rarely. Her position amongst and her relationships with these men, however, prove to be of vital importance for the history of science in this epoque.

Emilie du Châtelet and the Transformation of Metaphysics

Much has been written about Du Châtelet and her reversal from Newtonianism to Leibnizianism and possibly back. She was a straight, genuine, and independent thinker, by no means a disciple or an idolizer. From the very beginning of her writing she positions herself between the reasonable scholastics and the experience-based modern scientists. Her specific perspective is her power and her drive within the Leibniz-Newton debate. I have tried to sketch out how Du Châtelet was an attendant but self-determined philosopher from the very beginning of her entrance to the philosophical circles. Her philosophy shows clear differences when compared with Voltaire and her former teacher, Maupertuis. She was never conform and never subjected herself to the opinions of others. Yet, with Voltaire she found the philosophical environment that gave her the chance to articulate and to specify her own ideas within a friendship, as she documents in the introduction to her translation of the *Fable of the Bees*. She took this chance to position herself and her thoughts, criticizing her companion's idol, John Locke. She justifies the necessity to presuppose a priori principles and pronounces a sort of rational metaphysical approach from the very beginning. She defines a metaphysical conviction that formal principles constitute the basis of our knowledge. This kind of metaphysical approach can be found quite early in Du Châtelet's writings, and the origin and the root for her deeper understanding of Leibniz might be seen in this.

Du Châtelet criticizes Locke's philosophy in an elementary respect. She does not contest all of Locke's insights, but she emphasizes the necessity of the verification of knowledge through experience. She indicates the rational deficiencies of the empiricists as she points at the deficiencies of the rationalists and scholastics. Descartes filled the Book of Nature with "fables and reveries", Locke is "confused" and his idea of *thinking matter* is harshly rejected by her at an early stage. In her view the rationalists before her were as bad at metaphysics as the Newtonian scholars, as the former reasoned without considering history and experience, and the latter were not interested in finding out why and how the laws of attraction should be true. Du Châtelet criticized the philosophical traditions she faced and obviously her criticism was well known among the enlightened philosophers. Maupertuis' and La Mettrie's reference to Du Châtelet's deliberations on motion and free will, on *thinking matter* and numbers and on the way to do metaphysics indicate the importance of her reflections. She positions herself as a strong metaphysician, convinced of innate principles, free will and the human capacity to elaborate reliable knowledge. Mistaking the effects for the causes and grounding the truth in scientific or mathematical laws causes the truth to be twisted and unreachable. She rebuts the claim to finding truth by using mathematical laws, since the most convincing and beautiful mathematical law can be rich in evidence, but it remains a result, a consequence as Du Châtelet argues against Maupertuis.

Du Châtelet pronounces explicitly the hypothetical character of science and she is aware that a strong science needs strong metaphysics for its foundation. She indeed created a metaphysics of science that is in its method critical. Her method to establish scientific findings and justification by hypotheses and probabilities is outstanding in its systematic approach. Science and metaphysics belong together as science tests probabilities and justifies knowledge within this process, paying thus tribute to the importance of "history and experience" within her philosophy. Her metaphysics is based on the conviction of a strong rationalism as science is created by and an expression of free human beings. Against the arbitrariness of this approach and scepticism she holds a realistic metaphysics. Knowledge can partly indicate truth, however knowledge is never complete. It is the world which forces us to get to know it and there is no science and no method that guarantees its findings. Following, there is neither a scientist nor a mathematician nor a philosopher who has ever found or will ever find the whole truth. Knowledge and truth are much more the result of a constructive and heuristic undertaking to which everybody who is willing to dedicate her life to understand the theories and their relations is able to contribute.

Establishing a new way of hypothetical scientific reasoning that was situated between the scholastic minded rationalists and the modern phenomenalists, – leaving Descartes and rebuffing Locke – designates Emilie du Châtelet's intellectual achievement. She transforms equally metaphysics and experience based science, constituting metaphysics of science. From then on metaphysics had to pay attention to experience and science as well as experience based philosophy and science had to become aware of its hidden metaphysical roots. In this period metaphysics and science started to separate. In Du Châtelet we can learn how the two disciplines can be related to each other. This kind of transformation is also the essence of Du Châtelet's contribution to the Leibniz-Newton debate of her time. My contribution was written to position Du Châtelet within the context of her time. The discussion between Newtonians and Leibnizians was one of the most prospering disputes in the history of Western thought. It is important to show how deeply Du Châtelet was rooted in this discussion and her efforts to find a new way out of this fruitless confrontation. She created a new kind of reflection to reevaluate the empiricistic and rational arguments in her method of a critical metaphysics of science. The roots she put down grew into fruits that inspired a wide range of philosophers. Her influence on Voltaire, Maupertuis, Euler, Kant and La Mettrie are being discussed. Her inspiration for many others, such as Diderot, d'Alembert, Condorcet, D'Holbach, and Hume, will have to be analysed. A new kind of history of philosophy which includes women philosophers is taking its first steps.

References

Adelson, R. 2008. La belle Issé: Mme Du Châtelet musicienne. In *Émilie Du Châtelet: Éclairages & documents nouveaux*, ed. Ulla Kölving and Olivier Courcelle, 127–34. Ferney-Voltaire: Centre International d'Étude du XVIIIe Siècle.

Alexander, H. G. 1998. *The Leibniz-Clarke correspondence*. Manchester: Manchester University Press.

Arndt, H. W. 1983. Rationalismus und Empirismus in der Erkenntnislehre Christian Wolffs. In *Christian Wolff, 1679–1754. Interpretationen zu seiner Philosophie und deren Wirkung*, ed. Werner Schneiders, 31–48. Hamburg: Felix Meiner.

Baruzzi, A. 1968. *Aufklärung und Materialismus im Frankreich des 18. Jahrhunderts, La Mettrie - Helvétius, Diderot-Sade*. München: List

Bayle, P. ²1702. *Dictionnaire historique et critique*. Rotterdam.

Beeson. 1992. *Maupertuis: an intellectual biography*. Oxford: Voltaire Foundation.

Bellugou, H. 1962. *Voltaire et Frédéric au temps de la marquise du Châtelet. Un trio singulier*. Paris: Marcel Rivière.

Bernoulli, Johann. 1742. *Opera omnia tam antea sparsim edita, quam hactenus inedita*. Georg Olms.

Bessire, F. 2008. Mme Du Châtelet épistolière. In *Émilie Du Châtelet: Éclairages & documents nouveaux*, ed. Ulla Kölving and Olivier Courcelle, 25–35. Ferney-Voltaire: Centre International d'Étude du XVIIIe Siècle.

Biarnais, M.-F. 1982. *Les Principia de Newton. Genèse et structure des chapitres fondamentaux avec traduction nouvelle*. Paris: Société française d'histoire des sciences et des techniques.

Böttcher, F. 2008. La réception des *Institutions de physique* en Allemagne. In *Emilie Du Châtelet: Éclairages & documents nouveaux*, ed. Ulla Kölving and Olivier Courcelle, 243–54. Ferney-Voltaire: Centre International d'Étude du XVIIIe Siècle.

Brown, A., and U. Kölving. 2003. Qui est l'auteur du Traité de métaphysique? *Cahiers Voltaire* 2:85–94.

Brown, A., and U. Kölving. 2008. À la recherche des livres d'Émilie Du Châtelet. In *Émilie Du Châtelet: Éclairages & documents nouveaux*, ed. Ulla Kölving and Olivier Courcelle, 111–20. Ferney-Voltaire: Centre International d'Étude du XVIIIe Siècle.

Brucker, J. J. 1745. *Bilder-Sal heutigen Tages lebender und durch Gelahrtheit berühmter Schrifftsteller, in welchem derselbigen nach wahren Original-malereyen entworfene Bildnisse in schwartzer Kunst in natürlicher Aehnlichkeit vorgestellt, und ihre Lebensumstände, Verdienste um Wissenschafften, und Schriften aus glaubewürdigen Nachrichten erzählt werden*. Augsburg: Joh. Jacob Haid.

Brunet, P. 1929. *Maupertuis. Etude biographique*. Paris: A. Blanchard.

Carboncini, S. 1984. Lumière e Aufklärung – A proposito della presenza della filosofia di Christian Wolff nell'Encyclopédie. *Annali della Scuola Normale Superiore di Pisa* 14, no. 4:1297–301.

D'Alembert, J. B. 1743. *Traité de dynamique*. Paris.

Deschamps, J. 1742. Cours abrégé de la philosophie Wolfienne en formes de lettre. In *Gesammelte Werke*, ed. Christian Wolff. Hildesheim/Zürich/New York: Olms

Diderot, D. 1955 ff. *Correspondance*. 11 vols. Paris: Éditions de Minuit.

Douay-Soublin, F. 2008. Nouvel examen de la *Grammaire raisonée* de Mme Du Châtelet. In *Émilie Du Châtelet: Éclairages & documents nouveaux*, ed. Ulla Kölving and Olivier Courcelle, 173–196. Ferney-Voltaire: Centre International d'Étude du XVIIIe Siècle.

Droysen, H. 1910. Die Marquise du Châtelet, Voltaire und der Philosoph Christian Wolff. *Zeitschrift für französische Sprache und Literatur* 35:226–48.

Du Châtelet, E. 1738. Lettre sur les Eléments de la Philosophie de Newton. Amsterdam. *Journal des sçavans*:458–75.

Du Châtelet, E. 1947. L'Essai sur l'optique: Chapitre IV. De la formation des couleurs par Me du Chastellet. In *Studies on Voltaire. With some unpublished papers of Mme. du Châtelet*, ed. Ira O. Wade, 188–208. Princeton: Princeton University Press.

Du Châtelet, E. 1947. Mme. du Châtelet's Translation of the *Fable of the Bees*. In *Studies on Voltaire. With some unpublished papers of Mme. du Châtelet*, ed. Ira O. Wade, 131–87. Princeton: Princeton University Press.

Du Châtelet, E. 1961. *Discours sur le bonheur: introduction et notes de Robert Mauzi*. Paris: Les Belles-Lettres.

Du Châtelet, E. 2009. *Emilie Du Châtelet: Selected Philosophical and Scientific Writings*. Selected and edited by Judith P. Zinsser. Translated by Isabelle Bour and Judith P. Zinsser. Chicago: University of Chicago Press.

Dufrenoy, M.-L. 1963. Maupertuis et le progrès scientifique. *SVEC* 25: 519–587.

Eberhard, J. A. 1789. Über den wesentlichen Unterschied der Erkenntnis durch die Sinne und durch den Verstand. In *Philosophisches Magazin*. Vol. I. Halle: Johann Jacob Gebauer.

Edwards, Samuel. 1989. *Die göttliche Geliebte Voltaires. Das Leben der Emilie du Châtelet*. Engelhorn: Stuttgart.

Formey, J. H. S. 1789. *Souvenirs d'un citoyen*. 2 vols. Berlin: Lagarde.

Gandt, F. de, ed. 2001. *Cirey dans la vie intellectuelle: la réception de Newton en France*. Oxford: Voltaire Foundation.

Gardiner, L. J. 1982. Searching for the Metaphysics of Science: the Structure and composition of Madame du Châtelet's *Institutions de physique*, 1737–40. *SVEC* 201:85–113.

Gardiner, L. J. 2008. Mme Du Châtelet traductrice. In *Émilie Du Châtelet: Éclairages & documents nouveaux*, ed. Ulla Kölving and Olivier Courcelle, 167–72. Ferney-Voltaire: Centre International d'Étude du XVIIIe Siècle.

Gooch, G. P. 1961. Mme du Châtelet an her Lovers. *Contemporary Review* 200: 648–53.

Guyot, P. 2008. La pédagogie des *Institutions de physique*. In *Émilie Du Châtelet: Éclairages & documents nouveaux*, ed. Ulla Kölving and Olivier Courcelle, 267–81. Ferney-Voltaire: Centre International d'Étude du XVIIIe Siècle.

Hagengruber, R. 1999. Eine Metaphysik in Briefen. Emilie du Châtelet an Maupertuis. In *Pierre Louis Moreau de Maupertuis (1698–1759). Eine Bilanz nach 300 Jahren*, ed. Hartmut Hecht, 187–206. Berlin: Spitz-Verlag.

Hagengruber, R. 2002. Emilie du Châtelet. Gegen Rousseau und für die Physik. Wissenschaft im Zeitalter der Aufklärung. *Konsens* 3, no. 18:27–30.

Hagengruber, R. 2010. Das Glück der Vernunft - Emilie du Châtelets Reflexionen über die Moral. In *Von Diana zu Minerva. Philosophierende Aristokratinnen des 17. und 18. Jahrhunderts*, ed. Ruth Hagengruber and Ana Rodrigues, 109–28. Berlin: Akademie Verlag.

Hagengruber, R. 2010. Von Diana zu Minerva. Philosophierende Aristokratinnen des 17. und 18. Jahrhunderts. In *Von Diana zu Minerva. Philosophierende Aristokratinnen des 17. und 18. Jahrhunderts.*, ed. Ruth Hagengruber and Ana Rodrigues, 11–32. Berlin: Akademie-Verlag.

Hankins, T. L. 1970. *Jean D'Alembert. Science and the Enlightenment*. Oxford: Clarendon Press.

Häseler, J. 2005. Friedrich II. von Preußen – oder wie viel Wissenschaft verträgt höfische Kultur? In *Geist und Macht*, ed. Brunhilde Wehinger, 73–81. Berlin: Akademie Verlag.

Hecht, H., ed. 1991. *Gottfried Wilhelm Leibniz im philosophischen Diskurs über Geometrie und Erfahrung*. Berlin: Akademie Verlag.

Hecht, H., ed. 1999. *Pierre Louis Moreau de Maupertuis (1698–1759). Eine Bilanz nach 300 Jahren*. Berlin: Spitz-Verlag.

Hecht, H. 1994. Pierre Louis Moreau de Maupertuis oder die Schwierigkeit der Unterscheidung von Newtonianern und Leibnizianern. In *Leibniz und Europa: VI. Internationaler Leibniz-Kongreß*, ed. Herbert Breger, 331–38. Hannover: G.-W.-Leibniz-Gesellschaft.

Hecht, H. 2001. Leibniz'Concept of Possible Worlds. In *Between Leibniz, Newton, and Kant*, ed. Wolfgang Lefèvre, 27–45. Dordrecht/Boston/London: Kluwer.

Heimann, P. M. 1977. Geometry and Nature: Leibniz and Johann Bernoulli's Theory of Motion. *Centaurus* 21:1–26.

Hutton, S., and J. E. Force, eds. 2004. *Newton and Newtonianism: new studies*. Boston: Kluwer.

Iltis, C. 1977. Leibniz and the Vis viva Controversy. *Isis* 62:21–35.

Iltis, C. 1977. Madame du Châtelet's Metaphysics and Mechanics. *Studies in history and philosophy of science* 8, no. 1:29–48.

Iverson, J. R. 2006. A female member of the Republic of Letters: Du Châtelet's Portrait in Bilder-Sal (…) berühmter Schrifftsteller. In *Emilie Du Châtelet: rewriting Enlightenment philosophy and science*, ed. Judith P. Zinsser and Julie C. Hayes, 35–51. Oxford: Voltaire Foundation.

Iverson, J. 2008. Émilie Du Châtelet, Luise Gottsched et la Société des Alétophiles: une traduction allemande de l'échange au sujet des forces vives. In *Émilie Du Châtelet: Éclairages & documents nouveaux*, ed. Ulla Kölving and Olivier Courcelle, 283–99. Ferney-Voltaire: Centre International d'Étude du XVIIIe Siècle.

Jauch, U. P. 1998. *Jenseits der Maschine: Philosophie, Ironie und Ästhetik bei Julien Offray de La Mettrie (1709–1751)*. München: Hanser.

Kant, I. 1902. Gedanken von der wahren Schätzung der lebendigen Kräfte und Beurtheilung der Beweise, deren sich Herr von Leibniz und andere Mechaniker in dieser Streitsache bedient haben. In *Kants Werke. Akademie-Textausgabe*, vol I, 1–182. Berlin: Königl. Preußische Akademie der Wissenschaften.

Klens, U. 1994. *Mathematikerinnen im 18. Jahrhundert: Maria Gaetana Agnesi, Gabrielle-Emilie Du Châtelet, Sophie Germain: Fallstudien zur Wechselwirkung von Wissenschaft und Philosophie im Zeitalter der Aufklärung*. Pfaffenweiler: Centaurus.

Knobloch, E. 1995. Das große Spargesetz der Natur: Zur Tragikomödie zwischen. Euler, Voltaire und Maupertuis. *Mitteilungen der Deutschen Mathematikervereinigung* 3, 14–20.

Kölving, U., and O. Courcelle, eds. 2008. *Émilie Du Châtelet: Éclairages & documents nouveaux*. [… du Colloque du Tricentenaire de la Naissance d'Emilie Du Châtelet qui s'est tenu du 1er au 3 juin 2006 à la Bibliothèque Nationale de France et à l'ancienne Mairie de Sceaux]. Ferney-Voltaire: Centre International d'Étude du XVIIIe Siècle.

Kölving, U., and A. Brown. 2008. Émilie Du Châtelet, lectrice d'une Apologie d'Homère. In *Émilie Du Châtelet: Éclairages & documents nouveaux*, ed. Ulla Kölving and Olivier Courcelle, 135–65. Ferney-Voltaire: Centre International d'Étude du XVIIIe Siècle.

Kölving, U. 2008. Bibliographie chronologique d'Émilie Du Châtelet. In *Émilie Du Châtelet: Éclairages & documents nouveaux*, ed. Ulla Kölving and Olivier Courcelle, 341–85. Ferney-Voltaire: Centre International d'Étude du XVIIIe Siècle.

Kölving, U. 2008. Emilie Du Châtelet devant l'histoire. In *Émilie Du Châtelet: Éclairages & documents nouveaux*, ed. Ulla Kölving and Olivier Courcelle, 1–13. Ferney-Voltaire: Centre International d'Étude du XVIIIe Siècle.

La Mettrie, J. O. de. 1747. *Histoire naturelle de l'ame, traduite de l'Anglois de M. Charp, par feu M.H.*** de l'Académie des Sciences, et. Nouvelle Edition … & augmentée par la Lettre Critique de M. de la Mettrie à Madame la Marquise Du Châtelet*. Oxford.

La Mettrie, J. O. de. 1921. *Introduction et notes de Maurice Solovine*. Paris: Bossard.

La Mettrie, J. O. de. 1975. *Discours sur le bonheur*. Critical edition by John Falvey. Banbury/Oxfordshire: Voltaire Foundation.

La Mettrie, J. O. de. 1988. *Der Mensch als Maschine*. LSR-Verlag.

Lefèvre, W. 2001. *Between Leibniz, Newton, and Kant. Philosophy and Science in the Eighteenth Century*. Boston Studies in the Philosophy of Science. Kluwer: Dordrecht.

Leibniz, J. G. 1686. *Discours de métaphysique*.

Leibniz, G. W. 1887. *Nouveaux essais sur l'entendement humain*. Paris: J.H. Vrin.

Leibniz, G. W. 1998. *Monadologie*, ed. Hartmut Hecht. Reclam: Stuttgart.

Locke, J. 1975. *An Essay Concerning Human Understanding*, ed. Peter Nidditch. Oxford.

Maglo, K. 2008. Mme Du Châtelet, l'Encyclopédie, et la philosophie des sciences. In *Émilie Du Châtelet: Éclairages & documents nouveaux*, ed. Ulla Kölving and Olivier Courcelle, 255–66. Ferney-Voltaire: Centre International d'Étude du XVIIIe Siècle.

Mairan, J. J. D. de. 1741. *Dissertation sur l'estimation et la mesure des forces motrices des corps*. Paris: Charles-Antoine Jombert.

Mariotte, E. 1673. *Traité de la percussion ou choc des corps, dans lequel les principales règles du mouvement, contraires à celles que Mr. Descartes et quelques autres modernes ont voulu établir, sont démontrées par leurs véritables causes*.

Maupertuis, P. L. M. de. 1746. Les Loix du mouvement et du repos déduites d'un principe metaphysique. *Histoire e l'Académie Royale des Sciences et des Belles Lettres*: 1745–1769. Berlin: A. Haude, 1753. 267–94.

Maupertuis, P. L. M. de 1752. *Lettres de M. de Maupertuis*. Dresden: George Conrad Walther.
Maupertuis, P. L. M. de. 2010. *Essai de philosophie morale*. Whitefish: Kessinger.
Maupertuis, P. L. M. de. 1750. *Versuch in der moralischen Weltweisheit*. Halle: Gebauer
Maupertuis, P. L. M. de. 1752. *Lettres de Mr. De Maupertuis*. Dresden: Georg Conrad Walther.
Maupertuis, P. L. M. de 1768. *Oeuvres et lettres*. 4 vols. Lyon: Bruyset.
McMullin, Ernan. 2009. *Hypothesis in Early Modern Science*. Berlin/New York: De Gruyter.
Mozans, H. J. 1913. *Women in science: with an introductory chapter on on woman's long struggle for things of the mind by H.J. Mozans; with a preface by Cynthia Russett and an introduction by Thomas P. Gariepy*. New York/London: D. Appleton.
Newton, I. 1999. *Philosophiae Naturalis Principia Mathematica*, ed. Bernard Cohen and translated by Anne Whitman. University of California.
Otto, S. 1989. *Giambattisto Vico. Grundzüge seiner Philosophie*. Stuttgart: Kohlhammer.
Rey, A.-L. 2008. La figure du leibnizianisme dans les *Institutions de physique*. In *Émilie Du Châtelet: Éclairages & documents nouveaux,* ed. Ulla Kölving and Olivier Courcelle, 231–42. Ferney-Voltaire: Centre International d'Étude du XVIIIe Siècle.
Robinet, A. 1999. Les trois modèles cartésiens d'automates. 35–41. In *Matière pensante : études historiques sur les conceptions matérialistes en philosophie de l'esprit*, ed. Jean-Noel Missa. Paris: Vrin.
Rodrigues, A. 2010. Emilie du Châtelet, Julien Offray de La Mettrie und Pierre Louis Moreau de Maupertuis im Zwiegespräch über das Glück. In *Von Diana zu Minerva. Philosophierende Aristokratinnen des 17. und 18. Jahrhunderts*, ed. Ruth Hagengruber and Ana Rodrigues, 151–60. Berlin: Akademie Verlag.
Rousseau, J. J. 1978. *Emile oder über die Erziehung*. Paderborn: Klett-Cotta.
Schwarzbach, B. E. 2006. Mme du Châtelet's *Examens de la Bible* and Voltaire's *La Bible enfin expliquée*. In *Emilie Du Châtelet: rewriting Enlightenment philosophy and science*, ed. Judith P. Zinsser and Julie C. Hayes, 142–64. Oxford: Voltaire Foundation.
Schwarzbach, B. E. 2008. Mme Du Châtelet et la Bible. In *Émilie Du Châtelet: Éclairages & documents nouveaux,* ed. Ulla Kölving and Olivier Courcelle, 197–211. Ferney-Voltaire: Centre International d'Étude du XVIIIe Siècle.
Specht, R. 1989. *John Locke*. München: Beck.
Suisky, D. Forthcoming. Emilie du Châtelet und Leonhard Euler über die Rolle von Hypothesen. In *Emilie du Châtelet und die deutsche Aufklärung*, ed. Ruth Hagengruber and Hartmut Hecht. Hildesheim/New York: Olms.
Szabo Compare, I. 1997. Der philosophische Streit um das wahre Kraftmaß im 17. und 18. Jahrhundert. In *Geschichte der mechanischen Prinzipien und ihrer wichtigsten Anwendungen*, ed. István Szabo Compare, 47–85. Basel: Birkhäuser.
Terrall, M. 2002. *The man who flattened the Earth – Maupertuis and the sciences in the Enlightenment*. Chicago: University of Chicago Press.
Terrall, M. 2004. Vis viva revisited. *History of Science* 42, no. 2:189–209.
Tonelli, G. 1959. La nécessité des lois de la nature au XVIIIe siècle et chez Kant en 1762. *Revue d'histoire des sciences et de leurs applications* 12, no. 3: 225–241.
Vaillot, R. 1978. *Madame du Châtelet: Préface de René Pomeau*. Paris: Albin Michel.
Vartanian, A. 1960. *La Mettrie's L'Homme Machine: A Study in the Origins of an Idea*. Princeton: Princeton University Press.
Voltaire. 1736. Épître à madame la marquise du Châtelet. In *Alzire ou les Américains. Tragédie de M. de Voltaire*. Représentée pour la première foi le 27 janvier 1736, 5–12. Paris: Jean-Baptiste-Claude Bauche.
Voltaire. 1738. *Éléments de la philosophie de Newton*. Amsterdam: Etienne Ledet & Compagnie (ou) Jacques Desbordes.
Voltaire. 1877–1885. *Oeuvres complètes de Voltaire,* ed. Louis Moland. Paris: Garnier frères.
Voltaire. 1894. *Letters on England*. New York: Cassell.
Voltaire. 1968–1977. *Complete works of Voltaire. Les oeuvres complètes de Voltaire*, ed. Theodore Besterman. Vols 135. Genève: Institut et Musée Voltaire.

Voltaire. 1997. *Elemente der Philosophie Newtons. Verteidigung des Newtonianismus. Die Metaphysik des Neuton,* ed. Renate Wahsner and Horst-Heino v. Borzeszkowski. Berlin/New York: de Gruyter.

Wade, Ira O., ed. 1947. *Studies on Voltaire. With some unpublished papers of Mme. Du Châtelet.* Princeton: Princeton University Press.

Waithe, M. E., ed. ²1991. *A History of Women Philosophers (Vol. III). Modern Women Philosophers, 1600–1900.* Dordrecht/Boston/London: Kluwer Academic Publishers.

Walters, R. L. 2001. La querelle des *forces vives* et le rôle de Mme du Châtelet. In *Cirey dans la vie intellectuelle: la réception de Newton en France,* ed. François de Gandt, 198–211. Oxford: Voltaire Foundation.

Whitehead, B. 2006. The singularity of Mme du Châtelet: An analysis of the *Discours sur le bonheur.* In *Emilie Du Châtelet: rewriting Enlightenment philosophy and science,* ed. Judith P. Zinsser and Julie C. Hayes, 255–76. Oxford: Voltaire Foundation.

Zinsser, J. P., and O. Courcelle. 2003. A remarkable collaboration: the marquise du Châtelet and Alexis Clairaut. *SVEC* 12: 107–20.

Zinsser, J. P. 2005. *Men, women and the birthing of modern science.* DeKalb (IL): Northern Illinois University Press.

Zinsser, J. P. and J. C. Hayes, eds. 2006. *Emilie Du Châtelet: rewriting Enlightenment philosophy and science.* Oxford: Voltaire Foundation.

Zinsser, J. P. 2007. *Emilie du Châtelet: Daring Genius of the Enlightenment.* New York: Penguin.

Zinsser, J. P. 2008. Mme Du Châtelet: sa morale et sa métaphysique. In *Émilie du Châtelet: Éclairages & documents nouveaux,* ed. Ulla Kölving and Olivier Courcelle, 219–29. Ferney-Voltaire: Centre International d'Étude du XVIIIe Siècle.

In the Spirit of Leibniz – Two Approaches from 1742

Hartmut Hecht

In the 1740s, German scholars were slowly beginning to depart from their original focus on Leibniz' traditional contributions to metaphysics such as the theory of monads and the problem of theodicy to thus far unexplored territories of his work, thereby slowly uncovering new perspectives. The prize question of the *Académie des Sciences et Belles Lettres*, formerly known as *Societät der Wissenschaften*, in 1747 and the subsequent dispute over the actual priorities in formulating principle of least action quantity demonstrate that Leibniz' work was widely interpreted from a natural science point of view, and the academy's competition, *L'examen de l'hypothèse des Monades*, was consequently not merely addressed to philosophers but to a range of other fields and disciplines as well. Works focusing on the concept of monads and on the particular value it might have for scientific research in general were more than welcome. Similarly, Maupertuis' principle of least action might at first sight seem to be a primarily metaphysical one. Yet, as his overall work proves, to him, it proved to be less of a theoretical concept than it was an *esprit systématique* which contributed greatly and innovatively to all areas of his research.

It is an established fact that this new approach to Leibniz' work originated in the francophone parts of Europe and was first introduced to German academic circles by scholars such as Maupertuis and Euler. However, even before Euler could illustrate the way in which the *Monadologie* was compatible with the thoughts and theories of physics in the eighteenth century in his *Gedancken von den Elementen der Cörper* in 1746, Emilie du Châtelet had already presented a much more detailed and more extensive contribution to the field. Her *Institutions de physique* was published in 1740; it was followed by a second edition in 1742, and it was subsequently translated into German in 1743.

H. Hecht (✉)
Berlin Brandenburgische Akademie der Wissenschaften, Jägerstrasse 22/23, 10117 Berlin, Germany
e-mail: hecht@bbaw.de

It shows that the French influence on the Leibniz debate was considerable but by no means homogenous. While Maupertuis' and Euler's critical studies focussed on aspects of natural science primarily, Mme du Châtelet's *Institutions de physique* could be seen as a statement in favour of a systematic philosophy which was aimed at the breathtaking scientific developments of her own time. The German Enlightenment was therefore in a position to choose between two very different approaches. By introducing two relevant studies from 1742, the following now seeks to illustrate these in some more detail.

Maupertuis' *Lettre sur la comète*

Admittedly, Maupertuis' *Lettre sur la comète* is not the first thing that comes to mind when looking for early-eighteenth-century interpretations of Leibniz' work. However, although Leibniz' name is never mentioned in the *Lettre sur la comète*, Maupertuis' study may well serve to demonstrate the influence of Leibniz' work on scientific debates in the decades following his death. In his *Lettre*, Maupertuis explicitly refers to a phenomenon of astronomy which had been causing disputes since antique times, and it was in fact Emilie du Châtelet who had initiated the discussion.

Emilie was interested in Maupertuis' opinion regarding a comet which had been predicted to appear in March 1742 and that had already caused old superstitions and fears to re-emerge among her contemporaries. She was looking for some form of scientific enlightenment which could serve as a valid explanation for the phenomenon. Therefore, she not merely expected Maupertuis to inform her of the latest research developments and results; she also wanted to know how he would judge their relevance for the idea of reason and for the European Enlightenment in general. Maupertuis was more than happy to comply with her request. Du Châtelet received his reply on "ce 26 Mars 1742".[1]

In his letter, Maupertuis presents himself as a *philosophe*, and it takes him only 50 octavos to discuss a scientific problem that for generations had been the cause of numerous speculations rather than of serious scientific calculation. Maupertuis assumes a classicist approach to Emilie's question and quotes Horace's "Tu ne quaesieris, scire nefas"[2] on his title page. Combining both historical and contemporary research on the matter of comets, the verse thus propagates an understanding which is based on scientific fact rather than on superstition. By calculating the comet's orbits, Maupertuis successfully demonstrates how superstitions such as bad omens can be overcome via a more scientific approach to astronomy. Yet he also shows that there are more riddles to comets than could be solved by merely establishing their orbits, and he extends his subject matter to a scientific methodology which may serve to describe phenomena such as comets as more than mere 'objects' in Newton's sense. In other words, in his *Lettre*, Maupertuis seeks to develop a methodology which would acknowledge Newton's results while, at the same time,

---

[1] Maupertuis. 1768b. Lettre. 256.
[2] Maupertuis. 1768b. Lettre. 207.

could serve to problematise their metaphysical cause. And it is thus Leibniz' work which eventually inspires him to a natural philosophy *en miniature*.

It only takes three pages for Maupertuis to reveal an essential aspect of this particular understanding of Leibniz. The passage reads:

> Il est bien vrai qu'il y a une connexion universelle entre tout ce qui est dans la Nature, tant dans le physique que dans le moral: chaque événement lié à celui qui le précede, & à celui qui le suit, n'est qu'un des anneaux de la chaîne qui forme l'ordre & la succession des choses: s'il n'étoit pas placé comme il est, la chaîne seroit différente, & appartiendroit à un autre Univers.[3]

Maupertuis' approach is based on the concept of possible worlds here, and the passage provides ample proof that he had a profound knowledge of Leibniz' metaphysics and its implications. If an event were to occur in a different way than it actually does, there would still be nothing to prevent us from saying that this would be another possible universe which God has chosen, and it would in that case be truly another individual one. At the same time, however, Maupertuis' grasp of Leibniz' ideas also demonstrates that he sought to utilise the Leibniz' principles for a more general scientific purpose.

Stating that the notion of a universal harmony of things postulates any phenomenon to provide ample information on the nature of the change in others, Maupertuis proceeds to uncover a central principle in all of Leibniz' thoughts. But while he follows Leibniz' arguments to some great extent, he also draws attention to the fact that they can not be empirically verified, due to the limitations human cognitive capacity shows when trying to grasp temporal and spatial relations. As a consequence, he claims that, in accordance with reason, it is imperative to concentrate on those phenomena we can comprehend. His advice to Mme du Châtelet is that "[j]e ne vous parlerai que de celles qui sont à notre portée, & dont ont peut donner des raisons mathématiques ou physiques."[4]

In relation to Mme du Châtelet's question on the matter of comets, then, this means that their existence will have to be perceived as part of an overarching and systemic order, since they orbit the sun on elliptic paths not unlike those of the planets. Although there exist subtle differences between the two, namely the fact that comets have larger eccentricities and show larger orbital periods than the planets, Maupertuis eventually demonstrates how Newton's theory can be seen to integrate the comets as "nouvelles planetes"[5] into our system of the world.

From the start, Maupertuis' discussion of the cometary law of motion makes frequent mention of those philosophers who claimed that planets rotate around the earth on whirls, and he explains how imperative it is to uncover the true causes behind Newton's forces. Maupertuis not only examines the theory of comets from an empirical angle, he also seeks to implement empirical results into philosophical debates on the system of the world, its innate structure and the changes within. This he does in some detail, and it is illustrated by his theoretical

---

[3] Maupertuis. 1768b. Lettre. 211 f.

[4] Maupertuis. 1768b. Lettre. 213.

[5] Maupertuis. 1768b. Lettre. 224.

interest in the living conditions one might find on comets with elongated and elliptical orbits.

Maupertuis' comments here already bear witness of a very distinctive methodology. His assumptions are generally grounded in what is completely determined according to Leibniz, i.e. in a mathematical theory on cometary motion, and he goes on to integrate this theory into a world system defined by mathematical principles. As a *philosophe*, however, Maupertuis also realises that any notion of the world which can be reduced to mere mathematical facts is bound to have its flaws. Consequently, he seeks to adhere to a coherent finality in all of his thoughts which in many ways resembles Leibniz' own thinking. Where such finality can not be attained, he is not afraid to formulate hypotheses based on existing results which leave room for some scientific imagination.

Maupertuis' understanding of 'enlightenment' is therefore not devoid of imagination. He frequently sets the idea of enlightenment against common superstition when entering the current debate on comets. Yet he also claims that inspiration alone will inevitably have to fall short of explaining the laws of nature. The second part of his *Lettre sur la comète* illustrates this point further.

If one were to consider that the number of comets known to us only represents a mere fraction of the total comet population, and that it is only possible to anticipate the orbits of these comets due to insufficient data, it is perfectly conceivable that one of them could describe a movement so close to the earth that it might physically affect it, Maupertuis writes, and he turns out to be a master of scientific imagination when it comes to illustrating such events. Claiming that it could be possible for the axis of the Earth to change its relation to the eclipses, thereby turning prospering landscapes into polar wastelands, he subsequently goes on to discuss possible changes in the shape and position of the Earth's orbit and conceives of a comet that could knock out pieces which would then come to orbit it as moons. He also imagines a new moon which could be added to our planet, and he suggests it might be possible that the Earth itself could fall under the lead of one of the larger comets.

Maupertuis derives all of these ideas from speculation and simple variation of the cometary orbits and their respective parameters. In addition, he also considers the impact of the comets' tails. Once these spread the earth with poison gas, he claims, whole populations could either be poisoned or burned by their heat. And by finally referring to the ideas of William Whiston, he also states that it might be possible for a comet to cause the great catastrophes of the apocalypse, since God's wrath will have to be bound to a natural outlet.[6]

All of these ideas are, of course, greatly indebted to Leibniz. Based on his concept of pre-established harmony in § 88 of the *Monadology,* they find their expression in the following passage:

---

[6] Maupertuis. 1768b. Lettre. 238.

> Cette Harmonie fait qve les choses conduisent à la grace par les voyes mêmes de la Nature; et qve ce globe par exemple doit etre detruit et reparé par les voyes naturelles, dans les momens qve le demande le gouuernement des Esprits; pour le chatiment des uns, et la recompense des autres.[7]

In his *Protogaea*, Leibniz provides a more detailed insight into his theory by – amongst other things – providing a scientific explanation for the Flood.[8] Unlike Maupertuis, however, Leibniz generally dismissed the impact of comets, as he considered it to be too insubstantial. He never specified his reasons, but it is likely that accepting theories such as the one Maupertuis proposes would have introduced coincidence as a scientific variable and was thus not well-founded.

Even though Maupertuis' and Leibniz' concept have a lot in common, therefore, it is worth mentioning that there exist subtle but significant differences between their respective approaches to the subject. Leibniz' perception of the world is based on a universal order which proves to be absolute in all of its aspects and has to extend to all assumptions in relation to this world. It is precisely this order, however, which cannot be fulfilled, were one to describe the occurrence of comets as incidental events which only allow for arbitrary prognoses.

Maupertuis' explorations into the peripheral areas of scientific fact were certainly eloquent enough, yet he himself did not seem entirely confident as to their effect on Emilie du Châtelet when he wrote:

> Si vous n'êtes pas convaincue, Madame, que le Déluge & la *conflagration* de la Terre dépendent de la Comete, vous avouerez du moins, je crois, que sa rencontre pourroit causer des accidens assez semblables.[9]

In fact, there was ample reason to be sceptical, given that Maupertuis' *esprit systématique* must have seemed somewhat closer to *esprit* than it was *systématique* to Mme du Châtelet. However, her scepticism did not affect her admiration for the designated president of the Berlin Academy. In her *Institiutions de Physique*, she wrote:

> Mr. de Maupertuis est de tous les Philosophes Français, celui qui a poussé le plus loin ses recherches sur l'attraction; il donna en 1732 à l'Académie un Mémoire, dans lequel il recherche la raison de la préférence, que le Créateur a donnée à la loi d'attraction, en raison inverse du quarré des distances, qui a lieu dans les Phénomènes astronomiques, & dans la chute des corps, sur les autres loix possibles, qui semblent avoir eu un droit égal à être employées [...].[10]

And although her philosophical understanding later proves to somewhat depart from her mentor's view, she even states that "[l]e Mémoire de Mr. de Maupertuis, dont je viens de parler, est comme tout ce que fait ce Philosophe, plein de sagacité & de finesse de calcul [...]."[11]

---

[7] Leibniz. 1998. *Monadologie*. 60, 62.
[8] Leibniz. 1949. *Protogaea*. 24 ff.
[9] Maupertuis. 1768b. Lettre. 241.
[10] Inst1742Rep1988 §392.
[11] Inst1742Rep1988 §393.

Emilie du Châtelet's *Institutions de Physique*

The differences in Du Châtelet's and Maupertuis' approaches to philosophy and natural science become apparent right away when considering the sheer length of their respective studies. Maupertuis' works are usually decisively shorter and he presents his thoughts as reports on scientific progress or in the style of short essays for the Academy rather than as scientific treatises of a whole discipline or as a pedagogical introduction to a particular branch of science. As a further look at the differences in style and form betrays, out of the two authors, Pierre Louis Moreau de Maupertuis also shows a decisively different interest in his approach to science. While Du Châtelet subscribes to an empirical approach to philosophy in accordance with the common academic interests of her time, Maupertuis' work is primarily directed at the emerging scholarly circles of his fellow academics. Both authors, however, succeed in allowing mutual access to their discipline, and they both furthered a process which would eventually lead to a paradigm shift in the fields of philosophy and of the natural sciences alike. Leibniz' ideas decisively influenced both Du Châtelet and Maupertuis, although his thoughts obviously led to somewhat different interpretations in their respective works.

Maupertuis clearly emphasised the methodological value of Leibniz' concept of possible worlds; it eventually lead him to formulate his own *principe de la moindre quantité d'action*. "*Lorsqu'il arrive quelque changement dans la Nature, la quantité d'action employée pour ce changement est toujours la plus petite qu'il soit possible*",[12] Maupertuis writes in this, and applied to the concept of possible worlds, it means, of course, that all other conceivable changes belong to other possible worlds. The existence of such an *extremum* therefore indicates choice. Nature is set up in such a way that the principle can be fulfilled, and, therefore, it allows for the very idea of final causes.

Mme du Châtelet agrees with this particular relation between possible worlds and final causes. However, instead of drawing her own conclusions from it regarding her scientific methodology, she primarily discusses the issue in its relation to the problem of theodicy. This shows that her approach to legitimising natural science is obviously somewhat different. Maupertuis was under the impression that his actions had revealed the actual forces of nature. He was satisfied with his results because they allowed him to describe these forces scientifically without requiring further metaphysical insights. Mme du Châtelet, however, disagreed:

> Plusieurs vérités de physique, de métaphysique, & de géométrie sont évidemment liées entre elles, la Métaphysique est la faîte de l'Edifice, mais ce faîte est si élevé, que la vue en devient souvent un peu confuse.[13]

To her, what is required for scientific understanding is the very idea of reason, especially regarding a research methodology that could eventually serve as a conduit

---

[12] Maupertuis. 1768a. Essay de Cosmologie. 42.
[13] Inst1742Rep1988 XII.

In the Spirit of Leibniz – Two Approaches from 1742   67

to truth and verification. And, according to Du Châtelet, this can be found in Leibniz' work, because "il nous a fourni dans le principe de la raison suffisante une boussole capable de nous conduire dans les sables mouvans de cette science."[14]

Principles in Natural Philosophy

In accordance with their great significance, Du Châtelet discusses two fundamental principles of human understanding right at the beginning of her first chapter, namely the principle of sufficient reason and the principle of contradiction. The request for non-contradiction in philosophy, she claims in § 7, has been an established factor in philosophy at all times. This means that something cannot be for as long as it is contradictory, and Emilie du Châtelet adds that this principle suffices when it comes to verifying necessary truths. Still, she also argues that, the principle of contradiction generally needs to be completed by the principle of sufficient reason if there are contingent truths involved. This particular principle, she claims, was first introduced by Leibniz.

If the principle of contradiction can provide information on the possibility or non-possibility of things, it is therefore the principle of sufficient reason which establishes whether such things exist in the first place. §§ 13–17 consequently introduce the principle of continuity in some detail (it follows from the principle of sufficient reason) and they illustrate its geometric and physical relevance. Both paragraphs show how Du Châtelet sees philosophical principles rooted in science itself.

Figures 1 and 2 discuss geometric curves which seem to be in accordance with the principle of continuity, and by introducing Fig. 3, Du Châtelet subsequently provides an example which, according to her, does indeed not follow Leibniz' principle. As opposed to curves AC in Figs. 1 and 2, curve AC in Fig. 3 actually consists of two parts, AB and BC, which are based on different functions and, thus, negate continuous transition. Her argument seems convincing enough when considered in

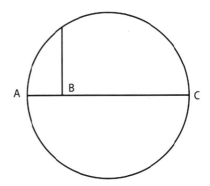

Fig. 1 Continuous curve

---

[14] Inst1742Rep1988 XII.

Fig. 2 Continuous curve consisting of two parts

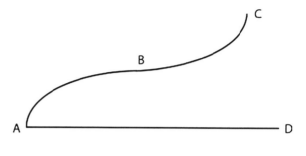

Fig. 3 Discontinuous curve

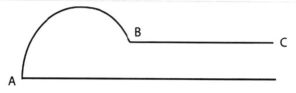

the light of the continuity of graphs. Yet is this kind of continuity really similar to Leibnition principle of continuity? And is this still in keeping with Leibniz' theory?

The *Discours de métaphysique* provides a similar passage. In it, Leibniz claims that, in accordance with the universal order of the world, there always exists a geometric line connecting any number of points arbitrarily thrown onto a piece of paper, and that it is always possible to generate a function that could describe such a phenomenon. He goes on to use an impressive image to illustrate his point. According to him, there exists not a single facial outline that would not be part of a geometric line, and he further suggests that any such outline could be drawn via a continuous movement based on a particular function. Eventually, he also addresses the problem Mme du Châtelet discusses. Even if someone were to draw a line which partly consisted of circles and partly of straight and curved lines, there would always exist a function describing the entire progression.[15]

Regardless of the similar wording Du Châtelet provides, however, it is quite obvious that her work is still a mere interpretation of Leibniz' thoughts, and this becomes even more apparent when considering her explorations of mathematics and physics. Like Leibniz, Mme du Châtelet seeks to demonstrate the principle of continuity for the field of physics. To that end, she discusses the reflection of light from a mirror, a process which can be rendered easily in geometric terms. At the point of reflection, a beam of light is thrown back at an angle equal to its angle of incidence. The reflection can thus be rendered geometrically. This also means that the path of the beam will always show a point, namely the peak point, where it is not continuous. To Du Châtelet, this, however, clearly indicates that geometric optics cannot properly describe the light's reflection. She states that the beam,

---

[15] Leibniz. 1999. Discours de métaphysique. 1538. The passage reads: "Et si quelcun traçoit tout d'une suite une ligne qui seroit tantost droite, tantost cercle, tantost d'une autre nature, il est possible de trouver une notion ou regle ou equation commune à tous les points de cette ligne, en vertu de la quelle ces mêmes changemens doivent arriver."

qui se réfléchit sur un miroir, ne rebrousse point subitement, & ne fait point un angle pointu au point de la réfléxion; mais il passe à la nouvelle direction qu'il prend en se réfléchissant par une petite courbe qui le conduit insensiblement & par tous les dégrés possibles qui sont entre les deux points extrêmes de l'incidence & de la réfléxion.[16]

Consequently, Emilie du Châtelet here suspends a fundamental principle of Leibniz' philosophy of nature. Although she generally agrees with Leibniz regarding his conviction that metaphysical principles are indispensable from science, she does not follow him when he excludes their application to physical details, and this is certainly not in agreement with his thoughts. From Leibniz' point of view, a solution to the problem will have to be found by physical means only, i.e. without the means of a metaphysical intervention. And in the *Tentamen anagogicum*, he demonstrates how this can be done.

In relation to the phenomenon Du Châtelet describes, Leibniz makes use of a new kind of geometry he calls *Characteristica geometrica* or *Analysis situs*. This particular invention enables him to distinguish between two attributes of physical space, namely quantity and order. Via the means of quantity and order, he argues, any movement can be described via discrete values and variables without questioning their continuity. His *Tentamen anagogicum*[17] therefore demonstrates that, by considering spatial relationships, in particular from the point of view of spatial order, it will be possible to define any one point out of a continuum of possible reflection points. This point he calls the 'most determined' ("le plus determine").[18]

Before defining lines as quantities, or, more precisely, paths as a special quantity, then, it has to be established that there exists this one point where everything is fully determined. Geometry in the ordinary sense, according to Leibniz, is generally concerned with quantities only. If one were to try to describe the reflection of light ray via this kind of geometric means only, however, one would undoubtedly face the problem Mme du Châtelet mentions. Whilst Du Châtelet thus seeks to describe the continuity of light propagation in geometry by bending the reflection point into a curve, Leibniz, quite to the contrary, embeds the aforementioned process into a continuum of spatial relationships described by his *analysis situs*. This way, order becomes inconceivable without quantity, just as much as quantity cannot be without order, and this is why Leibniz can geometrically describe the curve without worrying about its reflection point. It will always already be given in the world of rational order. The very reason for an application of the *Analysis situs* to a physical problem is therefore that all physical phenomena are based on metaphysical entities called monads. The universal order Leibniz supposes in his argument is of course a result of the activities of these monads, and his discussion of the light reflection from a mirror is thus not to be understood as a geometrical problem, as it is for Mme du Chatelet, but has to be understood as a dynamical one.

This particular difference in scientific methodology between Leibniz and Emilie du Châtelet is, of course, of a fundamental nature. It becomes most apparent in her

---

[16] See also: Inst1742Rep1988 §13.

[17] See also Hecht. 2006. Maupertuis und die Leibniz-Tradition an der Berliner Akademie.

[18] Leibniz. 1890. Tentamen Anagogicum. 275.

Institutions de Physique, and it appears especially in her discussions of current research issues. In the chapter "Du Repos, & de la Chute des Corps sur un plan incline", for instance, Du Châtelet's argument quickly progresses from causes of falls to particular discussions of advanced problems in physics, as will be shown in the following.

A Detour into Physics

The chapter opens by stating that the effects of gravity are always directed towards the geo centre. Every deviation from the rule therefore requires an explanation that would take external factors into consideration. These factors, however, can be of either an active or of a passive nature.

> Ces causes étrangères peuvent être actives ou passives; les causes actives sont celles qui impriment un nouveau mouvement aux corps, comme lorsque je jette une pierre qui seroit tombée par la seule force de sa gravité.
> Les causes passives sont celles qui n'impriment aucun nouveau mouvement au corps, mais qui changent seulement sa direction.[19]

In the case of a falling object hitting an inclined plane, then, its fall will be decelerated considerably. The forces at work here can be precisely as "gravité absolue"[20] and "gravité respective"[21] with absolue gravity describing an active force and respective gravity referring to the same force diminished by resistance. Resistance affects the object in its vertical fall, and it deflects it in a horizontal direction. From this, Du Châtelet concludes:

> Ainsi, les corps en descendant dans un plan incliné, n'ont d'autre mouvement que celui que la gravité leur imprime sans cesse pour arriver au centre de la terre.[22]

Emilie was well aware that Galileo's falling bodies experiment had proven that all objects fall at the same rate and that they also achieve the same speed, whether they fall vertically or are being decelerated via an inclined plane (see Fig. 4).[23] She was also aware of Galileo's discovery that an object rolling down an inclined plane would re-ascend to the same height it had started from, if one were subsequently to let it roll up an inclined plane of the same height, regardless of the plane's inclination.

In accordance with Varignon, Du Châtelet, however, reaches the conclusion that matters would be different if one were to combine the inclined planes in such a way that objects would have to pass parts such as AB and BC in Fig. 5.[24] These parts possess variable inclination angles in relation to the horizontal line, and in passing

---

[19] Inst1742Rep1988 §401.
[20] Inst1742Rep1988 §403.
[21] Inst1742Rep1988 §403.
[22] Inst1742Rep1988 §422.
[23] Inst1742Rep1988 §427 f.
[24] Inst1742Rep1988 §437.

In the Spirit of Leibniz – Two Approaches from 1742

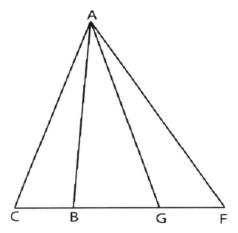

Fig. 4 Inclined plans

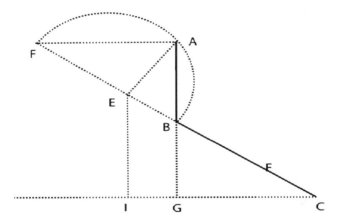

Fig. 5 Combination of inclined plans

point B, the object would encounter a deceleration which would stop it from re-ascending to its starting height. In an effort to clarify the matter, Du Châtelet refers to the principle of continuity and assumes that the objects in question have to be non-elastic, "[…] car on fait abstraction ici du ressort des corps […]".[25]

Mme du Châtelet's argument rests on two assumptions here: (1) it assumes Galileo's insight into the fact that an elastic object falling in a gravitational field will always gain a velocity that will allow for its re-ascension and (2) it also rests on the behaviour of non-elastic objects when passing inclined planes of variable angles. The former part invariably leads to Leibniz' *Principle of Conservation of Energy*, the latter would have to be considered an exercise in defining the one principle that could describe the movement of such non-elastic objects in general. And this is exactly how

---

[25] Inst1742Rep1988 §437.

Maupertuis perceives the matter, when he sees the achievement of his principle of action in its ability to describe a principle of motion which is able to regulate the occurrence and the disappearance of motion for all existing objects, be they elastic or entirely solid.[26]"Après tant de grands hommes qui ont travaillé sur cette matiere, je n'ose presque dire que j'ai découvert le principe universel sur lequel toutes ces loix sont fondées; qui s'étend également aux corps durs & aux corps élastiques […]."

Emilie du Châtelet disagrees. "Galilée", she claims,

> ést le prémier qui s'est mépris sur cet article, peut-être cette erreur de Galilée étoit-elle une erreur invincible pour ce Philosophe, qui ignoroit la Géométrie de l'infini, car il n'y a que la Géométrie de l'infini qui puisse éxpliquer pourquoi le Corps perd de sa vitesse dans les plans inclinés contigus, & n'en perd pas dans les courbes […].[27]

We are confronted here with the same problem as discussed above, namely the question of light's reflection. Du Châtelet's objection to Galileo is that his description of motion along two inclined planes does not take into account that there is one and only one point which interrupts the continuity of motion. She thus refers to Huygens' *Horologium oscillatorum*, which helps her to generalise Galileo's ideas, because he demonstrated that, regardless of how an object moves from a higher to a lower position within the gravitational field, in its lowest position, it will always possess a velocity that enables it to return to its initial starting point.[28]

In an attempt to settle the disagreement between Galileo and Huygens on the one and Varignon on the other hand, Mme du Châtelet reduces the movement of any object on an inclined plane to a mere geometric problem that can be solved by the help of infinitesimal mathematics. Not unlike earlier, when she tried to save continuity in the reflection of light beams via a reinterpretation of angles into curves, she underlines the necessity of metaphysics in science by firmly arguing from a perspective which was well in keeping with the scientific progress of her time. In fact, metaphysics play a valuable role here, for its principles allow for intervention whenever scientific dialogue is not (yet) in a position to offer any definite answers.

Gravity obviously remained one of these insoluble questions throughout the first half of the eighteenth century, and it caused numerous and far-reaching disputes in relation to its true nature. In query 31 of his *Opticks* – and contrary to his adepts – Newton unequivocally stated that attraction was not to be understood as being inherent to all matter.[29] Mme du Châtelet duly noticed this difference and accepted it as a starting point for her own research.

Du Châtelet initially sets out to discuss and evaluate a claim which Newton's followers made regarding the universality of attraction. As opposed to their assumptions that attraction could be understood as a quality which regulates movements in physics just as much as it influences chemical reactions and biological processes, Du Châtelet demonstrates that, in order to be valid, the universal law of gravitation

---

[26] Maupertuis. 1768a. Essay de Cosmologie. 42.

[27] Inst1742Rep1988 §437.

[28] Inst1742Rep1988 §437.

[29] Inst1742Rep1988 §389.

would have to be altered to such an extent that the universality of its reciprocal effects would then be called into question altogether.

In this context, then, she also mentions Maupertuis, who was the first to officially confirm the validity of Newton's law of gravity in a paper he presented to the Academy and which was later amended by a philosophical justification for his *Discours sur la différent figure des astres* in 1732. While Newton had deliberately excluded a metaphysical justification of the law in question via his well known dictum *hypotheses non fingo*, Maupertuis in fact turns it into the central theme of his argument. By comparing Newton's law of gravity to other principles, he found that the former was the only one to maintain its form under all circumstances. Consequently, there can exist only one law which proves to be significantly different to all other laws of attraction, namely Newton's $1/r^2$ law. And this particular law's validity in nature has to be more than a mere coincidence. Maupertuis concludes that, supposing God had wanted to assign a fundamental law to matter in the creation process, and given that this was not to be merely incidental, it had to be Newton's law of universal gravitation.

So much for Maupertuis' ideas. To Mme du Châtelet, however, the issue is far from being resolved. Her objection regarding Maupertuis' explanations aims at the idea that matters could be explained by stating God's reasons for his choice.[30] To her, it is imperative to formulate the reason for attraction itself, and she goes on to illustrate this via a fictitious example.

Imagining a world with only two mutually attractive objects in it in which each one is within a distance to the other that would allow attraction, she states that, according to Newton's followers, these objects would have to attract each other. But how exactly could this happen? Emilie du Châtelet considers three conceivable scenarios:

1. The bodies in question move of their own accord. This, however, can be ruled out easily, since the law of inertia clearly states that an object will only alter its state of motion via an external force.
2. An external impulse might be able to initiate the movement. This is logically inconceivable, for there are only two bodies in this particular world.
3. God could cause the attraction. This is also not an option, since the initial scenario did not provide any information on the bodies' velocity or the direction of their movement. And God will have to be given some kind of reason for his choice.

None of the above scenarios proves to be of any use as an explanation for the origin of gravitational force. In fact, the problem is that they are all ill-posed and that God cannot intervene, unless he was to be considered an arbitrary God; and this is clearly impossible in Mme du Châtelet's metaphysical thinking. The solution to the problem is therefore already implicit in the third scenario. This is, of course, the principle of sufficient reason. In order to be able to understand how physical motion is possible under the primacy of gravity, one has to state a cause, i.e. a mechanical procedure. Simply referring to the process of attraction will not do the trick.

---

[30] Inst1742Rep1988 §394.

To Emilie du Châtelet, it is certain that cause has to be of a mechanical nature, and in § 340 she states accordingly:

> Le principe de la raison suffisante, qu'il ne faut jamais perdre de vue, ne nous permet pas de douter qu'une matière qui ne pese point (§. 76.), & qui se meut tres rapidement, cause la chute des corps vers la terre [...].[31]

Her conclusion is indebted to Descartes, and it is certainly presented along those lines. Yet Du Châtelet's ideas also add a whole new dimension to his original hypotheses. In assuming that Newton's description of the oblateness of a celestial body might be a direct result of an antecedent Cartesian cosmos, this could also hold true for an explanation of the beginnings of attractive motion. Mme du Châtelet calls for the declaration of causes in order to resolve the problem of an ill-posed scenario. These causes, however, as § 378 clearly indicates, might well be of a historical nature,[32] and Newton's cosmos might well be the result of a precursory Cartesian one.

Conclusion

Considering that Emilie du Châtelet formulated her ideas on nature and attraction 15 years prior to the publication of Kant's *Allgemeine Naturgeschichte und Theorie des Himmels*, it will become apparent that her understanding of natural philosophy was well in keeping with the spirit of her time. Her scientific methodology proved highly successful in pinpointing philosophical implications of scientific problems in an exemplary way, yet it generally failed when it came to solving particular scientific dilemmas. Here, her writing differs somewhat from Maupertuis' because the metaphysical theses of the latter usually also allow for a scientific interpretation.

Maupertuis and Du Châtelet share a desire to further and to heuristically stimulate scientific progress via a normative methodology based on Leibniz' ideas and on his problem of choosing the best of all possible worlds. However, where Maupertuis turns Leibniz' dictum into a methodology of possible worlds, Emilie du Châtelet focuses on a different aspect of his thoughts, namely the principle of sufficient reason and its implications for a transition from possible to real worlds.

In both cases, Leibniz proves to be a vital starting point, and although each author selects different aspects of his theory, both approaches turn out to be mutually beneficial for an eventual redefinition of the fields of philosophy and natural sciences in the eighteenth century. While Mme du Châtelet favoured metaphysics, Maupertuis concentrated on the natural sciences. Both authors considered themselves to be philosophers.

---

[31] Inst1742Rep1988 §340.
[32] Inst1742Rep1988 §378.

References

Hecht, H. 2006. Maupertuis und die Leibniz-Tradition an der Berliner Akademie. In *Hofkultur und aufgeklärte Öffentlichkeit: Potsdam im 18. Jahrhundert im europäischen Kontext*, ed. Günther Lottes and Iwan D'Aprile. Berlin: Akademie Verlag.

Leibniz, G. W. 1999. Discours de métaphysique. In *Sämtliche Schriften und Briefe,* ed. Gottfried Wilhelm Leibniz, series VI, vol 4, Part B. Berlin: Akademie Verlag.

Leibniz, G. W. 1890. Tentamen Anagogicum. In *Die philosophischen Schriften*, ed. Carl Immanuel Gerhardt, vol 7. Berlin: Weidmann.

Leibniz, G. W. 1998. *Monadologie,* ed. Hartmut Hecht. Stuttgart: Reclam.

Leibniz, G. W. 1949. *Protogaea*. Transl. by W. v. Engelhardt. Stuttgart: Kohlhammer.

Maupertuis, P. L. M. de. 1768a. Essay de Cosmologie. In *OEuvres et lettres*, vol 1. Lyon: Bruyset.

Maupertuis, P. L. M. de. 1768b. Lettre sur la comète qui paroissoit en 1742. In *OEuvres et lettres*, vol 3. Lyon: Bruyset.

Between Newton and Leibniz: Emilie du Châtelet and Samuel Clarke

Sarah Hutton

As one of the first French champions of the philosophy of Leibniz and Wolff, and as one of France's most eminent Newtonians, Gabrielle-Emilie le Tonnelier de Breteuil, Marquise du Châtelet, is deservedly celebrated for her contribution to the scientific and philosophical debates of both the German and French Enlightenments. Her *Institutions de Physique* (1741) is framed according to Leibnizian principles, while her crowning achievement was her translation of Newton's *Principia mathematica* into French, which was completed just before she died.[1] Du Châtelet shared her interest in Newton with Voltaire, but she parted company with him in her emphasis on metaphysics and her receptivity to Leibniz.

Du Châtelet's interest in these opposing intellectual giants has been explained at different times in different ways. Voltaire accused her of changing her mind, of abandoning Newton in favour of Leibniz, only to return to Newton later. Latterly, scholars have been more inclined to take seriously her claim that she sought to provide a metaphysical basis for scientific enquiry, and that it was for this reason that

---

[1] Completed in 1749, but not published until 1756/9 it was published with the title *Principes mathématiques de la philosophie naturelle par feue Madame la Marquise du Chastellet*. On Du Châtelet's translation, see Cohen. 1968. The French Translation of Isaac Newton's Philosophiae naturalis principia mathematica. 261–280; Zinsser. 2001. Translating Newton's *Principia*: the Marquise Du Châtelet's Revisions and Additions for a French Audience. 227–45. Tâton. 1969. Madame du Châtelet, tradutrice de Newton. 185–210; idem 1970. Isaac Newton, *Principes mathématiques de la philosophie naturelle*, traduction de la marquise du Chastelet, augmentée des commentaries de Clairaut. 175–80. Also, Badinter and Muzerelle. 2006. *Madame du Châtelet. La femme des lumières*. Exhibition catalogue. For Du Châtelet's biography, see Zinsser. 2007. *La Dame d'Esprit. A Biography of Madame Du Châtelet*, and Tâton. 1971. Gabrielle-Émilie le Tonnier de Breteuil, Marquise du.

S. Hutton (✉)
Aberystwyth University

she turned to Leibnizianism.² However, her apparently wavering faith has done her intellectual reputation no service.³ Worse, rumours abounded at the time that she was simply a plagiarist who opportunistically published others people's work as her own. The source of these rumours was her some-time tutor, Samuel Koenig. Her criticism of aspects of Newtonian science notwithstanding, it is clear from the content of *Institutions* that she did not abandon Newtonianism, when she developed an enthusiasm for Leibnizian metaphysics Nevertheless, her subscription to *both* Newton and Leibniz remains problematic from a modern perspective, especially in view of the way Newton and Leibniz are nowadays treated as leading representatives of the empiricist-rationalist schools, and therefore of rival British and German philosophical traditions.⁴ This posthumous counter-posing of Newton and Leibniz is prefigured in the long-running quarrel between them, which soured relations in their lifetime and dominates the historical legacy of both, serving to emphasize their differences and to obscure any common scientific purpose. In its last phase, in the years 1715–1716, Newton's dispute with Leibniz was conducted on his behalf by his English admirer and populariser, Samuel Clarke. The Leibniz-Clarke dispute has since come to be viewed as emblematic of the irreconcilable opposition of Newton and Leibniz. Since Emilie du Châtelet's *Institutions de Physique*, assumes compatibility between Newtonian theories and Leibnizian metaphysics, it is reasonable to expect that her view of this controversy should hold clues about her understanding of the relationship of Leibniz's philosophy to Newton's. As it turns out, she does in fact make very little reference to Clarke in *Institutions de Physique*. But this paucity of reference is itself instructive. Furthermore, investigation shows that there was more common ground between Clarke and Madame du Châtelet than her brief, dismissive remarks on Clarke in her *Institutions* seem to suggest. This paper explores the common ground with Clarke and examines her comments on the Leibniz-Clarke correspondence in relation to her attempted reconciliation of Leibniz and Newton in her *Institutions*.

---

[2] Barber. 1967. Madame Du Châtelet and Leibnizianism: the Genesis of *Institutions de Physique*. 200–22, revised and reprinted in Zinsser and Julie Candler Hayes. 2006. *Emilie Du Châtelet. rewriting Enlightenment philosophy and science*. 5–23.; Gardiner. 1982. Searching for the Metaphysics of Science: the Structure and Composition of Madame du Châtelet's *Institutions de physique*, 1737–1740. 85–113; Zinsser. 2007, *La Dame d'Esprit*. chapter 4. The claim by Janik, and, following her, Paul Veach Moriarty, that the Leibniz-Clarke correspondence was a primary source for her knowledge of Leibnizian metaphysics is, in my view, unsustainable. Moriarty. 2006. The Principle of Sufficient Reason in Du Châtelet's *Institutions*. 203–225.

[3] Ira O.Wade deemed her *Institutions de Physique*, 'essentially derivative', believing that she was unlikely to have influenced Voltaire who, he claimed, had 'a deeper understanding of Newton', Wade. 1947. *Studies on Voltaire*. 221.

[4] Stuart Brown discusses the adverse impact of this stereotype for Leibniz's reputation as a philosopher: Brown. 1984. *Leibniz*. 204.

Background to *Institutions*

Madame du Châtelet had already embarked on an extensive study of Newtonianism, before she published her *Institutions de Physique* in 1741. She was certainly thoroughly conversant with Newton's *Opticks* (as she told Maupertuis in 1738, 'je sais presque par coeur l'optique de M. Newton').[5] And she had already embarked on a study of his *Principia mathematica* by this time. She was also conversant with the literature relating to Newton: in 1738 she sent her bookseller, Laurent-François Prault, a shopping list of books that she wished to acquire, including runs of *Philosophical Transactions* of the Royal Society, *Nouvelles de la république de letters* and 'tous les livres de physique que vous trouverez sur votre chemin'.[6] She also asked him for 'une belle edition' of *Principia mathematica philosophiae naturalis* and the 'recueil de letters de Leibnits et de Clarke'.[7] The letter also lists Newtoniana already in her possession, including Whiston, Maupertuis, 'sGravensande, Musschenbroek. Her *Institutions de Physique* confirms her acquaintance with these and other British and European Newtonians, for example, Keill, Raphson and Pemberton, as well as the Leibniz-Clarke correspondence. As I have argued elsewhere, notwithstanding its debt to Leibniz and Wolff, *Institutions de physique* was conceived as part of the same campaign to promote Newtonianism with which Voltaire and Algarotti were engaged in *Elémens de la philosophie de Newton* and *Neutonianismo per le dame*.[8] Madame du Châtelet's *Institutions* is also the product of the period when she and Voltaire were exposed most directly to the thought of Leibniz's disciple Wolff, thanks to the presence at Cirey of the Wolffian, Samuel Koenig, who had been introduced there by the Newtonian, Pierre-Louis Moreau de Maupertuis. Koenig's arrival at Cirey came in the wake of the contact between Voltaire and Crown Prince Frederick of Prussia, whose enthusiasm for Wolff could not be ignored in the Cirey circle. Where Voltaire responded with courteous coolness about Wolff, Madame du Châtelet's curiosity was aroused, and she took steps to acquaint herself with his philosophy. It seems that Frederick encouraged her by arranging for a French translation of Wolff's metaphysics, which he sent to her.[9] This seems to have inspired her to write a work of Wolffian philosophy in French,

---

[5] Du Châtelet to Maupertuis, 1st December 1738, LetChBI 152, 273.

[6] Du Châtelet to Prault, 16th February 1739, LetChBI 186, 329.

[7] Ibid. She had access to *Principia mathematica* prior to this in Voltaire's library. The reference Clarke and Leibniz must be to the Clarke-Leibniz letters published in *Recueil de diverses Pièces sur la Philosophie, la Religion Naturelle, & autres Autheurs célèbres*, vol.1 (Amsterdam, 1720), and not to separate editions of their letters as Besterman's note suggests.

[8] Hutton. 2004. Emilie du Châtelet's *Institutions de physique* as a Document in the History of French Newtonianism. 151–63. See also, Gandt. 2001. *Cirey dans la vie intellectuelle. La réception de Newton en France*; Le Ru. 2005. *Voltaire newtonien. Le Combat d'un philosophe pour une science*.

[9] Du Châtelet to Johann Bernouilli, LetChBII 241, 18. cf. also Du Châtelet to Frederick of Prussia, LetChBII 237, 13.

as she told Frederick, 'J'ai le desire de donner en français une philosophie entière dans le gout de mr. Wolf, mais avec une sauce française'.[10] This plan never came to fruition. But she did incorporate her new-found interest in her *Institutions de Physique*.

Samuel Clarke

Samuel Clarke (1675–1729) played a significant role in the dissemination and defence of Newtonianism.[11] Not only did he champion Newton against his most formidable critic, Gottfried Wilhelm Leibniz, but he was an informed populariser of Newtonian science. He was personally acquainted with Sir Isaac Newton and a friend of Newton's disciple, William Whiston. Educated at the University of Cambridge, he was a prominent member of the liberal wing of the Church of England, dubbed Latitudinarian for their more inclusive and tolerant attitude to articles of faith and religious observance. His ecclesiastical career received a boost when he was appointed royal chaplain by Queen Anne in 1706. However, his *The Scripture Doctrine of the Trinity* (1712) provoked questions about his theological orthodoxy and resulted in his being branded a heretic by his own church.[12] His promotion within the Church of England was effectively halted, and he was obliged to resign his position as royal chaplain. But he continued to hold the prestigious living of St James, Westminster, and to reach a wide public through publication of his sermons—he was in fact one of the most widely read sermon writers of the period. After the Hanoverian succession in 1714, Clarke enjoyed good relations with the German Princess of Wales, Caroline of Anspach, wife of the future King George II, and dedicatee of the London edition of Voltaire's *Henriade*. And it was through Princess Caroline that Clarke became involved in the controversy with Leibniz.

Clarke's role in the history of Newtonianism was far wider than his championship of Newton against Leibniz. He was twice invited to deliver the prestigious Boyle lectures.[13] These were sermons instituted by the will of the scientist, Robert Boyle for the express purpose of combating perceived threats to the Christian religion.

---

[10] Du Châtelet, LetChBII 244, 24. This work of Wolffian metaphysics was, presumably, planned for a future date. She was already engaged on writing *Institutions de physique*, of which she had already sent Frederick a copy the first part. See LetChBII 237 and 240.

[11] Clarke is overdue for re-appraisal. But see Ferguson. 1974. *The Philosophy of Samuel Clarke and its Critics*; idem. 1976. Samuel Clarke: an Eighteenth-century Heretic. Also Vailati. 1997. *Leibniz and Clarke. A Study of their Correspondence*; idem. 1998. Introduction to his edition of Clarke's *A Demonstration of the Being and Attributes of God and Other Writings*.

[12] Clarke's book was condemned by the Convocation of Canterbury in 1714. Clarke shared his Arian view of the Trinity with Whiston and, of course, with Newton – though Newton was careful to dissemble his view. His reputation for heterodoxy can only have been confirmed by Voltaire's description of him as 'le plus ferme patron de la doctrine arienne' in the seventh of his *Lettres Philosophiques*.

[13] On the Boyle Lectures see Jacob. 1976. *The Newtonians and the English Revolution, 1689–1720*.

Clarke's sermons were published as *A Demonstration of the Being and Attributes of God*, in 1704, and *A Discourse concerning the unchangeable obligations of natural religion and the Certainty of the Christian Revelation*, in 1705. These sermons—along with many others by him—were many times reprinted throughout the eighteenth century. Clarke's Boyle Lectures established him as promoter and defender of Newtonian Science in the Anglican tradition instigated by Richard Bentley, who, when he was Boyle Lecturer, had led the way in invoking Newtonian natural philosophy in defence of religious faith against atheists and other enemies of religion. Clarke therefore contributed to the image of Newtonianism as the natural ally of godliness in general and English Protestantism in particular. Clarke's credentials for this role were firmly based on his detailed knowledge of contemporary science in general, and Newtonian physics in particular. In 1706, with the approval of Newton, he translated *Opticks* into Latin. Clarke's *Optice: sive de reflexionibus, refractionibus, inflexionibus & coloribus lucis* (London,1706) is based on the first (1704) edition of *Opticks*.[14] It doesn't print the two mathematical treatises printed at the end of *Opticks*. But it does add seven additional queries. Clarke also played a role in disseminating Newtonianism through the footnotes he supplied to his Latin translation of Jacques Rohault's Cartesian *Traité de physique* (originally published in French in 1671). These notes amount to Newtonian rebuttals of various aspects of Rohault's text.[15] Clarke's role as mediator of Newtonianism extended beyond his compatriots. Several of his writings were translated into French,[16] but, most importantly, his ideas were mediated by none other than Voltaire. When Voltaire was in London in 1727, he became personally acquainted with Clarke.[17] As Voltaire's references to him show, their topics of conversation included Newton's theories, as well as Clarke's theological opinions. Voltaire expresses high regard for Clarke in his published writings and

---

[14] Newton. 1704. *Opticks, or a Treatise of the Reflexions, Refractions, Inflexions and Colours of Light*. Further editions with additional queries were published in 1717, 1718, 1721, 1730.

[15] This translation was first published in 1702. The annotated version appeared in 1710. This was republished in 1718. It also appeared in Europe (Cologne, 1713, Louvain 1729 and 1739). See Clair. 1978. *Jacques Rohault (1618–1672). Bio-bibliographie*. One of Clarke's reasons for undertaking the translation was in order to correct the shortcomings of Theophile Bonet's Latin translation of 1674. See Barber. 1979. *Voltaire and Samuel Clarke*. 47–61. His notes were incorporated in European reprints of Rohault, and in his brother, John's English translation, Rohault 1723 *Rohault's System of Natural Philosophy illustrated with Dr. Samuel Clarke's Notes taken mostly out of Sir Isaac Newton's Philosophy*.

[16] These included his Boyle lecture *Demonstration of the Being and Attributes of God*, which was translated by Pierre Ricotier in 1717 (*De l'existence et des attributs de Dieu*), which saw several editions; his notes, which were included in European editions of his Rohault translations, and his dispute with Leibniz, published in Desmaizeaux's, *Recueil*. Clarke's Latin translation of *Rohault*, which incorporated his notes, was also published in Europe.

[17] According to Barber. 1979. *Voltaire and Samuel Clarke*. Voltaire's contact with Clarke was enhanced by the fact that Clarke had a fluent command of French. This gave Clarke an advantage in relations with Princess Caroline who was not an English speaker but was francophone. That was also an advantage for Voltaire when he was introduced to her circle.

treats him as an important authority, especially in matters of metaphysics. Clarke undoubtedly helped to mould Voltaire's view of Newton. When asserting his credentials as author of his *Elémens*, Voltaire made much of his direct contact with English Newtonians, as well as his knowledge of Newton's writings. The only person named among the former is Clarke, whose dispute with Leibniz is also mentioned:

> Je dirai fidèlement, soit ce que je recueillis en Angleterre de la bouche de ses disciples, et particulièrement du philosophe Clarke, soit ce que j'ai puisé dans les écrits de Neuton, et dans la fameuse dispute de Clarke et de Leibnits...[18]

Voltaire's debt to Clarke means that there is an important sense in which, through Voltaire, Clarke contributed to the French view of Newtonianism. It also means that Clarke was part of the Newtonian backdrop to Emilie du Châtelet's introduction to Newton. Even while Du Châtelet was preparing her *Institutions* for publication, Voltaire was working on his *La métaphysique de Neuton*, which he published in Amsterdam in 1740 with the subtitle, *parallèle des sentiments de Neuton et de Leibnitz*. The following year he published this as part I of his revised *Elemens de la philosophie de Neuton*.[19] This makes plain his respect for Clarke, especially in matters of metaphysics. Chapters 1–5 discuss the nature of God, time, space, free will and natural religion by reference to Clarke's arguments in the Leibniz-Clarke controversy.

The Leibniz-Clarke Debate

The debate between Clarke and Leibniz was mediated through the good offices of Princess Caroline, sometime pupil of Leibniz. Although called the Leibniz-Clarke controversy, this was really a Leibniz-Newton controversy. It was conducted during the last years of Leibniz's life (1715–1716), and constitutes the final episode of ongoing hostilities between Leibniz and Newton which dated from at least 1705. The Leibniz-Clarke letters were published in 1717 as *A Collection of Papers which passed between the late learned Mr Leibnitz and Dr Clarke in the years 1715 and 1716*.[20] There is no space here to elaborate the history of this

---

[18] Voltaire. 1992. Elémens de la Philosophie de Neuton. 193.

[19] On the various editions of Voltaire's *Elémens* and their composition, see the Introduction to Voltaire. 1992. Elémens de la Philosophie de Neuton. This includes a list of eighteenth-century editions complied by Andrew Brown, pp. 141-68.

[20] They were also published in volume 1 of Pierre Desmaizeaux's *Recueil* (1720), which printed facing-page French and English versions of the letters, in which Leibniz's French was translated by Clarke, and Clarke's English translated by Michel de la Roche. The Demaizeaux *Recueil* saw several reprintings. There was also a German edition of the Leibniz-Clarke letters, with a forward by Wolff. 1729. *Merckwürdige Schriften welche ... zwischen dem Herrn Baron von Leibniz und dem Herrn D. Clarke uber besondere Materien der naûrlichen Religion in Franzôs. und Englisher Sprache gewechselt und ... in teutscher Sprache herausgegeben worden von Heinrich Köhler*. For a modern English edition, see Alexander. 1956. *The Leibniz-Clarke Correspondence*, which prints relevant excerpts from *Principia mathematica*, *Opticks* and Leibniz's letters.

controversy,[21] but suffice to say that the key topics around which it turned were key topics of Leibnizian metaphysics: the relation of God to the universe, space, time, the Principle of Sufficient Reason, the Principle of the Identity of Indiscernibles. They also debated the relative merits of positing the existence a Vacuum as against a Plenum, Newton's theory of gravitation, and the highly topical related issue of the measurement of force. Leibniz's first charge against Newton concerns his account of space as the *sensorium* of God. He also denied the existence of absolute time and space, and the atomic structure of matter, and charged Newton with re-introducing scholastic occult qualities in his theory of gravitation. It should not be forgotten that, in the main, Leibniz had a high regard for Newton, especially *Opticks*, and he praised Newton's experimental method. Also, his critical remarks were based on the first printing of *Opticks*. Later editions of the work silently incorporate changes, which were probably made in direct response to his criticisms.[22] Nevertheless Clarke concedes nothing in his responses, and makes no reference to the changes.

Emilie du Châtelet's Knowledge of Clarke

There are not many references to Clarke in the writings of Emilie du Châtelet. But she certainly knew the Clarke-Leibniz controversy, both indirectly through Voltaire's references to it in the *Elémens* and directly through possession of the published text. As the dedicatee of Voltaire's *Elémens*, whose contribution to the book was acknowledged by its author, she could not but have been aware of Clarke's Newtonian objections to Leibniz. To be introduced to Newton by an admirer of Clarke was to encounter a strong predisposition towards Newton against Leibniz, as was the case with Voltaire. This would seem to be borne out by a letter to Maupertuis in 1738 regarding Mairan's *Mémoire* on *forces*. Her comments suggest that at this time she agreed with Clarke on all points, except for the matter of *forces vives*:

> Le docteur Clarke dont mr de Mairan a rapporté tous les raisons dans son mémoire traite de mr de Leibnitz avec autant de mépris sur la force des corps, que sur le plein, et les monades, mais il [a] grand tort a mon gré car un homme peut être dans l'erreur sur plusieurs clefs, et avoir raison dans le reste. Mr Leibnitz à la verité n'a guerre raison que sur les forces vives, mais enfin il les a decouvertes, et c'est avoir deviné un des secrets du créatur.[23]

As already mentioned, Emilie du Châtelet was well read in the literature of Newtonianism. Her letter to Prault of 1738 mentions that she owned the Leibniz-Clarke correspondence—no edition is named, but the term 'recueil' suggests that it was, in all likelihood the collection of philosophical papers translated and edited by

---

[21] But see H.G. Alexander's Introduction to *Leibniz-Clarke Correspondence*; for recent studies, the Bibliography of Clarke 1998 A Demonstration.

[22] *Leibniz-Clarke Correspondence*, p. xviii. Also Koyré and Cohen. 1961. The Case of the Missing Tanquam. Leibniz, Newton and Clarke. 555–66.

[23] Du Châtelet to Maupertuis, 10th February,1738, LetChBII 120, 217.

Pierre Desmaizeaux, and published as *Recueil de diverses pièces sur la Philosophie, la Religion Naturelle, & autres Autheurs celebres* (Amsterdam, 1720).[24]

Besides the Leibniz-Clarke debate, another possible point of contact between Du Châtelet and Clarke is the fact that his Latin translation may have been a source for her knowledge of Newton's *Opticks*. This was first published as *Optice* in 1706 and reprinted in Europe in 1740 (Geneva) and 1749 (Leiden). We can't be sure about this, since her command of English appears to have been sufficient for her to be able to read *Opticks* in English. She may well have read it in the Coste translation of 1721.[25] At all events, she appears to have made a special study of *Opticks*, which was the work of Newton's that she knew best prior to the appearance of *Institutions de physique*. This is scarcely surprising, since *Opticks* was easily the most widely known of Newton's works in the eighteenth century, far more so than *Principia mathematica*.[26] After all, as Madame du Châtelet herself acknowledged, the *Principia* presupposed a level of attainment in mathematics that was far beyond the average reader.[27] By contrast Newton's *Opticks* was an accessible and practical text. Newton himself claimed of the experiments it describes were so straightforward that, 'A Novice may the more easily try them'.[28]

Another work of Clarke's with which Du Châtelet was certainly familiar was Clarke's version of Rohault's Cartesian *Traité de physique*. The significance of Clarke's translation is that he updated his Latin translation by adding extensive footnotes which provided Newtonian rebuttals of various aspects of the text.[29] In 1710 Clarke added further footnotes from *Optice*, his Latin translation of Newton's *Opticks*. These notes were incorporated in European reprints of Rohault, and in the English translation by Clarke's brother, John Clarke, which was published in 1723 with the unwieldy title, *Rohault's System of Natural Philosophy illustrated with Dr. Samuel Clarke's Notes taken mostly out of Sir Isaac Newton's Philosophy*. In the letter to Prault, mentioned above, Madame du Châtelet tells him

---

[24] This is the version on which Voltaire draws in his *La métaphysique de Neuton*. See Elémens, ed. Besterman, no.1, 201.

[25] Newton. 1720. *Traité d'optique*. This is based on second English Edition, 1717. A second French edition, with corrections, was published in 1722.

[26] Hall. 1993. All was *Light. An Introduction to Newton's 'Opticks'*.

[27] Newton's philosophy she wrote, bristles with so many calculations and algebra, that it is a kind of mystery, is only accessible to intiates ('sa Philosophie hérissée de calculs & d'algèbre, étoit une spece de mistere auqel les seuls initiés avoient droit de participer') Du Châtelet. 1738. Lettre sur les *Eléments de la Philosophie de Newton*. Lady Mary Wortley Montagu held the same view. See Grundy. 1999. *Lady Mary Wortley Montagu, Comet of the Enlightenment*. 524. As Paolo Cassini notes, the *Principia* was challenging to non-mathematicians, such as Locke, Halley, and the *eruditi* of Leipzig, and its reputation 'was based on the authority of a few competent readers'. See Casini. 1988. Newton's *Principia* and the Philosophies of the Enlightenment. 35–52, 42. See also Fellman. The *Principia* and Continental Mathematicians, ed. ibid., 13–34.

[28] Newton. 1952. *Opticks*. 25. In his study of *Opticks*, Rupert Hall attributes the success of *Opticks* in England to its being written in the vernacular, its experimentalism, and the speculative openness of the Queries. Hall. 1993. *All was Light*. 183.

[29] Hoskins. 1961. Mining all within. Clarke's Notes to Rohault's *Traité de physique*. 353–63.

that she owned a copy of 'Rohault commenté par Clark'. Whether this was the English or Latin version is not clear from the letter.

Aside from any use she may have made of Clarke's Rohault, there is another respect in which Clarke's up-dated Rohault is point of connection with Du Châtelet.[30] And that is that *Institutions de Physique*, like Clarke's Rohault, had an educational purpose. Not only that, both texts expressly sought to provide an up-to-date perspective on natural philosophy. Her *Institutions* was, as Madame du Châtelet explained to Prince Frederick of Prussia, written for the instruction of her son, to supply the lack of a comprehensive manual of modern physics suitable for a young man of his age,

> Ce livre n'était destiné que pour l'éducation d'un fils unique que j'ai, ... & voulant lui apprendre les elements de la physique, j'ai été oblige d'en composer une, n'y ayant point en français de physique complète, ni qui soit à la portée de son âge ...[31]

In the Avant-Propos to the *Institutions*, she is specific that it was intended to replace Rohault's *Traité de physique* (1671) which was now out of date. If Du Châtelet consciously thought of her *Institutions* as a successor to Rohault, it is hard to judge whether her treatment of Newtonian topics in *Institutions de physique*, is indebted to Clarke's notes on Rohault. For example, a topic copiously annotated by Clarke is optics, but this is not a topic covered in *Institutions de physique* (although it is possible that she proposed to do so in the planned later volumes). Unlike Clarke's Rohault, which was designed for the use of university students, Du Châtelet's book was, as the title indicates, designed to lay early foundations of the study of natural philosophy. It was intended to meet the needs of a young reader, who had only basic knowledge of mathematics, and was not yet conversant with the intricacies of higher mathematics, or what she calls 'algèbre'. As she tells her son in the 'Avant-Propos':

> je tâcherai dans cet Ouvrage, de mettre cette Science à votre portée, & de la dégager de cet art admirable, qu'on nomme l'Algébre, lequel séparant les choses des images, se dérobe aux sense, & ne parle qu'à l'entendement: vous n'étes pas encore à la portée de cette Langue, qui paroît plûtot celle des Intelligences des Hommes, elle est reserve pour faire l'étude des années de votre vie qui suivront celle où vous étes.[32]

Accordingly, she sought to provide an account of physics appropriate to the youth and educational attainments of her reader, namely someone without an advanced knowledge of mathematics. As she tells him, 'je tâcherai lui donner ici celle qui peut convenir à votre age, & de ne vous parler que des choses qui peuvent se comprendre avec le seul sécours de la Géometrie comme que vous avez étudiée.[33] Her reason for choosing a rationally-based methodology was, therefore, practical as

---

[30] Of course, it is important not to overstate the parallels between Clarke's Rouhault and Du Châtelet's *Institutions* and thereby lose sight of the differences—differences both of scale and conception.

[31] LetChBII 257. The original purpose of *Institutions* is highlighted in the title of the 1742 edition, and in the German and Italian translations: *Institutions physiques de Madame le marquise Du Chastellet addresses a son fils* (Inst1742Rep1988) 1742; *Der Frau Marquisinn von Chastellet Naturlehre an ihren Sohn*, translated by Wolf Balthasar Adolph von Steinwehr (Naturlehre1743); *Istituzioni di fisica di Madama la Marchesa Du Chastellet indiritte a suo figliuolo* (Venice 1743).

[32] Inst1740 Avant-propos I.

[33] Ibid.

much as anything. Leibnizianism as mediated via Wolff, furnished her with a set of rational principles for discussing and evaluating the natural philosophy contained in her book. These are set out systematically in the second chapter of the book. The attraction of Leibnizianism as a non-mathematical mode of philosopizing is confirmed in the letter to Prince Frederick, in which she told him that she wished to write a work of Wolffian philosophy in French:

> ... je suis persuade que mes compatriotes goûteront cette façon precise & sévere de raisonner, quand on aura soin de ne les point effrayer par les mots de lemmes, de théorèmes, et de démonstrations, qui nous semblent hors sphere, quand on les emploie hors de la géometrie.[34]

Emilie du Châtelet, therefore, had a practical reason for her recourse to Leibniz and Wolff, determined by the fact that *Institutions de physique* was conceived as an educational manual, and therefore designed for readers as yet unversed in 'algebre'.

Both Clarke's Rohault and Du Châtelet's *Institutions* are, in a sense, hybrid texts. Each entails the conjunction of two very different philosophies—Newton and Leibniz in the case of Du Châtelet, Newton and Descartes in the case of Clarke.[35] However, they are hybrids of very diverse kinds: Clarke's Newtonian intervention in Rohault's text, juxtaposed Newton to Descartes, without any attempt at mediation. Madame du Châtelet, by contrast is careful to lay the ground for integrating her diverse source materials. First of all she sets out the rational principles on which she bases her arguments at the start of the book—in this respect, she proceeds, one might say, *en bonne newtonienne*, by setting out the rules by which she intends to proceed. Secondly, as already noted, but she presents the contents of her book as hypotheses to be confirmed by evidence. The importance of this chapter is not just her acceptance of hypothesis *per se*, but, in marked contrast to Clarke's application of Newtonianism to Cartesianism, Du Châtelet pays attention to how scientific and theoretical theory should to be applied. In this chapter Du Châtelet exhibits an acute understanding of probabilism in scientific enquiry. She points out that hypotheses have an essential function in the process of scientific investigation, comparing them to the scaffolding which supports a building under construction.[36] Hypotheses, she argues, are useful, to the extent that they can help explain phenomena as yet undemonstrated by experiment. Their 'certainty' is, therefore, at best provisional, but may be increased by cumulative demonstration through observation. For Du Châtelet, to rely on hypotheses does not exclude observation and experiment. On the contrary, the watchword of her *Institutions* is 'expérience', which she calls 'le bâton que la Nature a donné aux nous autres aveugles'. Du Châtelet was too good a mathematician to eschew hypotheses and too thorough a student of Newton to accept that his theories were not in need of demonstration. She correctly reminds her readers

---

[34] Du Châtelet. LetChBII 244, 24.

[35] Marina Frasca-Spada discusses Clarke and the practice of publishing through footnotes in Compendious Footnotes. In Frasca-Spada and Jardine. 2000. *Books and the Sciences in History*. 171–189.

[36] On Du Châtelet's use of hypothesis, see Karen Detlefsen's forthcoming paper in the collection edited by Eileen O'Neill. A large extract from Du Châtelet's discussion was incorporated in the *Encyclopédie*.

that hypotheses were not peculiar to Cartesianism, and Newton himself, she reminds her readers, employed hypotheses.[37] And many aspects of Newtonianism remained hypothetical, because non-proven (e.g. his positing that the earth is an oblate spheroid in shape). She, therefore, castigates the repudiation of hypotheses by 'quelque Philosophes de ce tems' (by which she doubtless meant contemporary Newtonians). What distinguishes good from poor hypothesising is taking care not to exceed the available evidence and not to take hypothesis for truth. Furthermore, she displays a very modern sense of the inter-dependence of scientific theories, of the debt of more up-to-date theories to the predecessors they render obsolete. Rather than presenting the moderns as destroyers of the old, she recognises the dependence of the moderns on their predecessors—of Descartes on Galileo, Newton on Kepler. As she explained to her son, nobody had a monopoly on the truth—'la verité peut emprunter différentes formes'. Even in her recourse to Leibniz as the basis for a metaphysical foundation for physics, Madame du Châtelet suspended her judgement as to its truth value. As she told Frederick, she turned to Leibniz as the best available metaphysics.

> Je suis persuadée que la physique ne peut se passer de la métaphysique, sur laquelle elle est fondée j'ai voulu lui donner une idée de la métaphysique de mr de Leibnits, que j'avoue est le seule qui m'ait satisfaite, quoiq'il me reste encore bien des doutes.[38]

Her acknowledgement of doubts about Leibniz's system, indicates that Du Châtelet did not subscribe to his metaphysics dogmatically, but accepted his philosophical ideas provisionally as one would an hypothesis.

Institutions de Physique and the Leibniz-Clarke Debate

There are only two explicit references to the Leibniz-Clarke dispute in *Institutions de Physique*. These occur in Chap. 4, on the subject of space, and in Chap. 5 on the subject of time—though there is considerably more to Du Châtelet's discussion of space and time than the discussion of the Leibniz-Clarke debate: she is not just concerned with the ontological question of the nature of space and time, but also with the epistemological issue of how we arrive at knowledge of them. Given her Newtonian credentials and her earlier comments to Maupertuis, she might have been expected to side with Clarke. However, this is not the case. It transpires that she had apparently reversed the position she took when she had declared to Maupertuis in 1738, Leibniz was only right on *forces vives*. That *Institutions de*

---

[37] E.g. *Principia* II, proposition X, is followed by 'Hypothesis I. That the centre of the system of the world is immovable. In her translation of the General Scholium of Newton's *Principia*, Du Châtelet renders his famous statement, 'Hypotheses non fingo', as 'Je n'imagine pas d'hypothese', a translation which indicates that Newton's Latin retains the possibility of hypothesis, but excludes fanciful ones ('fingo' is the root of the English word 'fiction').

[38] Du Châtelet. LetChBII 257, 12. cf. John Locke, who recommended reading contemporary systems of natural philosophy as hypotheses rather than as purveyors of truth and fact Locke. 1989. *Some Thoughts concerning Education*. 247.

physique should include a critique of Newton's theories of time and space is consistent with the criticisms of his theories which she makes elsewhere in the book.[39] Her pro-Leibnizian stance on space and time is also consistent with her systematic application of the principle of sufficient reason throughout *Institutions de physique*. The fact that her references to the Clarke-Leibniz debate are confined to the topics, space and time, would suggest that the immediate context for her inclusion of it was most likely Voltaire's treatment of it in his *Métaphysique de Neuton* (subsequently incorporated into his *Elémens*).[40]

Du Châtelet's discussion of the Leibniz-Clarke debate focuses on what she considered to be the irrationality and absurd consequences of Clarke's arguments, absurdities which have theological consequences. She follows Leibniz in regarding absolute space and time as illusions of the imagination, rather than the product of true reasoning. In her critique of Newton's concept of space, Du Châtelet, argues that Clarke is unable to offer a rational explanation why the universe should occupy one part of space and not another. Instead he is forced to appeal to the arbitrary will of God, so contradicting the nature of God, and undermining his argument for void space. Clarke's defence of Newton,

> fait ecrouler son opinion & découvre le foible de sa cause, car Dieu ne sauroit agir sans des raisons prises dans son Entendement, & sa volonté doit toujours se déterminer avec raison. Ainsi être obligé de recourir à une volonté arbitraire de Dieu …c'est être réduit à l'absurde, la raison de la place de l'Univers dans l'Espace, & celles du limite de l'étendue n'étant donc ni dans les choses mêmes, ni dans la volonté de Dieu, on doit conclure que l'hypothèse du vuide est fausse et qu'il n'y eu point dans la nature.[41]

Also like Leibniz, Du Châtelet focuses on Newton's comparison of space to the *sensorium* of God, which she expounds as, 'ce, par le moyen de quoi, Dieu est prèsent à tout les choses'.[42] She picks up on Leibniz's essentially theological objection, that Newton's conception of space as the *sensorium* of God implies that God is a material being, if not a creature. She expatiates on this point, citing the example of Joseph Raphson's reification of space as an attribute of God as an instance of how Newton's conception of space has resulted undesirable theological consequences.

---

[39] These chapters are incorporated in the articles 'Espace' and 'Tems' in the *Encyclopédie*.

[40] Voltaire had already sent Frederick II a copy of *La métaphysique de Neuton* before Madame du Châtelet had completed her *Institutions*. LetChBII 237, 13. Judith Zinsser interprets this context in terms of antagonism between Voltaire and Du Châtelet, arguing that Voltaire set out to undermine Du Châtelet, and adducing manuscript changes in support of her interpretation. While I defer to her superior knowledge of the sources, it does seem to me that the printed texts suggest disagreement on interpretation, but not necessarily presented as deliberate attempt to undermine her. She herself explained their differences of view to Frederick of Prussia, as good evidence of their amity: citing Montaigne, she says 'Il me semble meme que notre amitié est plus respectable et plus sûre puisque même la diversité de l'opinion ne la pu altétérer'. LetChBII 237, 14.

[41] Inst1740 §74.

[42] Inst1740 §73.

> Il y a encore une grande absurdité à dévorer dans l'opinion de l'espace absolu, c'est que tous les attributs de Dieux lui conviennent; car cet Espace, s'il étoit possible, seroit réellement infini, immuable, incrée, nécessaire, incorporeal, présent par tout. C'est an partant de cette supposition que M. Raphson à voulu démontrer géometriquement que l'Espace est un attribut de Dieu, & il exprime son essence infinie & illimitée & c'est effectivement ce que suit très naturellement de la supposition de l'Espace absolu, quand on l'a une fois admises.[43]

Du Châtelet was no more impressed by Clarke's attempt to defend Newton's hypothesis (as she calls it) of absolute time. Like Leibniz, she objects that to posit absolute time is to make time a necessary being, like God, eternal, immutable and self-subsisting—that is, he endows time with the attributes of God. Absolute time, like absolute space is an imaginary concept. Section 96 focuses on what she calls Clarke's 'fameuse question', namely, why did not God create the world sooner— 'pourquoi Dieu n'avoit pas créé l'univers 6 mille ans plûtôt ou plus tard?' In her reply she argues if we follow our reason, and define time as the successive order of things, the question never arises, since 'le Tems n'est qu'un Etre abstrait, qui n'est rien hors des choses'.

Other aspects of Newtonianism which are covered in the Leibniz-Clarke letters are discussed elsewhere in *Institutions*. But these make no reference to their debate and can't be directly linked to it in the sense that the Leibniz-Clarke letters were her point of reference. Although, generally speaking, Du Châtelet takes a pro-Leibnizian stance, her criticisms do not merely reproduce Leibniz's objections and they do not amount to an outright rejection of Newtonianism. For example, in the only chapters specifically devoted to Newtonian topics, she takes a relatively agnostic position as to the explanatory value of Newton's theories of gravitation and attraction (Chapter XV, 'Des Decouvertes de M. Newton sur la pesanteur', and Chapter XVI, 'De l'Attraction Neutonienne'), while affirming that Newton's theory of universal attraction accounts for a large number of phenomena more satisfactorily than Cartesian physics (e.g. the fall of bodies, tidal movement, the rotation of the earth, irregularities in the path of the moon, comets, etc.).[44] Her criticisms are directed not so much at Newton, as at his followers, such as Keill and Freind. In the main, her criticisms are in line with the kind of criticisms that were being advanced in both France and Germany at that time. For example, her agnosticism in respect of Newtonian attraction was shared by Maupertuis, whom she praises as one of the best students and expositors of Newton.

---

[43] Inst1740 §75. In Inst1742Am, she replaces this paragraph with the simple statement, 'Ainsi le raisonnement de Mr. De Leibnitz contre l'Espace absolu est sans replique, & l'on est forcé d'abandoner cet Espace, ou renoncer le principe de la raison suffisante, c'est à dire, au fondement de toute verité, (§ 74, pp. 98-9). On Raphson's concept of Space, see Copenhaver. 1980. Jewish Theologies of Space in the Scientific Revolution: Henry More, Joseph Raphson, Isaac Newton and Their Predecessors. 515 ff.

[44] 'mais je ne m'engagerai pas ici dans le détail des Phénoménes & de leurs causes méchaniques; mon but étant seulement de vous faire voir en générale, comment les Newtoniens prétendent expliquer ces Phénoménes par l'attraction, & quelles sont les raisons qui doivent faire rejetter cette attraction, lorsqu'on la donne pour cause'. Inst1740 §389.

Reconciling Leibniz and Newton: *forces vives*

Du Châtelet's siding with Leibniz against Clarke in *Institutions de physique* does not mean that she was anti-Newton on all counts. Her pro-Leibnizian stance notwithstanding, it is clear that this did not involve abandoning Newtonianism. On the contrary, the *Institutions de physiques*, contains plenty of evidence of her abiding interest in Newton's theories, the problems they solve and the questions they raise—the question of the shape of the earth is a case in point.[45] This is a good example of a hypothesis confirmed by observation. There are, furthermore, places where she is explicit about the compatibility of Newton and Leibniz. This is particularly evident in her account of force, especially *force vives,* as discussed in her correspondence with Dortous de Mairan, which she added by way of appendix the 1742 edition of her *Institutions* (which has the modified title, *Institutions physiques*).[46] In her reply to Mairan, Du Châtelet plays down the opposition between Newton and Leibniz on theories of force, arguing that *forces vives* solve a Newtonian problem—the problem of the conservation of force, contrary to what the laws of Newtonian physics logically entail. This is a problem identified by Newton in *Opticks*, but one for which he had not produced an answer, beyond proposing that force must be conserved by divine intervention to re-calibrate the universe. Du Châtelet went so far as to suggest that in the last queries of his *Opticks*, Newton had put forward an alternative solution by proposing that forces are proportional to their velocity. And this solution, Du Châtelet argued, *anticipated* Leibniz's *forces vives*. However, he never set out the proofs. Had he done so, she claims, he would have arrived at a similar position to Leibniz, who explained the proportionality of force to the quantity of motion in terms of *forces vives*.

> [Newton] croyait la force des corps proportionelle à leur simple vitesse; mais comme il n'en parle que dans les questions quit sont à la fin de son Optique, & que nous n'avions aucun Ouvrage de lui, qui nous fasse voir qu'il ait discuté les preuves, que l'on apporte en faveur des forces vives, on peut peut-être raisonablement douter de quelle opinion Mr Newton auroit été s'il les avoit discutées, car il étoit assez grand-homme pour embrasser une opinion dont Mr. de Leibnits étoit l'Auteur, s'il l'avoit jugé véritable.[47]

As I have argued elsewhere, the significance of this passage is not what it tells us about Du Châtelet's lack of awareness of Newton's hostility towards Leibniz.[48] Its significance lies, rather, in what it tells us about the interpretative context for Newtonianism in eighteenth-century France, the context within which Du Châtelet studied Newton. She was highly conscious of a number of unresolved issues in

---

[45] I deal with this more fully in my article, 'Emilie du Châtelet's *Institutions de physique* as a document in the history of Newtonianism'.

[46] *Lettre de M. Mairan à la Marquise du Chastelet, sur la question des Forces Vives* and Madame du Châtelet's reply (originally sent 26th March 1741).

[47] Inst1742Am. 541.

[48] Hutton. 2004. Emilie du Châtelet's *Institutions*.

Newtonianism, and the discussions surrounding them—*forces vives* being a case in point. Her acceptance of the Leibnizian position, moreover, did not derive from a prejudice in favour of metaphysics: she had experimental grounds for her view, especially the experimental findings of Jakob Bernouill's pupil, Jacob Hermann, which she considered 'decisive ... en faveur de forces vives'. As she saw it, this was a case where the physics concurred with metaphysics:

> Toutes les expériences concourent à prouver les forces vives, mais la méthaphysique (*sic*) parle presque aussi fortement que la physique en leur faveur.[49]

Madame du Châtelet's attempt to reconcile Newton and Leibniz on the matter of *forces vives*, is sure evidence of her commitment to Newtonianism over an issue where she believed that Leibniz and Newton could be reconciled—not because Leibniz was wrong, but because she believed, on the strength of the late Queries appended to *Opticks*, that Newton was already working towards a position cognate to that of Leibniz on this issue. Other Newtonians, like 'sGravensande and Hermann, had conceded on *forces vives*.[50] She certainly was not the first *savant* to read Newton through Leibnizian spectacles.[51] In one sense, her use of the Leibniz-Clarke debate is a classic example of Madame du Châtelet doing just that. Her siding with Leibniz is predictable, given the method of argument she brought to the case.

Conclusion

When evaluating Du Châtelet's position in relation to the Leibniz-Clarke debate, it is important to remember that her views were contingent upon changes and developments in both her own thinking and the wider intellectual world. They therefore must be understood as part of an ongoing process of scientific discovery and not a static picture. Although her engagement with the Clarke-Leibniz debate was relatively insignificant within her work as a whole, there are a number of conclusions to be drawn from her engagement with it.

First of all it underlines her reliance in *Institutions de physique* not on Newtonian methodology or mathematics, but on Leibnizian principles, notably the principle of sufficient reason—this is consistent both with her aim of providing a metaphysical foundation for scientific theory, and with her aim to lay the foundations of a scientific

---

[49] Inst1742Am. 460. Jacob Hermann (1678–1733) is an example of a an admirer of Newton, who like Madame du Châtelet, was also a respecter of Leibniz, to whom he dedicated his *Pharonomia* (Amsterdam, 1716).

[50] See Hankins. 1985. *Science and the Enlightenment*. 31–3. Before Du Châtelet s'Gravensande was the first Newtonian to see *forces vives* as an answer to a problem in Newton in 1722.

[51] Johann Bernouilli and others used Leibniz's contributions to mathematics in their study of Newtonian mechanics. (Greenberg 1986). Maupertuis, for example, learned Leibnizian analysis from Bernouilli and applied it to solving problems raised in the *Principia* (Terrall. 2002. *The Man who Flattened the Earth. Maupertuis and the Sciences*, 7.)

education for readers without an advanced level of mathematical expertise.[52] The emphasis Du Châtelet places on the principle of sufficient reason is a feature of the *Institutions* as a whole, not just of her discussion of space, time and attraction. Leibniz himself had lamented that Clarke did not accord value to this principle. Had he done so, he told Princess Caroline, Newton would surely have been unable to resist the force of his (Leibniz's) arguments.

In the second place, the aspects of the Leibniz-Clarke debate to which Du Châtelet gives most attention, space and time, are ones which foreground theological issues—the objection that the Newtonian concepts of absolute space and time are too much like the attributes of God, and that Newton's description of space as God's *sensorium* imputes creaturely properties to the deity. The primary motivation for Leibniz's critique of Clarke was his concern about the religious implications of Newtonianism–or, at least, it was theological concerns that he put first among his objections. In his first letter to Princess Caroline, he expressed the fear that Newtonianism and other theories being put forward by some Englishmen (e.g. Hobbes) could lead to atheism. Madame du Châtelet expresses no such fears, but her critique of Newton's theory of absolute space echoes his theological concerns.[53] Her receptivity towards the metaphysical approach of Wolff and Leibniz may indicate the influence of Pierre Louis Moreau de Maupertuis who brought a theological and metaphysical dimension to his interest in Newtonian physics. That influence did not last, however. The changes she makes in the 1742 *Institutions physiques* show that she placed less emphasis on theological arguments. This is true of her discussion of gravitation, as also her discussion of space and time. For example, the revised version of her discussion of space omits the passage where she accuses Joseph Raphson of making space and attribute of God. It must be said that theological considerations were never extensive or primary features of her *Institutions*, and the changes in question are relatively small. But the changes in the 1742 edition which involved the removal or reduction of theologically-based objections, suggest that, as far as Newtonianism was concerned, theological issues were no longer so pressing. To the extent that revisions to her chapters on space and time are of this order, they encapsulate the drift of her thinking on Newtonianism generally. In her later work, Du Châtelet makes no reference to the Leibniz-Clarke debate. It is not mentioned in her *Abregée* appended to her translation of the *Principia*. Nor is there is any mention of Maupertuis in the *Abregée*. Instead, the hero of the hour is Claude Alexis Clairaut, the French Newtonian who accompanied Maupertuis to the pole, and who helped prepare Du Châtelet's *Principes* for publication. Why there should be no reference

---

[52] Du Châtelet's reliance on the Leibnizian principle of sufficient reason, is also consistent with the probability that, at this time Madame du Châtelet's own competence in mathematics were not yet fully accomplished, and that her knowledge of Newtonianism was largely derived from *Opticks*, rather than *Principia mathematica*.

[53] I discuss the impact of theological considerations on Du Châtelet's earlier discussions of Newtonianism in my article: Hutton. 2004. Emilie du Châtelet's *Institutions de physique* as a Document in the History of Newtonianism.

to Maupertuis when Clairaut is named, is not at all clear. But this does throw into relief the fact that while she was writing *Institutions de physique*, Du Châtelet came under the influence of Maupertuis. The silence about Maupertuis in *Principes* suggest that Du Châtelet's views had, by that time, diverged significantly from his.

Limited though it is in the context of her *oeuvre* as a whole, Du Châtelet's engagement with the Leibniz-Clarke debate is useful because it throws into relief her changing relationship with Newtonianism which was part of an on-going evaluation of Newtonian natural philosophy. Where, in 1738, she had taken a Newtonian position against Leibniz on all the issues in the debate except *forces vives*, in her *Institutions* she modifies that position. She also modifies her position on *forces vives*, to the extent that she now argues that Newton was closer to Leibniz than she had previously acknowledged. What is important about this—though unexpected perhaps in a self-declared admirer of Leibniz—is that her position now is not one of straightforward counter-positioning of Leibniz to Newton, but one where she saw the possibility for an accommodation between the two. The brevity of Du Châtelet's engagement with the Leibniz-Clarke debate is perhaps indicative of a desire not to be side-tracked by partisan issues. Emilie du Châtelet's admiration for both Newton and Leibniz was not, therefore, a *prima facie* inconsistency, but explicable in terms of the historical moment, and her aims when writing her *Institutions de physique*—among which was the educational purpose which she shared with Samuel Clarke.

References

Alexander, H. G. ed. 1956. *The Leibniz-Clarke Correspondence*. Manchester: Manchester University Press 1956.
Badinter, E., and D. Muzerelle, eds. 2006. *Madame du Châtelet. La femme des Lumières*. Paris: Bibliothèque Nationale de France.
Barber, W. H. 1979. *Voltaire and Samuel Clarke. SVEC* 179: 47–61.
Barber, W. H. 1967. Mme du Châtelet and Leibnizianism: the genesis of the *Institutions de physique*. In *The Age of Enlightenment: Studies presented to Theodore Besterman*, ed. William H. Barber, J.H Brumfitt, R.A Leigh, R. Shakleton, and S.S.B Taylor, 200–22. Edinburgh/London: Oliver and Boyd.
Brown, S. 1984. *Leibniz*. Brighton: Harvester Press.
Casini. 1988. Newton's *Principia* and the Philosophies of the Enlightenment. In *Newton's Principia*, eds. Hall and King-Hele 1988, 35–52. The Royal Society.
Clair, P. 1978. *Jacques Rohault (1618–1672): Bio-bibliographie, avec l'Édition critique des Entretiens sur la Philosophie*. Paris: Editions de Centre National du la Recherche Scientifique (CNRS).
Clarke, S. 1998. *A Demonstration of the Being and Attributes of God and Other Writings*, ed. Vailati. Cambridge: Cambridge University Press.
Cohen, I. B. 1968. The French Translation of Isaac Newton's *Philosophiae Naturalis Principia Mathematica* (1756, 1759, 1966). *Archives internationales d'histoire des sciences* 21:261–90.
Copenhaver, B. P. 1980. Jewish Theologies of Space in the Scientific Revolution: Henry More, Joseph Raphson, Isaac Newton and Their Predecessors. *Annals of Science* 37: 515 ff.
Du Châtelet, E. 1738. Lettre sur les *Eléments de la philosophie de Newton*. Paris. *Journal des sçavans*:434–41.
Ferguson, J. P. 1976. *Samuel Clarke: an Eighteenth-century Heretic*. Kinetown: Roundwood Press.

Ferguson, J. P. 1974. *The Philosophy of Samuel Clarke and its Critics*. New York: Vantage Press.
Frasca-Spada, M. and Jardine, N. 2000. *Books and the Sciences in History*. Cambridge: Cambridge University Press.
Gandt, F. de. 2001. *Cirey dans la vie intellectuelle: la réception de Newton en France*. Oxford: Voltaire Foundation.
Gardiner, L. J. 1982. Searching for the Metaphysics of Science: the Structure and composition of Madame du Châtelet's *Institutions de physique*, 1737–40. *SVEC* 201:85–113.
Grundy, I. 1999. *Lady Mary Wortley Montagu, Comet of the Enlightenment*. Oxford: Oxford University Press.
Hall, A. R. 1993. *All was Light. An Introduction to Newton's 'Opticks'*. Oxford: Clarendon Press.
Hall, A. R., and King Hele, D. G. 1988. *Newton's Principia and its legacy*. London: Royal Society.
Hankins, T. L. 1985. *Science and the Enlightenment*. Cambridge: Cambridge University Press.
Hoskins, M. 1961. Mining all within. Clarke's Notes to Rohault's *Traité de physique*. *The Thomist* 24: 353–63.
Hutton, S. 2004. Emilie du Châtelet's *Institutions de physique* as a Document in the History of French Newtonianism. *Studies in history and philosophy of science* 35 A, no. 3:515–31.
Jacob, M. 1967. *The Newtonians and the English Revolution, 1689–1720*. Hassocks: Harvester Press.
Koyré A., and Cohen, I. B. 1961. The Case of the Missing Tanquam. Leibniz, Newton and Clarke. *Isis* 52. 555–66.
Le Ru, V. 2005. *Voltaire newtonien. Le Combat d'un philosophe pour une science*. Paris: Vuibert.
Locke, J. 1989. *Some Thoughts concerning Education*, ed. John W. and Jean S. Yolton. Oxford: Clarendon Press.
Moriarty, P. V. 2006. The principle of sufficient reason in Du Châtelet's *Institutions*. In *Emilie Du Châtelet: rewriting Enlightenment philosophy and science*, ed. Judith P. Zinsser and Julie C. Hayes, 203–25. Oxford: Voltaire Foundation.
Newton, I. 1704. *Opticks, or a Treatise of the Reflexions, Refractions, Inflexions and Colours of Light*. Smith and Walford.
Newton, I. 1720. *Traité d'optique*. Transl. by Pierre Coste. Amsterdam.
Rohault, J. 1723. *Rohault's System of Natural Philosophy, illustrated with Dr. Samuel Clarke's Notes taken mostly out of Sir Isaac Newton's Philosophy*. Done into English by John Clarke, printed for James Knapton. London.
Samuel, C. 1998. Introduction. In *A Demonstration of the Being and Attributes of God and Other Writings*, ed. Vailati, E. Cambridge: Cambridge University Press.
Taton, R. 1971. Châtelet, Gabrielle-Émilie le Tonnier de Breteuil, Marquise du. In *Dictionary of scientific biography*, ed. Charles Coulston Gillispie, 215–17. New York: Charles Scribner's Sons.
Taton, R. 1970. Isaac Newton. *Principes mathématiques de la philosophie naturelle*, trad. de la marquise du Chastellet, augmentée des commentaires de Clairaut. *Revue d'histoire des sciences et de leurs applications* 23:175–80.
Taton, R. 1969. Madame du Châtelet, traductrice de Newton. *Archives internationales d'histoire des sciences* 22: 185–210.
Terrall, M. 2002. *The Man who Flattened the Earth. Maupertuis and the Sciences*. Chicago: Chicago University Press.
Vailati, E. 1997. *Leibniz and Clarke. A Study of their Correspondence*. Oxford: Oxford University Press.
Voltaire. 1992. *Eléments de la philosophie de Neuton*, 1738. In *The Complete Works of Voltaire*, ed. R. L. Walters and W. H. Barber. Oxford: Voltaire Foundation.
Wade, I. O. 1947. *Studies on Voltaire. With Some Unpublished Papers of Mme du Châtelet*. Princeton: Princeton University Press.
Wolff, C. 1729. *Merckwürdige Schriften welche ... zwischen dem Herrn Baron von Leibniz und dem Herrn D. Clarke uber besondere Materien der naûrlichen Religion in Franzôs. und Englisher Sprache gewechselt und ... in teutscher Sprache herausgegeben worden von Heinrich Köhler*. Frankfurt/Leipzig.

Zinsser, J. P. and Julie Candler Hayes, eds. 2006. *Emilie Du Châtelet: rewriting Enlightenment philosophy and science.* Oxford: Voltaire Foundation.

Zinsser, J. 2001. Translating Newton's 'Principia': The Marquise du Châtelet's Revisions and Additions for a French Audience. *Notes and Records of the Royal Society of London* 55, no. 2:227–45.

Zinsser, J. P. 2007. *La Dame d'Esprit. A Biography of Madame Du Châtelet.* New York and London: Viking.

"Sancti Bernoulli orate pro nobis". Emilie du Châtelet's Rediscovered *Essai sur l'optique* and Her Relation to the Mathematicians from Basel

Fritz Nagel

The quick prayer to which the title of this article refers is taken from a letter of Emilie du Châtelet written to Johann II Bernoulli on 28th April 1739. Along with her prayer, Emilie du Châtelet reports on her recently erupted dispute with Dortous de Mairan over the evaluation of the living forces. The prayer itself was addressed to mathematicians of the Bernoulli family active in Basel at that time, namely to Johann I Bernoulli (1667–1748), his sons Daniel I (1700–1782) and Johann II (1710–1790) as well as their cousin Nicolaus I Bernoulli (1687–1759). This quick prayer illustrates how close Emilie du Châtelet was to these mathematicians, and demonstrates at the same time her extraordinarily high regard for them. Two questions arise from this:

1. What was the base of Emilie du Châtelet's relation to the mathematicians from the Bernoulli family?
2. Which subjects did she discuss with them?

Regarding the former, I have to restrict myself to the most essential aspect, and regarding the second question, I will explore it from one particular angle only. This, however, will be done by analysing a text by Emilie du Châtelet which had until now been considered to be lost. It is her *Essai sur l'optique*, of which I found a complete version in Basel in 2006.

Before I can go into further detail, I first have to draw attention to the question of the primary sources. The public library of the University of Basel is in the enviable position of having a large collection of relevant documents on the Bernoulli correspondence at their disposal. Within this collection we find 43 original letters from

---

F. Nagel (✉)
Bernoulli-Forschungsstelle, Universitaetsbibliothek,
Schoenbeinstrasse 18/20, 4056 Basel, UB, Switzerland
e-mail: fritz.nagel@unibas.ch

Emilie du Châtelet to Johann II Bernoulli.[1] In addition, the library holds eight letters Voltaire sent to Johann II Bernoulli dating from the time when Emilie du Châtelet used to correspond with Bernoulli.[2] The collection furthermore includes copies of three of Du Châtelet's letters to Pierre Louis Moreau de Maupertuis, letters which are closely connected to her correspondence with Johann II Bernoulli.[3] The library furthermore holds excerpts of Cardinal de Tencin's correspondence with his sister, relating to the relationship between Mme du Châtelet and Voltaire[4] and also has some excerpts from the *Poesies choisies de Voltaire*, Leipzig 1797, some of the verses of which were directly addressed to Emilie du Châtelet.[5] The excerpts were edited by Johann III Bernoulli, the son of Johann II, who added them – together with some other sources – to a set of documents which is now part of the volume with the shelf mark UB Basel Ms L I a 684. Maupertuis' correspondence with Johann I and Johann II Bernoulli, which is documented by 98 and 174 letters respectively,[6] provides another important source here. Finally, the library also holds a handwritten *Mémoire* by Voltaire on Du Châtelet's *Dissertation sur la nature et la propagation du feu*.[7]

The fact that we only have letters addressed to Johann II Bernoulli confronts us with some problems regarding the interpretation of these sources. Unlike his father, Johann II Bernoulli did not draw up copies of his own letters. Therefore, the actual location of his letters is mostly unknown. Unfortunately, this also applies to those letters which were addressed to Emilie du Châtelet. The content of his letters can thus only be reconstructed by examining her replies. But let me now turn to the actual theme of this article.

It was Maupertuis himself who arranged the acquaintance of Emilie du Châtelet with the mathematicians from Basel. He had moved to Basel in autumn 1729 and had been welcomed by Johann I Bernoulli in order to be introduced by the famous mathematician to new developments in the fields of mathematics and physics and to perfect his own knowledge. To that end, he had also enrolled with the University of Basel.[8] At the time, Johann I Bernoulli was uncontestedly the best mentor Maupertuis could find to teach him the latest and most elaborate mathematical

---

[1] UB Basel Ms L I a 648, 349–541. These letters were edited rather poorly by Theodore Bestermann with several misreadings and unsatisfactory and faulty comments. LetChB I and II.

[2] UB Basel Ms L I a 726, fo. 170–180. Voltaire. 1968–1977. *Complete works of Voltaire. Les oeuvres complètes de Voltaire*. Vol. 90.

[3] UB Basel Ms L I a 684, 555–562. These copies were sent some years later to Johann II Bernoulli by La Condamine (La Condamine to Johann II Bernoulli, 09 July 1761, UB Basel Ms L I a 685, 407–414).

[4] UB Basel Ms L I a 648, 543–548.

[5] Ibid, p. 549–552.

[6] UB Basel Ms L I a 662 and L I a 708.

[7] UB Basel Ms L I a 726, fol. 184r-187v. This *Mémoire* can be found in Voltaire. 1879. *Oeuvres complètes*, ed. Moland. Vol. 23, 65.

[8] Maupertuis enrolled with the University of Basel on 30 September 1729. See *Matrikel Basel*. 1980. Vol. V, No 135.

methods. Apart from his mathematical teachings, however, Bernoulli also introduced Maupertuis to Leibniz' fundamental ideas of dynamics. It was during this first stay in Basel that Maupertuis became close friends with Johann II Bernoulli, the third son of Johann I Bernoulli.[9]

When in 1731 Maupertuis was elected *"pensionaire géomètre"* of the *Académie des Sciences* shortly after he had returned from Basel to Paris, Johann I Bernoulli hoped that he had found a supporter who could help him to introduce the Leibnizian dynamics into the academy. As it turned out, however, Maupertuis took on a rather passive role and concentrated on making his stand on the controversy over the cosmological ideas between the Cartesian and Newtonian parties. This distance to the Leibnizian dynamics naturally changed his relation with Johann I Bernoulli completely, and when Maupertuis returned to Basel in autumn 1734, he and his companion Alexis-Claude Clairaut, who was hoping to be taught by Johann I Bernoulli, could feel the alienation with Johann I Bernoulli. In the end, however, they again could profit from the mathematical skill of Maupertuis' former mentor, although they no longer shared the same convictions regarding Cartesian cosmology and Leibniz' dynamics. When Johann I Bernoulli accused Maupertuis that his prepossession for the oblateness of the earth towards its poles would have affected his scientific results obtained during his famous Lapland expedition, their friendship was finally damaged beyond repair. Despite Maupertuis' attempts to reconcile their friendship and to keep up his correspondence with Johann I Bernoulli until his death, their former cordiality could never be reestablished. Maupertuis' friendship with Johann II on the other hand would last a whole lifetime.

It was Maupertuis' third visit to Basel which plays an important role in our context. Travelling from Cirey, Maupertuis arrived in Basel on 20th January 1739, where he then stayed until mid-March.[10] His visit not only allowed for a meeting with his friend Johann II Bernoulli, but Maupertuis was also introduced to Samuel König, who at the time was a student of Johann Bernoulli senior. On his arrival in Basel, Maupertuis received a letter from Emilie du Châtelet stating that he had inspired her to engage shortly before their parting more closely with geometry and the differential calculus. In the meantime, it seems, she had realized that her research work would require her to seek further instruction by an adequate teacher.

Thus she wrote:

> Si vous pouvez déterminer un de m.^{rs} Bernoüilli à apporter la lumière dans mes ténèbres j'espère qu'il sera content de la docilité, de l'application, et de la reconnaissance de son écolière. Je ne puis répondre que de cela, je sens avec douleur que je me donne autant de peine que si j'apprenais le calcul, et que je n'avance point, parce que je manque de guide.[11]

---

[9] For Maupertuis and his relations to Emilie du Châtelet and the Bernoulli cf. Beeson. 1992. *Maupertuis: An intellectual biography*, and Terrall. 2002. *The Man who flattened the Earth. Maupertuis and the Enlightenment.*

[10] La Condamine to Johann II Bernoulli, 18 June 1761, 09 July 1761 and 26 July 1761, UB Basel Ms L I a 685.

[11] Emilie du Châtelet to Maupertuis, 20 [January] 1739, LetChBI 175.

Maupertuis tried to comply with her wishes. He arranged that both his friend Johann II Bernoulli and Samuel König would accompany him back to Cirey the following March.

At the time, Johann II Bernoulli was 29 years old.[12] He had studied law in Basel, had written his doctoral thesis "De compensationibus" in 1729 and had obtained his "Doctor utriusque juris" on 27th March 1732.[13] With his father, he had also intensively studied mathematics and a long stay in Vevey had helped him to polish his language skills in French. In 1732 he had gone on a long journey via Frankfurt, Marburg, Hamburg and Lübeck to St. Petersburg, where he had visited his brother Daniel, always hoping to secure an adequate academic position. As it turned out, he had hoped for a post in vain and he eventually had to relinquish his plans. He, accompanied by his brother, travelled back to Basel in 1733. Their journey took the two brothers to Danzig, Holland, Paris and Strasbourg, where Johann II found time to attend the wedding of his third brother Jacob. In a short autobiographical fragment, Johann proudly mentions how Maupertuis introduced him and his brother to the academic circles of Paris. The Bernoullis were admitted for example to a meeting of the *Académie des Sciences* and were lucky enough to witness a reading of the

Fig. 1 Johann II Bernoulli, Basel, Historisches Museum, Inv. Nr. 1991.157

---

[12] Figure 1.
[13] See *Matrikel Basel*. 1975. Vol. IV, Nr. 2836.

titles and devices of articles that had been submitted in that particular year to the scientific contest of the Royal Academy. Beneath them were two articles, one written by their father, the other by Daniel Bernoulli. Johann II Bernoulli notes:

> Da sich nun die anwesenden einbildeten, es würde uns etwan eine von diesen Schrifften nicht unbekannt seyn, so waren gleichsam aller Augen auff uns gerichtet um zu sehen, ob unser Angesicht solches bey ablesung der überschrifft und der Devise nicht verrathen wurde. In der that hatte es sich just gefügt, dass damahls über die nemliche question mein Vatter eine Schrifft von hier auss und mein Bruder von Petersburg auss eine andere nacher Pariss geschickt hatten, und zwar so hatten nach der Hand diese Schrifften beyde das Glück, dass der vorgesetzte Preiss (welcher das vorige mahl niemand ware zuerkannt worden und also dissmahl verdoppelt ware) unter sie getheilet wurde ...[14]

After his return in 1734, Johann II Bernoulli unsuccessfully applied for the chair of Ethics, Natural and International Law at the University of Basel. This rejection, however, allowed him to accompany the ruling Margrave of Baden-Durlach, who had asked him for his companionship, from his exile in Basel to Langensteinbach, a spa near Karlsruhe. From there he returned to Basel the following autumn in attendance on the Margrave himself. Several "Noms de plusieurs personnes de différentes qualités et sexe..." with whom he was friendly may serve to illustrate just how important Johann II considered his associations with aristocratic circles. Among others, he proudly lists the Marquis de Poié, the Baron Ponicau, the Comte de Tressan, the Prince Eduard [Stuart], the Marquis de Condorcet, the Margrave of Baden, the Princesse regnante d'Anhalt-Zerbst, the Duc de la Rochefoucauld, the Marechal de Ligny and a certain "M.e la marquise du Chatelet, qui m'a honoré de ses lettres jusqu'au temps de ses dernierès couches dans lesquelles elle est morte.[15]

His educational journeys together with his association with aristocratic circles as well as the aristocratic environment of a court and estates might all have been responsible for the cosmopolitan attitude this young scholar had already developed and which eventually led Maupertuis to introduce him to Emilie du Châtelet's and Voltaire's circle in Cirey. Unfortunately, Johann II Bernoulli himself tells us only very little about his visit. In his autobiography he merely notes:

> Ao. 1739 begleitete ich M. de Maupertuis, welcher uns nach seiner bekannten lapländischen Reyse heimgesucht hatte, wieder zurück biss halbwegs Pariss nacher Cirey, einem der Marquise du Châtelet zugehörigen Lustschloss, allwo wir uns noch einige Zeit mit einand auffhiellten und ich also Gelegenheit bekam, mit dieser verständigen und gelehrten Dame, wie auch mit dem berühmten Poeten Mr. de Voltaire, welcher sich gleichfalls allda befande, in Bekanntschafft zu gerathen, welche Bekanntschafft ich seitdeme durch Briefwechsel biss zu dem Absterben dieser Dame unterhalten habe.[16]

---

[14] Taken from Bernoulli. 1922. *Gedenkbuch der Familie Bernoulli zum 300. Jahrestag ihrer Aufnahme in das Basler Bürgerrecht*. 115–116. (*Gedenkbuch* in the following). Johann II Bernoulli refers here to both prizewinning works of Johann I Bernoulli (Op. CXLVI) and of Daniel Bernoulli (St. 24) on *l'Inclination des orbites des planets*.

[15] Bernoulli. 1922. *Gedenkbuch*. 127–129.

[16] Bernoulli. 1922. *Gedenkbuch*. 118.

Johann II Bernoulli only spent 2 weeks in Cirey, but his visit must have left Emilie du Châtelet deeply impressed. Shortly after his departure she wrote:

> Je saisis avec empressement cette occasion de vous dire combien on vous regrette à Cirey. Je me trouverais bien heureuse si le petit séjour que vous y avez fait vous avait inspiré un peu d'amitié pour moi. Je le désir infiniment. Vous êtes bien sûr de l'estime de tous les gens qui vous conaissent, mais soyez persuadé qu'il n'y a aucun endroit au monde où l'on fasse plus de cas de votre commerce et de votre amitié qu'à Cirey.[17]

Her subsequent letters to Johann II Bernoulli all seem to have one agenda only: Emilie du Châtelet tried to convince Johann II Bernoulli to return to Cirey in order to assist her in her research work. Du Châtelet's wish to employ Johann II Bernoulli as a tutor at that time coincided with an argument she had had with her former mentor Samuel König, who, needless to say, was a follower of Christian Wolff's philosophy. Emilie du Châtelet goes as far as to promise Johann II Bernoulli everything she could think of to secure his assistance. She even urged her husband to arrange for Bernoulli's salary to be guaranteed in case of her or her husband's death. In addition and along with various apartments in Cirey, Brussels and Paris, she also offered Johann a private servant, free rein to travel to Basel any time and a complimentary carriage for the journey. Eventually she saw to it that even Voltaire – who openly admitted that "... je seroy sans compliment Bernoullien toute ma vie"[18]- made an effort to lure Johann II to Cirey.

Emilie du Châtelet's letters of that period (1739–1749) show an fervent and urging style which she has applied also in other cases where she wanted to convince people to realize her plans. In the end, Johann II Bernoulli made a promise to return to Cirey for a longer stay, which he later withdrew at short notice and much to Emilie du Châtelet's chagrin. The Marquise suspected an intrigue by Maupertuis behind Bernoulli's sudden withdrawal, and as it turns out, it was indeed Maupertuis who had intervened in the last moment and who strongly advised Bernoulli to turn down her generous offer. On 12th January 1740 he wrote to his friend in Basel:

> Quant au conseil que vous me demandés je me donne bien de garde de vous conseiller d'y aller et quoique Me du C. en meure d'envie et que je voulusse luy rendre service en toute autre chose mon amitié pour vous me deffend de la servir en cecy outre que je crois qu'elle a grand tort avec K[önig], apres toutes ces histoires la place pour vous ne me paroit pas honeste: cette place de plus a encor d'autres inconveniens c'est qu'on ne scauroit y etre avec quelque agrement de la part de Me du Chast. qu'on ne soit tout d'un coup insupportable à Voltaire, et Voltaire aura toujours le dessus, et on luy sacrifiera tout; on ne peut presque plus luy doner aucun sujet de Mecontentement, par plus d'une raison. Tout cela posé, le Mathematicien de M.^{e} du Ch. joue un assez mauvais personage.[19]

The actual reasons for Maupertuis' intervention, however, have to remain of a speculative nature. Whether it was plain jealousy that made him advise against Du Châtelet's offer or whether the reason can be found within a certain weakness in her character which Maupertuis used to refer to frequently cannot be satisfactorily

---

[17] Emilie du Châtelet to Johann II Bernoulli, 30 March 1739, LetChBI 203, 352.

[18] Voltaire to Johann II Bernoulli, Cirey 08 May 1739, UB Basel Ms L I a 726, fo. 169.

[19] UB Basel Ms L I a 708, fo. 40v-41r.

elucidated here. Whatever the reason for his intervention has been, Emilie du Châtelet was fully aware of the extent to which her former tutor was involved in the affair. She wrote to Johann II Bernoulli thus:

> Ce n'est point votre lettre, monsieur, qui m'a fait croire que m^r de Maupertuis vous avait empêché de venire chez moi, je le savais avant d'avoir reçu votre réponse, et je vous l'ai mandé. C'est donc à lui seul que je m'en prends, et ce n'est que de lui dont j'ai à me plaindre dans cette affaire, car on n'a véritablement à se plaindre que des gens sur qui l'on a des droits, et j'en avais sur son amitié. Je regarde Koenig comme un homme à qui la tête a tourné, et dont j'aurais dû me défaire plus tôt, mais je vous avoue que je ne m'attendais pas à la façon don't m^r de Maupertuis s'est conduit dans cette affaire, et comme je suis infiniment sensible au bien, et au mal, il m'est difficile de l'oublier.[20]

Although Johann II Bernoulli had rejected her offer, Emilie du Châtelet never felt any personal resentment towards him. In fact, she kept corresponding with him until shortly before her death.

What are the subjects of this correspondence? We learn for example that Emilie du Châtelet owned some printed works of the Bernoulli[21] and that she even came into possession of manuscripts of Johann I Bernoulli's correspondence with Leibniz and Montmort.[22] We learn about their opinion in the *vis viva* debate[23] or that her biography for Brucker's *Pinacoteca scriptorum*[24] was composed by Johann II Bernoulli, based on material Emilie du Châtelet had relinquished to him.[25] But in this context here I shall focus on another subject, namely Emilie du Châtelet's *Essai sur l'optique*, an essay which until now was considered lost and which is preserved in the archive of Johann II Bernoulli.

We were informed of Du Châtelet's *Essai sur l'optique* by the existence of a fragment of this *Essai*, which only comprises a version of the fourth chapter ("De la formation des couleurs"). The copy is to be found in Voltaire's legacy which is now being held at the National Library in St. Petersburg and which was edited by Ira O. Wade in 1947.[26] In the correspondence of Emilie du Châtelet her *Essai* is first mentioned on 30th March 1739 when she writes to Johann II Bernoulli shortly after his departure from Cirey:

> J'espère bien que vous me direz votre avis du petit essai d'optique que vous avez bien voulu emporter, car comme je ne l'ai encore montré à personne je ne sais s'il est bon ou mauvais, et vous êtes mon *criterium* de la vérité. Je compte autant sur votre indulgence que sur vos lumières, et j'ai également besoin de l'une et de l'autre.[27]

---

[20] Emilie du Châtelet to Johann II Bernoulli, 24 February 1740, LetChBII 234.
[21] Emilie du Châtelet was in possession of Jacob Bernoulli's *Ars conjectandi*, Basel 1713, bound together with Nicolaus I Bernoulli's *Dissertatio de usu artis conjectandi in jure*.
[22] LetChBI 203; 211; 214; 220.
[23] Cf. ex. gr. Walters. 2001. La querelle des forces vives et le rôle de Mme du Châtelet. 198–211.
[24] Brucker. 1744. *Pinacoteca scriptorum nostra aetate literis illustrium*, Decas III (no page numbers).
[25] A detailed description is given by Iverson. 2006. A female member of the Republic of letters. Du Châtelet's portrait in the Bilder-Sal […] berühmter Schriftsteller. 35–64.
[26] Wade. 1947. *Studies on Voltaire. With Some Unpublished Papers of Mme du Châtelet*. 188–208.
[27] LetChBI 203, 352.

In another letter dating from 28th April 1738 she kept pressuring him about his opinion on her *Essai* by writing:

> Je voudrais bien que vous me mandiez avec bonté ce que vous pensez du petit ouvrage que j'ai eu la hardiesse [de] vous donner. Je l'ai relu depuis et je me suis trouvée bien téméraire. Je vous supplie de ne le montrer à personne, car je suis bien loin d'en être contente et je ne la serai véritablement que quand je l'aurai corrigé sur vos avis.[28]

According to the above, she was still working on her *Essai* on optics whereas Johann II Bernoulli was proofreading it on his way from Cirey to Basel. Apparently he soon gave his advices for amendments to Emilie du Châtelet because on 3 August she wrote:

> En vous remerciant de l'attention que vous avez bien voulu faire à mon petit essai sur l'optique, dont je suis encore bien éloignée d'être contente, et je profiterai assurément des avis que vous voulez bien me donner sur cela dans votre lettre. Vous me donnez le désir et le courage d'en faire quelque chose puisqu'il vous a plu.[29]

Unfortunately, this particular letter from Johann II Bernoulli is missing, which means that we can gain no further insight into his proposals concerning the *Essai sur l'optique*. But fortunately the loss of Bernoulli's letter is more than compensated by the discovery of the complete text of Du Châtelet's *Essai* he had read and commented on.

Whilst preparing for the Potsdam symposium on Du Châtelet in 2006, I went through Bernoulli's legacy at the public library in Basel and accidentally came across a complete version of the *Essai sur l'optique* by Emilie du Châtelet.[30] The manuscript consists of 36 sheets stitched together in a 23,5×18,5 cm format with gold-plated edges. Johann III Bernoulli (1744–1807), the first son of Johann II, kept this convolute in a folded waste sheet which has now been made part of the manuscript with the shelf mark UB Basel Ms L I a 755, (fo. 230–265). The front page bears a note in a yet to be identified handwriting which says "Auteur de cet Essai est M.e la Marquise du Châtelet". The last page bears a note in pencil saying "Ma. Me du Ch. donna ce Msct à mon Père, lorsqu'il quitta Cirey, p.r le revoir voulant le faire imprimer – mais ensuite l'impression n'eut pas lieue, & le Mspt demeura à mon Père. DB". "DB" are, of course, the initials of Daniel II Bernoulli (1751–1834), third son of Johann II, who was professor of eloquence in Basel and also went on to act as a substitute for his uncle Daniel I Bernoulli, who was professor of physics at the same university. Eventually, however, Daniel II Bernoulli resigned from his duties at the university and started working as a so-called "Domprobsteischaffner", a position which entailed administrating monastic goods which had been confiscated during the Reformation.[31] We can assume that his copy of the *Essai sur l'optique* is indeed the manuscript Bernoulli had proofread on his

---

[28] LetChBI 211.
[29] LetChBI 220.
[30] "Du Châtelet. Essai" in the following.
[31] See *Matrikel Basel*. Vol. V, Nr. 1254.

way to Basel, since Johann II seems to be the unidentified editor who wrote his comments and notes in the margins. The text itself is written by a hand which is not that of Emilie du Châtelet. Further there are notes and comments in the text in a third handwriting. The margins show short summaries of individual paragraphs.

Du Châtelet opens her *Essai sur l'optique* with a reference to Newton's *Opticks*,[32] which, she argues, had contributed more to the progress of science than any other branch of physics. In fact, Emilie du Châtelet seems to have read Newton's work thoroughly and several times over. A letter she wrote to Maupertuis dating from 1739 may serve to illustrate that:

> [j]e sais presque par coeur l'optique de m^r Neuton et je vous avoue que je ne croyais pas qu'on pût révoquer en doute ses expériences sur la réfrangibilité. Ce sont celles qu'il a faites avec le plus de soin ...[33]

It is to be assumed that Du Châtelet herself has conducted most of the experiments on optics she mentions. At least her cabinet de physique at Cirey contained some of the relevant instruments.[34] A complaint regarding some repair work on her "chambre obscure", which Abbé Nollet had carried out poorly, may provide evidence for her expertise as well as her appeal to Algarotti to execute together some of those experiments the author had described in his *Dialogues sur la lumière*.[35] But nevertheless she calls, Algarotti "un singe de Fontenelle" and later she dismissed his ideas in the *Dialogues* as "frivole". Her real master was and always had been Newton. Du Châtelet believed that it was not in her to review his work critically, but rather to develop a theory which would outline the results reached irrevocably by experiments from a superior point of view. The second paragraph of her preface consequently states for example "....que plus on veut aprofondir les causes de l'opacité et de la transparence et plus il faut étudier le traité d'optique de ce grand homme."[36]

The *Essai sur l'optique* is divided into four chapters, all of which are explicitly listed in the *Introduction*. The first chapter deals with light in general, the second chapter evolves the causes of the transmission and refraction of light in transparent objects and the third chapter focuses on the causes of the reflexion by nontransparent objects. Finally, the fourth chapter turns to the question why certain objects reflect different colours than others.

The first chapter of the actual *Essai* sets out to establish four definitions, the first of which assumes that light is a property of any substantial being, i.e. a property we would usually refer to as fire. The second definition states that light is composed of

---

[32] It is to be assumed that Emilie du Châtelet would have read the French translation of Newton's *Opticks* (1704), which was published in Amsterdam under the title 'Traité d'optique sur les reflexions, refractions, inflexions et couleurs de la lumière' in 1720.

[33] LetChBI 152.

[34] Cf. Gauvin. 2006. Le cabinet de physique du château de Cirey et la physique naturelle de Mme du Châtelet et Voltaire. 165–202.

[35] Emilie du Châtelet to Algarotti, 20 April 1736, LetChBI 65.

[36] Du Châtelet. Essai. Introduction, 1.

luminous particles expanding with finite velocity. Emilie du Châtelet assumes that light travels at a speed that would take it approximately 7–8 min to travel from the sun to the earth, thereby accepting a result of Ole Roemer, which Newton had also been aware of.[37] She further argues that light is made up of extremely subtle particles, leaving us with a considerable problem regarding the measuring of these particles. Eventually, she states that a beam of light consists of a combination of seven differently coloured components, namely red, yellow, orange, green, blue, indigo and purple. Thus, Emilie du Châtelet remains well within Newton's ideas and theories on optics and closely follows his patterns in explaining optic phenomena.

Having established these definitions, Du Châtelet goes on both to discuss and to dismiss Aristotle's and Descartes' theories on optics. She discards Aristotle's definition of light as an "accidens" and refers to it as an "acte du transparent en tant que transparent" in a "jargon inintelligible qu'on a pris pendant plus de deux milles ans pour la verité".[38] Du Châtelet appreciated that Descartes had rejected Aristotle's concept of light and consequently Aristotle's view of the world. Yet she also had her doubts about Descartes' own theory. Descartes described light as a substance which, according to him, consisted of tiny globes. In keeping with this theory, visual perception would then have to be accomplished by shattering these globes, something for which Emilie du Châtelet could find little evidence. Instead of providing an explanation, Descartes' theory led her to the conclusion that "….le seul profit que les hommes ayent tiré des ouvrages de Descartes, tout grand homme qu'il etoit, c'est d'apprendre à se tromper avec methode".[39] Hence and in accordance with Newton, whom she admired greatly, Emilie du Châtelet appears to have believed in emission theory and ostensibly objected to the assumption that the properties of light could be altered by passing an object.

In her second chapter, Emilie du Châtelet then explores why transparent objects transmit light when they are vertically hit by it. She seeks to explain why refraction occurs when light passes obliquely from one medium to another and she particularly contradicts the idea that light transmission is dependent on the pore size of the object. By thoroughly observing her every-day environment, Emilie du Châtelet thereby produced valid evidence for her own theories. However, it will be her theoretical explanations of refraction, reflection and dispersion that will be of further interest in the following, rather than her views on Descartes and Aristotle.

Here too, Emilie du Châtelet decided to follow Newton's approach and tried to explain the aforementioned optic phenomena by a special interaction between light and the objects in question. Paragraph six of her second chapter reads:

> Les corps agissent sur la lumiere, puis qu'ils la forcent à prendre un nouveau chemin dans la refraction, et dans la reflexion et qu'ils luy font perdre son mouvement, en l'absorbant dans leur substance. Mais outre tous ces effets que les corps font sur la lumiere qui les ateint, ils exercent encor une action sur elle, avant qu'elle soit parvenüe à leur premiere surface.[40]

---

[37] Du Châtelet already dealt with optical principles and relevant scientific observations and results in her *Dissertation sur la nature et la propagation du feu*, drafted almost at the same time, for which she received the prix de l'Académie Royale (Paris 1739). We find there some remarks on light for example in the sections on attraction and dispersion. (pp. 6-7).

[38] Du Châtelet. Essai. Chap. 1, 4.

[39] Du Châtelet. Essai. Chap. 1, 5.

It is quite obvious that Emilie du Châtelet's hypotheses were inspired by Newton's *Principia*[41] and its notion of gravity in either case. Yet neither Newton nor Emilie du Châtelet was willing or able to provide any further details of their notions of these particular interactions. The nature of interaction between objects and light in optics, they both state, has to remain as obscure as the direct impulse and the notion of contactless gravitation when it comes to mechanics.

In Du Châtelet's own words,

> [t]ous les effets que les corps operent sur la lumiere, sont egallement incomprehensibles. On croit cependant mieux concevoir les effets qu'ils produisent sur la lumiere qui les atteint, que ceux qu'ils font sur elle en eloignement. Tout ce qui porte un air d'impulsion, en paroit plus aisé à comprendre; cependant si on examine ses idées avec severité, on trouvera peut-estre qu'on ne conçoit gueres mieux, comment un corps peut communiquer son mouvement à un autre corps en le poussant, que comment un corps peut agir sur un autre corps, sans le toucher. Le principe de ces deux effets nous est caché. L'impulsion tombe plus sous nos sens, mais nostre esprit n'en a peutestre pas une idée plus claire.[42]

It follows from this that Emilie du Châtelet assumed that a beam of light CD[43] passing a body of glass would have to be deflected from its linear course due to the gravity of the glass, even before it had hit surface FF from an oblique angle. To this

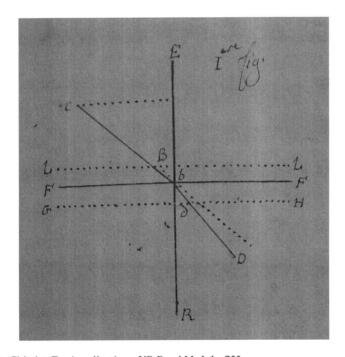

Fig. 2 Du Châtelet, Essai sur l'optique, UB Basel Ms L I a 755

---

[40] Du Châtelet. Essai. Chap. 2, 7.
[41] Newton 1687.
[42] Du Châtelet. Essai. Chap. 2, 7.
[43] See Figure 2 ("Iere fig.").

end, she postulates a "sphère d'activité" *LL* in the glass, inside of which gravity becomes noticeable. Subsequently, she also goes on to point out that any beam of light in this "sphère d'activité" would have to follow a slightly deflected course *Bb* even before breaking through the glass and progressing linearly within it. Following Newton's theory further, Emilie du Châtelet finally seeks to explain optical diffraction, a process she had been able to observe on a knife blade in her "chambre obscure". Her conclusion that it was imperative to acknowledge the gravity between objects and light[44] as established in this way, however, turned out to have been somewhat hasty.

Emilie du Châtelet then turns to the laws of gravitational force. She states that "Cette attraction paroit en plusieurs occasions, suivre d'autres loix que l'attraction que les corps exercent les uns sur les autres[45]," and subsequently establishes three laws.

(1). This aforementioned "attraction" affects light rays from separate colours differently and can thus be seen as the cause of their dispersion. (2). Compared to other objects, Du Châtelet repeatedly states that "corps sulphureux" affect light more strongly than other objects. (3). Just as Newton's gravitational force between bodies not in direct contact is effective even at a distance, the attraction between light and its respective object becomes noticeable even before the light actually hits the object. This particular law of "attraction", however, differs from the law of universal gravitation, as its force is not proportional to the product of the two masses and inversely proportional to the square of the distances between its respective objects. Consequently, Emilie du Châtelet ventures the question whether it should not be possible for this particular "attraction" to follow a different law altogether. Even though such a law would not yet have been formulated, she does not see any contradiction in proposing a different law to Newton's law of universal gravitation in the case of light attraction.

> La lumiere paroit un etre à part, qui n'est analogue à aucun de ceux que nous connoissons, et je ne verrois nulle contradiction a supposer qu'il y a pour elle, dans de certaines circonstances, d'autres loix d'attraction, que pour les corps, il y a tant de loix du mouvement produit par l'impulsion, pourquoy ne pouroit-il pas y en avoir plusieurs du mouvement produit par l'attraction? Je ne vois ny impossibilité, ny contradiction, à admettre sur cela, ce que les phenomenes nous découvriront.[46]

Even though the true nature of this "attraction" remains unknown to her, Emilie du Châtelet applies those physical methods to her research that were available to her. Assuming that a beam of light, when hit by an object, will be affected by two forces mainly and that, in such a case, one force will cause linear movement and the other one "attraction", the law of decomposition and composition on these two forces will also allow for the assumption of a law of refraction and reflection. Needless to say, some of these assumptions were quite inaccurate. Firstly, no force can have any direct effect on light travelling both at constant motion and according

---

[44] "Il est impossible, quand on considere avec attention tous les effets, de ne pas reconnoitre une attraction entre les corps, et la lumiere." Du Châtelet. Essai. Chap. 2, 8.

[45] Ibid.

[46] Du Châtelet. Essai. Chap. 2, 9.

to its direction of movement, and, secondly, the physical quantity of this assumed force of attraction lacks any basic definition. Furthermore, if velocity vectors had been added, the velocity vector of light would greatly differ from the velocity vector of "attraction", so that its parameters could not have led to the basic law of refraction, of which, it has to be said, Du Châtelet was well aware. If one were to follow her assumption regarding the effect of "attraction" on a beam of light here, one could easily see in it a forerunner of Einstein's assumptions regarding the bending of light in the presence of a gravitational field. Yet it was not until the solar eclipse of 1919 that this minimal effect could be experimentally verified via modern technology and refined methods. In other words, it had nothing to do with the law of refraction. Quite apart from the fact that Du Châtelet herself considered "attraction of light" and "gravitation of bodies" to be two completely separate forces.

Assuming that there was indeed an "attraction" between objects and incidental light, Du Châtelet approaches all subsequent phenomena in the *Essai* from this particular angle of Newton's hypotheses. She even goes as far as to hold the lack of knowledge regarding the "principe actif de la nature" responsible for Descartes' dispute with Fermat. She furthermore blames Fermat for the fact that his minimization principle (the path taken by a ray of light is always the path that can be traversed in the least of time) only served to defend a good cause very poorly, "car un principe moral, une cause finale ne peut contrebalancer un principe geometrique tel que celuy de la composition des mouvemens."[47]

In the third chapter Emilie du Châtelet tries among other things to find a valid explanation for the fact that light reflects off opaque objects. She also claims that, in the process of reflection, a fracture of light is always already being thrown back even before hitting the surface of an object. The cause for this phenomenon, she claims, has still to be found and even Newton "a eu la modestie et la bonne foy d'avoüer, qu'il en ignoroit la cause".[48] Newton had assumed that the phenomenon was caused by an extremely subtle and very elastic matter coating the object's surface. Emilie du Châtelet on her part suggests "la matiere electrique" to be such matter and hopes that "l'exactitude avec laquelle on examine à present les phenomenes de l'electricité, éclaircira quelque jour le mystere de la reflexion …"[49]

Finally, the fourth chapter of the *Essai* is committed to examining the dispersion of light into different colours. Since this chapter has already been edited by Wade,[50] I will limit myself to one fundamental principle by which Emilie du Châtelet seeks to explain the phenomenon of dispersion. It will come as no surprise that it is again the phenomenon of "attraction" which has to serve as a cause here. Colours, Du Châtelet claims, can neither be based on the modification of a medium which is able to propagate light nor on the diffusion of an object interacting with this light. It is rather in accordance with Newton's theory that white light is composed of spectrum

---

[47] Du Châtelet. Essai. Chap. 2, 13.
[48] Du Châtelet. Essai. Chap. 3, 6.
[49] Du Châtelet. Essai. Chap. 3, 7.
[50] Cf. footnote 28.

colours. Emilie du Châtelet's theory here states that the spectrum of a white beam's component colours will be slowed down at different frequencies when it hits a prism. Thus, each component passes through the prism at various frequencies. Here she utilizes the parallelogram of forces by applying it to each component, thereby receiving separate refraction angles for different colours due to various vectors of velocity. Based on these results and assumptions, Du Châtelet continues to explore a variety of phenomena such as reflection, refraction, dispersion and absorption. This variety and its proximity to Newton's ideas, however, would exceed the limited scope of this article.

In conclusion I shall simply summarize the following:

1. Emilie du Châtelet's *Essai sur l'optique* provides ample evidence of her extraordinary familiarity with the scientific debates of her time regarding the field of optics. This applies to both geometric optics and the phenomena of dispersion and optical physics.
2. In the field of optics, Emilie du Châtelet shared Newton's positions entirely. Not unlike many other scientists of her time, she obviously utilized Newton's experimental results and his explanation of dispersion especially. As she objects to the theory that a material medium transmitting light has to be hypothesized, she turns out to be a proponent of emission theory. However, contrary to Newton, Emilie du Châtelet failed to discuss her position critically and did not search for an alternative or for more subtle explanations regarding the sources of optical phenomena, something Newton did regularly for the "Queries" of his various *Opticks* editions.[51]
3. By combining various experimental facts under one integrative aspect of optics and using only a few common fundamental principles in order to prove her theory, Emilie du Châtelet's attempts are in line with a series of other contemporaries who tried to explain physical phenomena like gravity or heat via mechanical procedures. Her attempts are consistent and follow the ideal of an empirical method in natural sciences. As was the case with the research of several of her contemporaries, her attempts were nonetheless bound to fail. Just like other researchers of her time, who were trying to falsify Newton's notion of "action at a distance" by developing several working hypotheses in order to explain gravitational force by traditional mechanical principles based on direct contact of matter, Emilie du Châtelet can hardly be blamed for straying from Newton's path.
4. After all, one has to bear in mind what an extraordinary achievement it was for a woman living in the first half of the eighteenth century to acquire such detailed knowledge of optics. Knowledge, which enabled her to write an essay taken seriously by her male colleagues, first to mention Johann II Bernoulli.

It therefore would now be appropriate to compare Emilie du Châtelet's *Essai* with the work of contemporaries such as Huygens, `sGravesande, Musschenbroeck,

---

[51] Newton. 1704. Opticks. Newton. 1706. Optice. Newton 1720. *Traité d'optique*.

Malebranche, Mairan and others.[52] In this context, it would be particularly worthwhile to compare the *Essai* with Johann II Bernoulli's theses on optics in order to be able to draw some conclusions as to his long lost comments on Du Châtelet's work. Bernoulli's remarks would especially be interesting, because in his prizewinning dissertation on the propagation of light[53] he has developed a sophisticated model of oscillating light particles in a space filling medium which apparently is closer to Cartisian concepts than to Newtonian ideas. This, however, would far exceed the scope of this paper and will have to be dealt with elsewhere.

I would like to conclude here by stating that Emilie du Châtelet perhaps would have frowned upon my presentation, given that she did not consider the *Essai* to be worthy of publication yet. However, maybe she will kindly accept my offer of a printed edition of her work[54] as a present for her 300th birthday. As different as this might be to her original intentions when uttering the quick prayer "Sancti Bernoulli orate pro nobis" I referred to in the title of this article, she perhaps could feel that it has been answered in a certain unexpected way.Du Châtelet, Essai sur l'optique, UB Basel M s L I a 755

References

Beeson, D. 1992. *Maupertuis. An intellectual biography.* Oxford: Voltaire Foundation.
Bernoulli, C. A. 1922. *Gedenkbuch der Familie Bernoulli zum 300. Jahrestag ihrer Aufnahme in das Basler Bürgerrecht.* Basel: Helbing u. Lichtenhahn Bernoulli, J. 1713, Ars conjectandi, Basel.
Bernoulli, J. II. 1752. Recherches physiques et géométriques sur la question: Comment se fait la propagation de la lumière, proposée par l'académie royale des sciences pour le sujet du prix de l'année 1736. *Receuil des pieces qui ont remporté les prix de l'académie royale de sciences* Vol. 3. Paris.
Brucker, J. 1744. *Pinacoteca scriptorum nostra aetate literis illustrium, Decas* III. Augustae Vindelicorum (Augsburg): Haid.
Basel, M. 1975. *Die Matrikel der Universität Basel*, Bd. IV, eds. H.G. Wackernagel, M. Triet, P. Marrer, Basel.
Basel, M. 1980. *Die Matrikel der Universität Basel*, Bd. V, eds. Max Triet, Pius Marrer, Hans Rindlisbacher. Basel.
Gauvin, J.-F. 2006. Le cabinet de physique du château de Cirey et la philosophie naturelle de Mme Du Châtelet et Voltaire. In *Emilie Du Châtelet: rewriting Enlightenment philosophy and science*, ed. Judith P. Zinsser and Julie C. Hayes, 165-202. Oxford: Voltaire Foundation.

---

[52] For a more detailed overview see Casper Hakfoort's wonderful book 'Optics in the age of Euler, Conceptions of the nature of light, *1700-1795*'. 1995.

[53] Bernoulli. 1752. Recherches physiques et géométriques sur la question: Comment se fait la propagation de la lumière, proposée par l'Académie Royale des Sciences pour le sujet du prix de l'année 1736. Vol. 3.

[54] This edition is planned for 2011 as a volume of the series "Travaux sur la Suisse des lumières" edited by the Swiss Society for 18th century studies (Edition Slatkine, Genève).

Hakfoort, C. 1995. *Optics in the age of Euler, Conceptions of the nature of light, 1700–1795*. Cambridge: Cambridge University Press.

Iverson, J. R. 2006. A female member of the Republic of Letters: du Châtelet's Portrait in the Bilder-Sal (...) berümhter Schrifftsteller. In *Emilie Du Châtelet: rewriting Enlightenment philosophy and science*, ed. Judith P. Zinsser and Julie C. Hayes, 35–51. Oxford: Voltaire Foundation.

Newton, I. 1687. *Philosophiae naturalis principia mathematica*. London.

Newton, I. 1704. *Opticks: or a Treatise of Reflexions, Refractions, Inflexions and Colours of Light*. London. (Second edition, London 1718; third edition, London 1721; fourth edition, London 1730).

Newton, I. 1706. *Optice: Sive de reflexionibus, refractionibus, inflexionibus et coloribus lucis*, transl. by Samuel Clarke. London: Smith and Walford.

Newton, I. 1720. *Traité d'optique sur les reflexions, refractions, inflexions et couleurs de la lumière*, Traduit de l'anglois par M. Coste sur la seconde édition augm. par l'auteur. Amsterdam.

Terrall, M. 2002. *The Man who flattened the earth: Maupertuis and the sciences in the Enlightenment*. Chicago: University of Chicago Press.

Voltaire. 1879. *Oeuvres complètes: avec notices, préfaces, variantes, table analytique, les notes de tous les commentateurs et des notes nouvelles: conforme pour le texte à l'édition de Beuchot enrichie des découvertes les plus récentes et mise au courant des travaux qui ont paru jusqu'à ce jour; précédée de la vie de Voltaire par Condorcet et d'autres études biographiques; ornée d'un portrait en pied d'après la statue de foyer de la Comédie Française*. Paris: Garnier.

Voltaire. 1968–1977. Complete works of Voltaire, *Les oeuvres complètes de Voltaire*, ed. Theodore Besterman. Vols 135. Genève: Institut et Musée Voltaire.

Wade, I. O., ed. 1947. *Studies on Voltaire. With some unpublished papers of Mme. du Châtelet*. Princeton: Princeton University Press.

Walters, R. L. 2001. La querelle des forces vives et le rôle de Mme du Châtelet. In *Cirey dans la vie intellectuelle: la réception de Newton en France*, ed. François de Gandt, 198–211. Oxford: Voltaire Foundation.

Leonhard Euler and Emilie du Châtelet. On the Post-Newtonian Development of Mechanics

Dieter Suisky

Introduction

In view of the long history of science, Leonhard Euler (1707–1783) and Emilie du Châtelet (1706–1749) wrote their first comprehensive treatises on the basic principles of mechanics almost simultaneously in the very short period of the fourth decade of the eighteenth century. It was the time after Leibniz and Newton who dominated the discussions of almost all problems in mathematics, physics and philosophy by far not ever since, but mostly since Leibniz published the principles of the calculus[1] in 1684 and Newton the *Principia*[2] in 1687.

Euler becomes famous as the leading mathematician of the eighteenth century whereas Du Châtelet is known for the translation of Newton's *Principia* into French.[3] Du Châtelet's *Institutions de physique* had been published in 1740 [1740Inst] almost simultaneously with Maupertuis' *Loi du repos de corps*,[4] after Euler's *Mechanica*[5] in 1736, but before d'Alembert's *Traité*[6] in 1743 and Maupertuis' general principle for motion and rest[7] in 1746. Du Châtelet's book was rapidly translated into German [Naturlehre1743] whereas Euler's *Mechanica* had to wait for a German translation

---

[1] Leibniz. *Nova methodus pro maximis et minimis, itemque tangentibus, quae nec fractas nec rationales quantitatis moratur, et singulae pro illis calculis genus.*
[2] Newton. 1687. *Philosophiae naturalis principia mathematica.*
[3] PrincChat1756.
[4] Maupertuis P. L. M. Lois du repos de corps.
[5] Euler. 1736. *Mechanica sive motus scientia analytice exposita.*
[6] d'Alembert. 1743. *Traité de Dynamique.*
[7] Maupertuis P. L. M. Les lois du mouvement et du repos déduites d'un principe métaphysique.

D. Suisky (✉)
Humboldt-Universität zu Berlin, Institut für Physik, Newton str. 15, 12489, Berlin, Germany
e-mail: dsuisky@physik.hu-berlin.de

until the middle of the nineteenth century.[8] Reading the *Institutions*, Euler did not only appreciate Du Châtelet's idea and intentions in writing the *Institutions*, but was probably really challenged by her presentation and surpassed Du Châtelet only with the *Lettres à une princesse d'Allemagne*.[9]

In the 1730s, the scientific community was involved in controversial debates about basic concepts of mechanics like the nature of space and time and the true measure of living forces, the priority in the invention of the calculus and the application of the calculus to mechanics.

Why to read Euler in the twenty-first century? Why to read Du Châtelet? The surprisingly short, but nevertheless precise answer is that Euler, Du Châtelet and their contemporaries were involved in the same business we are dealing with still today. The question to be answered in the eighteenth century and today concerns the legacy of Descartes (1596–1650), Newton (1643–1727) and Leibniz (1646–1716). Nowadays, it has to be completed by the inclusion of the legacy of Euler, Du Châtelet and their contemporaries.[10] Therefore, reading today Du Châtelet's treatise as a commentary on the development of mechanics since Descartes and the later summaries by Kästner[11] and Gehler[12] we obtain a lively mirroring of the influence of Newton and Leibniz and, moreover, also of Euler in the first half of the eighteenth century which is not obscured by later interpretations.

Between 1734 and 1765,[13] Euler constructed a consistent theory of classical mechanics based on the mechanical, mathematical and methodological principles developed by Descartes, Newton and Leibniz. Between 1735 and 1749, Du Châtelet presented a *methodologically* and *historically* based summary of the controversial debates on the foundation of mechanics which completes advantageously Euler's *systematic* presentation in the treatises *Mechanica*,[14] *Anleitung*[15] and

---

[8] Euler. 1848. *Mechanik oder analytische Darstellung der Wissenschaft von der Bewegung.*

[9] "En lisant vos Institutions Physiques, j'ai également admire la clarté, avec laquelle Vous traitez cette science, qui la facilité, avec laquelle Vous expliquez les choses les plus difficiles sur le mouvement, qui sont même assez embarassantes, quand il est permis de se servir du calcul." Euler. 1963. Письма к ученым. Pis'ma k učenym (Letters to scholars). 275. Euler. 1760–1762. *Lettres à une princesse d'Allemagne sur divers sujets de physique & de philosophie.*

[10] For the present state of affairs compare Euler. 2006. *Leonhard Euler. Life, Work and Legacy.*

[11] Kästner. 1766. *Anfangsgründe der höheren Mechanik.* 213.

[12] Gehler. 1787. *Physikalisches Wörterbuch oder Versuch einer Erklärung der vornehmsten Begriffe und Kunstwörter der Naturlehre mit kurzen Nachrichten von der Geschichte der Erfindungen und Beschreibungen der Werkzeuge begleitet in alphabetischer Ordnung* (Kraft).

[13] Euler's works published after 1749 are to be included into the analysis because Euler discussed more in detail those topics which had been only presented in an abridge version in earlier papers like the rejection of the force of inertia Euler. 1736. *Mechanica sive motus scientia analytice exposita.* Praefatio Euler. 1765. *Theoria motus corporum solidorum seu rigidorum.* §§ 88 to 95.

[14] Euler. 1736. *Mechanica sive motus scientia analytice exposita.*

[15] Probably, Euler felt the need of such completion of the *Mechanica* by an essay on methodology. In goal and spirit, the *Anleitung zur Naturlehre* (Instructions for natural science) (Euler. 1746b). *Anleitung zur Naturlehre*) is just the book to fill the gap. Although the *Anleitung* had been only published in the nineteenth century and it was available neither to Du Châtelet nor to other contemporaries of Euler, it will be included into the analysis. The *Anleitung* was written shortly after the *Institutions* that makes it appropriate for a direct comparison between Du Châtelet and Euler.

Theoria.¹⁶ Du Châtelet's translation of the *Principia* is to be included into the analysis since her comments on motion and forces in the *Institutions* are advantageously completed by the modifications of Newton's propositions in the translated version of the *Principia*.¹⁷

Although Du Châtelet did not obtain a satisfactory solution of the problems being a matter of the debate on the Leibnizian theory of living forces, she prepared this matter for a solution by the same simultaneously performed two step procedure which had been also successfully applied by Euler, i.e. a reception and reinterpretation of the *Newtonian* and *Leibnizian* principles by inclusion of the *Cartesian* legacy and a consequent reference to the science of the ancients. For that reason, Euler and Du Châtelet revealed the hidden relationship between Newton and Leibniz which is mainly caused by their critical reading and development of Descartes. Most of the concepts in mechanics appeared in different representations being correlated either with Newtonian or Leibnizian principles. Nevertheless, there is a correspondence[18] between the conceptual frames that was mostly carefully hidden and obscured by the great authors themselves who, on the contrary, accentuated the differences.[19] Du Châtelet was open to the influences from both sides and introduced Leibnizian "adjustments" to Newtonian concepts, e.g. substituted the name of an "impressed moving force" with the name of "moving force"[20], and Newtonian "adjustments" to Leibnizian concepts. Decisively, both procedures are discussed with regard to those *Cartesian* concepts that were rigorously rejected by Newton and Leibniz, i.e. the theory of relative motion and the inertia of bodies, respectively.

Euler invented a similar procedure. The only, but striking difference is that Euler completed the system of *physical* concepts axioms and propositions with a complete system of *algorithms* formulated in the language of the calculus. The adaptation of mechanics to these mathematically defined algorithms changed considerably the shape of the science of motion.[21] Defending the Leibnizian measure, Du Châtelet promoted this transformation. The indispensability of both

---

[16] Euler. 1765. Theoria motus corporum solidorum seu rigidorum.

[17] PrincChat1756 (compare Sect. 1.3).

[18] The origin of this similarity can be traced back to the author's early years when Newton (Westfall. 1983. Never at Rest. 378 to 380) and Leibniz 1982b. *Specimen Dynamicum*. I (10) were not only in favour of Descartes, but created almost simultaneously the calculus that was published with a considerable delay. Newton. 1740. *La Méthode des Fluxions et de suites infinies*. Leibniz. *Nova methodus pro maximis et minimis, itemque tangentibus, quae nec fractas nec rationales quantitatis moratur, et singulae pro illis calculis genus*.

[19] A Collection of Papers which passed between the late Learned Mr. Leibniz and Dr. Clarke, in the Years 1715 and 1716, with an Appendix by Samuel Clarke.

[20] Inst1740 §229.

[21] "The attempts in this direction by Newton, Johann Bernoulli, Varignon, and Hermann are less successful than this new approach that we owe to Euler, and the *Mechanica* remains, as Lagrange put it in his *Mécanique analytique* (1788) 'The first great work where Analysis has been applied to the science of movement'." Euler. 1736. *Mechanica sive motus scientia analytice exposita*. Opera Omnia II, 1. Preface (Stäckel).

measures that had been previously only demanded could now be rigorously demonstrated by Euler.[22]

Reading the *Institutions* it can be concluded that Du Châtelet's attempts to construct "Newtonian corrections" of Leibniz's theory and "Leibnizian corrections" of Newton's theory were not sufficient to terminate the debate on the true measure of living forces. This is demonstrated by the unpleasant result of her correspondence with Mairan.[23] The hidden fault being the main source of misinterpretations and confusions is, that Du Châtelet discussed, following Leibniz,[24] the preservation of the Leibnizian *living forces* and the Cartesian *quantity of motion* preferentially in terms of the sums whose terms are formed by products of masses and velocities $m \cdot v \cdot v$ and $m \cdot v$, respectively, but not also in terms of the *increments* of these quantities, i.e. $m \cdot v \cdot dv$ and $m \cdot dv$, expressed explicitly in terms of *differentials* of the velocity dv employed by Euler.[25] The name of the *force* was assigned to the Leibnizian quantities instead of being correlated to the increments of them.[26] Thus, Euler's solution was not only obtained by the introduction of a new theory on the *origin of forces* due to the impenetrability, but also by a *new method* to represent mechanical relations making consequently use of the *differential* and *integral* calculus. The later progress in physics is based on this foundation that was never questioned, but maintained until now. Almost all basic laws in physics are represented in terms of differential equations.

The importance of Du Châtelet's writings is by no means diminished by Euler's discoveries. On the contrary, Du Châtelet explains to us how cumbersome the way was the scholars had to go in that time.

The paper is organized as follows. In Sect. 1 common and different principles in Euler and Du Châtelet are analyzed. It will be argued that Euler and Du Châtelet reinterpreted the legacies of Descartes, Newton and Leibniz putting aside the opposition between the partisans of their schools in the eighteenth century. In Sect. 2 Euler's and Du Châtelet's treatment of hypotheses and models for the demonstration of basic principles of mechanics will be discussed. In Sect. 3 the importance of the distinction between internal and external principles of mechanics will be analyzed. It will be examined how the different interpretations of these principles of bodies by Euler and Du Châtelet and other authors promoted or hampered the comprehension and development of the legacy of Newton and Leibniz. In Sect. 4, the analysis of the basic principles of mechanics will be completed by the comparison of Euler's and Du Châtelets concepts of relative motion. Finally, a summary is given which includes comments on the developments until the twentieth century physics appeared.

---

[22] Euler. 1736. *Mechanica sive motus scientia analytice exposita.* §§ 151 and 152. Euler. 1746b. Anleitung zur Naturlehre. Chap. 7.

[23] Naturlehre1743.

[24] Leibniz. *Essay de Dynamique sur les loix du mouvement.* 215.

[25] Euler. 1736. *Mechanica sive motus scientia analytice exposita.* §§ 151 and 152.

[26] Euler. 1746b. *Anleitung zur Naturlehre.* §§ 31 to 34.

Common and Different Principles in Euler and Du Châtelet

> Il n'y a eu jusqu'ici que deux systèmes de Physique, qui ayent fait grand bruit, & partagé les opinions des Physiciens: L'un est (...) introduit par M. DESCARTES; l'autre est celui de M. NEWTON (...). L'un & l'autre de ces deux systèmes est très bien imaginé,(...) mais aussi faut-il convenir qu'il y a de part & d'autre de grand défauts, & des difficultés que personne n'a encore entierement levées (...). De cette maniere, j'ai tâché de concilier ensemble les deux systèmes par leur beau côté, pour en former un nouveau.[27]

> Les sistêmes de Descartes & de Newton partagent aujourd'hui le monde pensant, ainsi il est nécessaire que vous connaissiez l'un & l'autre (...).[28]

The scientific life of scholars who were working in first decades of the eighteenth century was dominated by the reception of the legacy of Descartes, Newton and Leibniz. Neither Newton nor Leibniz nor their contemporaries gave a review of the state of art in mathematics, physics and philosophy where they took a position which was beyond the quarrels between the different schools.[29] Du Châtelet's intention was to make such a complete re-examination possible. Moreover, the chosen form of an instruction in natural science addressed to her son is as useful as the formerly used form of dialogues, since it allows for digressions to treat topics that are usually not included in the mathematical and mechanical textbooks, although they may belong to the foundation of mathematics and physics. Du Châtelet's *Institutions* may be considered as forerunner of Euler's later comprehensive review entitled *Lettres à une princesse d'Allemagne*[30] that caused an enthusiastic response by the contemporary readers.[31] Like Du Châtelet, Euler discussed almost all hot topics and reviewed all the quarrels appearing between different schools. Though Du Châtelet's *Institutions* and Euler's *Lettres* are addressed to the common reader they aimed to set, nevertheless, in all details the same standards of science that are found in the textbooks.

It will be argued that the discovery of common principles in the work of Newton and Leibniz is not only due to their followers. The same procedure had been already

---

[27] Bernoulli. 1735. La nouvelle *Physique céleste*. II and VIII.

[28] Inst1740 Avant-propos VI.

[29] An advanced study concerning the different measures entitled *Essai de Dynamique* had been given by Leibniz (G. W. *Essay de Dynamique sur les loix du mouvement*. 215) in the 1690s which was only surpassed later by Euler. Leibniz did not include his results in the published version, but only in the draft of the first unpublished version of the *Specimen Dynamicum*. Leibniz. 1982b.

[30] Euler. 1760–1762. *Lettres à une princesse d'Allemagne sur divers sujets de physique & de philosophie*.

[31] "The Letters of Euler to a German Princess have acquired, over all Europe, a celebrity, to which the reputation of the Author, the choice and importance of several subjects, and the clearness of elucidation, justly entitle them. They have deservedly been considered as a treasure of science, adapted to the purposes of every common seminar of learning. (...) they convey accurate ideas respecting a variety of objects, highly interesting in themselves, or calculated to excite a laudable curiosity; they inspire a proper taste for the sciences, and for that sound philosophy which, supported by science, and never loosing sight of her cautions, steady, methodical advances, runs no risk of perplexing or misleading the attentive student." Euler. 1823. *Letters on different subjects in Natural Philosophy addressed to a German Princess*. Preface by Condorcet.

performed by Newton and Leibniz themselves. There a two striking items to confirm this supposition. In the *Principia*, Newton[32] commented on Leibniz's publication of the calculus in 1684[33] and accentuated the *equivalence* between the different representations of the calculus, secondly, in 1695 Leibniz commented on Newton model of the excitation of the force of inertia by an impressed moving force[34] in terms of the relation between an *internal* and an *external* force.[35] Du Châtelet responded to an essential challenge in science that is related to the inherent, but hidden relationship between different representations of basically the same physical content. It is very likely that she was stimulated by Johann Bernoulli's attempt to merge the "best of Descartes" and the "best of Newton".[36] In the twentieth century, this challenge appeared after the invention of the *action parameter* or the *quantum of action* by Planck in 1900.[37] In 1925 and 1926, Heisenberg[38] and Schrödinger[39] developed quite different representations of quantum theory whose unexpected internal relationship had been readily demonstrated by Schrödinger.[40]

The Legacy of Descartes, Newton and Leibniz

The origin of main features of the currently accepted images of Newton and Leibniz as scientists is found in Mach's analysis of the historical development of mechanics written in the second half of the nineteenth century. Mach renewed Voltaire's

---

[32] "In literis quae mihi cum Geometra peritissimo G. G. Leibnitio annis abhinc decem intercedebant (...) rescripsit Vir Clarissimus se quoq; in ejusmodi methodum incidisse, & methodum suam communicavit a mea vix abludentem praeter quam in verborum & notarum formulis." Newton. 1687. *Philosophiae naturalis principia mathematica.* Book II, Section II, Lemma II, Scholion. 253 and 254. In 1687, Newton confirmed the equivalence of the approaches without emphasizing the differences. This had been done later in the famous debate on the priority in the invention of the calculus.

[33] Leibniz *Nova methodus pro maximis et minimis, itemque tangentibus, quae nec fractas nec rationales quantitatis moratur, et singulae pro illis calculis genus.*

[34] Newton. 1687. *Philosophiae naturalis principia mathematica.* Definitione.

[35] "Ex dictis illud quoque mirabile sequitur *quod omnis* corporis *passio sit spontanea seu oriatur a vi interna, licet occasione externi.*" Leibniz. 1982b. *Specimen Dynamicum.* II (5). Translated into Newton's terminology, it follows: "But a body exerts this force only, when another force, impressed upon it, endeavours to change its condition; (...)." Newton. 1845. *The mathematical principles of natural philosophy.* Definition III.

[36] Bernoulli. 1735. *La nouvelle Physique céleste.* VIII.

[37] Planck. 1913. *Vorlesungen über die Theorie der Wärmestrahlung.* §§ 144 to147.

[38] Heisenberg. 1925. Über quantentheoretische Umdeutung kinematischer und mechanischer Beziehungen.

[39] Schrödinger. 1926a. Quantisierung als Eigenwertproblem.

[40] Both approaches confirmed, moreover, an earlier result that Planck obtained in 1913 by purely thermodynamic calculations of the ground state energy of a system of Planckian harmonic oscillators which is, contrary to the classical theory, in thermodynamic equilibrium different from zero for absolute temperature T equal to zero. Planck. 1913. *Vorlesungen über die Theorie der Wärmestrahlung.* §§ 137 to 140. Schrödinger. 1926b. Über das Verhältnis der Heisenberg-Born-Jordanschen Quantenmechanik zu der meinen.

anti-metaphysical attitude in condemning Leibniz.[41] In this time, Newton was considered as physicist whereas Leibniz was regarded to be preferentially a theologian and the creator of one of the great metaphysical systems of the seventeenth century. Leibniz shared this destiny with Descartes whose contributions to mathematics and physics were forgotten or, as far as physics is concerned were only remembered as being disproved by Newton. The interpretation of Descartes' legacy becomes the domain of the philosophers and scholars investigating the history of philosophy. Descartes' contribution to mathematics and physics were overshadowed by the dominance of the mind-body dualism.[42]

In the eighteenth century, such an interpretation of Descartes', Newton's and Leibniz's legacy would be very unlikely, first of all because all schools intended to present the works of their fathers as complete as possible, but never in an abridged version. None of the founders was reduced to his contributions only to mathematics or only to physics or only to philosophy.

Voltaire and Du Châtelet give a typical example for the variety of approaches to treat the legacy of the seventeenth century science in that period, in this case of quite opposite intentions. Voltaire was defending Newtonianism by diminishing the relevance of Leibniz's work. The alternative example of the reception and development of the same subject is due Euler and Du Châtelet who, contrary to most of their contemporaries, were looking for common principles in Newton and Leibniz.

The origin of these common principles is not at all by chance, but due of the similarity in the Newton's and Leibniz's treatment of two common sources, first, their esteem of ancient science and, second, their opposition to Descartes. The main difference between Newton and Leibniz is due to their different recognition of Descartes' merits.[43]

The main difference between Euler and Du Châtelet is due to their different reference to Descartes and Newton. This will be demonstrated for the interpretation (i) of Newton's force of inertia and (ii) of Descartes' model of relative translation.

From Inherent and Impressed Forces to Internal and External Principles

The invention of the absolute space by Newton and the extraordinary role of forces being established by Leibniz result from their opposition to Descartes and the rejection of essential parts of the Cartesian frame of concepts. The young Newton as well as the young Leibniz was in favour of Descartes. Later, Newton turned back

---

[41] Mach. 1991. *Die Mechanik in ihrer Entwicklung.* 288 and 431.

[42] Vorländer. 1908. *Geschichte der Philosophie.* Falckenberg. 1892. *Geschichte der neueren Philosophie.* Windelband. 1892. *Geschichte der Philosophie.*

[43] Leibniz. 1982a. *Monadologie.* § 80. Westfall. 1993. *The life of Isaac Newton.* 165.

to Euclid[44] and Leibniz recovered the entelechies or substantial forms.[45] Thus, starting with Descartes, Euler and Du Châtelet were in good conditions to construct another continuation by referring also to the ancients, preferentially to Archimedes, and preserving those parts in Descartes that had been removed by Newton and Leibniz and substituted by other constructions. The Leibniz-Clarke correspondence[46] may by read as a review of the problems resulting from this incomplete and unfinished substitution.[47]

Euler[48] replaced the Cartesian *extension* consequently with the *impenetrability* and *steadfastness* (inertia) of bodies and the Leibnizian relational definition of time and space with the consequent relational concepts of *rest* and *motion* preserving Newton's idea of the properties of space as a physical magnitude. Du Châtelet maintained Descartes' notion of extension and constructed the notion of body by adding the Newtonian *force of inertia* and the Leibnizian *moving force* as *internal* principles to the theory of bodies as extended things. The *external* principles are represented by the relation between *dead* and *living* forces. This classification of the mechanical principles in terms of internal and external ones is due to Newton[49] and Euler,[50] but Newton referred, as Leibniz later also did, to the distinction between

---

[44] Westfall. 1983. *Never at Rest.* 377.

[45] Leibniz. 1982b. *Specimen Dynamicum.* I (1) to (5).

[46] A Collection of Papers which passed between the late Learned Mr. Leibniz and Dr. Clarke, in the Years 1715 and 1716.

[47] The intricacy of the problems may be estimated by the duration of the debates until a commonly accepted solution had been constructed. Inaugurated by Leibniz in 1686 (Leibniz *Brevis demonstratio erroris memorabilis Cartesii et aliorum circa legem naturalem*) and promoted by the publication of the *Principia* in 1687, the debate on the *true measure of living forces* was running over a period of 50 years until the *Mechanica* was issued 1736. Euler. 1736. *Mechanica sive motus scientia analytice exposita.* In the twentieth century, a similarly long-lasting debate between 1900 and 1935 was after the invention of the quantum hypothesis by Planck in 1900.

[48] Following Descartes, the properties of bodies and the specific relation between bodies are explicitly distinguished from the properties of spirits by Euler. Euler excluded those changes in the motion of bodies which are caused by spirits or by God. "Hier werden diejenigen Veränderungen mit Fleiss ausgeschlossen, welche unmittelbar von Gott oder einem Geiste hervorgebracht werden. Wenn wir also in der Welt nichts als Körper betrachten, so ist klar, dass ein jeder Körper so lange in seinem Zustande verbleiben muss, als sich von aussen keine Ursache ereignet, welche vermögend ist, in demselben eine Veränderung zu wirken. So lange aber die Körper von einander entfernt, so verhindert keiner, dass die Uebrigen nicht in ihrem Zustande, verharren könnten." Euler. 1746b. *Anleitung zur Naturlehre.* § 49. Obviously, Euler presupposed the whole set of Descartes' substances to define the subject of mechanics. Euler maintained this procedure in correlating the "impenetrability of bodies and the liberty of spirits" Euler. 1760–1762. *Lettres à une princesse d'Allemagne sur divers sujets de physique & de philosophie.* LXXXV.

[49] "Definition 5. Force is the causal principle of motion and rest. And it is either an external one that generates or destroys or otherwise changes impressed motion in some body; or it is an internal principle by which existing motion or rest is conserved in a body, and by which any being endeavours to continue in its state and opposes resistance." Newton. 1988. *De gravitatione et aequipondio fluidorum.*

[50] Euler. 1736. *Mechanica sive motus scientia analytice exposita.* Chapter I and II. Euler. 1765. *Theoria motus corporum solidorum seu rigidorum.* Introduction, Chapter II and III.

internal and external *forces*. Nevertheless, in its general form, i.e. by the rigorous *exclusion of forces* from internal principles, *De internis motus principiis* and *De causis motus externis seu viribus*,[51] it may be considered as a genuine Eulerian classification.

Excluding all kinds of forces from a world governed only by *internal* principles, especially the force of inertia, Euler constructed a consistent theory of relative rest and motion.[52] Moreover, the bodies form either (i) a world being free of forces and performing relative motions or being at rest or (ii) form a world of interacting bodies where the bodies change mutually their states or (iii) form a mixture of both kinds of unperturbed and perturbed motions.[53] The *external* principles govern a world where forces exist. Forces are related to the internal principles by their origin due to the impenetrability of bodies.[54]

Making use of this classification of principles, the main features of Du Châtelet's *Institutions* may be interpreted as follows: Du Châtelet constructed the *internal* principles by joining the ideas of the extension and the active and the passive forces[55] being of Cartesian, Leibnizian and Newtonian origin, respectively. The *external* principles concern the relations between dead and living forces and the transformation of dead into living forces.[56]

Referring to Descartes, Du Châtelet demonstrated that the criticism of the Cartesian basic concepts, especially Newton's rejection of *relativism* and Leibniz's rejection of *inertia*, ends up with an incomplete substitution of Descartes' theory. The great prospective, i.e. the substitution of the ancient science of *equilibrium* with the modern science of *motion* was not finally achieved. Leibniz intended to join theses two sciences by the correlation established between dead and living forces.[57] For that reason, Leibniz introduced the notion of living forces, but, unfortunately, he combined this progress with the almost complete exclusion of the Cartesian measure from dynamics.[58] Following Leibniz, Du Châtelet proposed the same substitution and defended the Leibnizian measure.[59] As a consequence, the transition from rest

---

[51] Euler. 1765. *Theoria motus corporum solidorum seu rigidorum*. Introduction, Chapter II and III.

[52] Euler. 1736. *Mechanica sive motus scientia analytice exposita*. §§ 7 to 20, 80. Euler. 1746b. *Anleitung zur Naturlehre*. Chapter 10 (compare Sect. 4.).

[53] Euler. 1746b. *Anleitung zur Naturlehre*. Chapter 7 to 9.

[54] Euler. 1746b. *Anleitung zur Naturlehre*. Chapter 6.

[55] Inst1740 Chapter V to XII.

[56] Inst1740 Chapter XIII, XX and XXI.

[57] Leibniz. 1982b. *Specimen Dynamicum*. I (8).

[58] Later, Leibniz accepted the Cartesian measure as terms of which an analytical expression is composed that describes the conservation of the centre of gravity. "II. *Equation plane*, qui exprime la conservation du progrès commun ou total des deux corps $av + by = ax + bz$. J'appelle *progrès* icy la quantité de mouvement qui va du costé du centre de gravité (…)." Leibniz *Essay de Dynamique sur les loix du mouvement*. 227. Here, a and b label the masses of bodies and v, y and x, z the velocities before and after the impact, respectively. Leibniz *Brevis demonstratio erroris memorabilis Cartesii et aliorum circa legem naturalem*.

[59] Inst1740 Chapter XXI.

to motion was brought in the fore by Leibniz and Du Châtelet whereas the as essential *change of motion* was emphasized by Newton and Euler.

The old and the new concepts in statics or the science of equilibrium and mechanics or the science of motion, however, were not clearly separated from each other and consistently formulated.[60] Newton himself preserved the principles of statics, i.e. the equilibrium between forces or weights, as a model for dynamics and constructed the science of motion by a pair of forces.

> The crux of Newton's dynamics lay in the relation of inherent force and impressed moving force, what he later called (as he struggled to clarify them) 'the inherent, innate and essential force of a body' and 'the force brought to bear or impressed on a body'. The continuing development of his dynamics hinged on the two concepts.[61]

Euler and Du Châtelet had to tackle the same problem. Looking through the eye-glasses of Leibnizian type, Du Châtelet saw the "force of inertia" or the "resisting force", whereas Euler, looking through the eye-glasses of Cartesian type, saw only the "inertia" being completely independent of forces.

Euler's and Du Châtelet's Interpretation of Newton's Axioms

The 1730s may be considered as a *transitional* period in the development of science being the first essential occurrence in the post-Newtonian and post-Leibnizian epoch. Moreover, in such periods it is just the time for *hypotheses* because the final solution of the new problems is not known and the former scheme of interpretation is still in power. In the first half of the eighteenth century, Newtonianism was only reluctantly accepted on the continent, especially as far as the theory of gravitation was concerned.[62] It is well-known that Johann Bernoulli was in favour of Descartes' theory of vortices.[63]

In the twentieth century, the *transitional* period was that between classical mechanics and quantum theory. In 1900, Planck inaugurated a new epoch by the quantum of action.[64] The typical features observed for the development of mechanics in the 1730s may be also assigned to the so-called "old quantum theory" that appeared between 1910 and 1925 where the classical theory was improved by adding of "quantum corrections". Finally, a theory of appropriate shape was constructed by Heisenberg and Schrödinger in 1925 and 1926, respectively.

---

[60] "Das Pleonastische, Tautologische, Abundante der Newtonschen Aufstellungen wird übrigens psychologisch verständlich, wenn man sich einen Forscher vorstellt, der, von den ihm geläufigen Vorstellungen der Statik ausgehend, im Begriff ist, die Grundsätze der Dynamik aufzustellen." Mach. 1991. *Die Mechanik in ihrer Entwicklung*. 241.

[61] Westfall. 1993. *The life of Isaac Newton*. 166.

[62] Nick. 2001. *Kontinentale Gegenmodelle zu Newtons Gravitationstheorie*.

[63] Nick. 2001. *Kontinentale Gegenmodelle zu Newtons Gravitationstheorie*. 42 to 77, 129 to 187.

[64] Planck. 1913. *Vorlesungen über die Theorie der Wärmestrahlung*. §§ 113 to 125.

In the eighteenth century, the new shape of classical mechanics had been presented by Euler in 1736.[65]

It is the merit of Du Châtelet to have discovered that there is a *hidden* common basis in Newton and Leibniz by repeating the procedure that was already observed by the Newton and Leibniz themselves. Leibniz tested the applicability of his own approach by an alternative demonstration of some of Newton's theorems. In 1695, Leibniz reinterpreted Newton's model of the relation between the inherent force of inertia and the externally impressed moving force.[66] Newton explained that this force is only activated in the presence of the "impressed moving force", i.e. in occasion of the *change of the state* whereas in non-interacting bodies this force is sleeping. The resulting partial equivalence of the Newtonian and Leibnizian constructions is readily demonstrated by the comparison of the following propositions.

> Ex dictis illud quouque mirabile sequitur quod omnis corporis passio sit spontanea seu oriatur a vi interna, licet occasione externi.[67]

> Exercet vero corpus hanc vim solum modo in mutatione status sui per vim aliam in se impressam facta; estque exercitium ejus sub diverso respectu & Resistentia & Impetus.[68]

In 1740, Du Châtelet did not only repeat, but recovered (this part of the Specimen[69] had not been published by Leibniz) the procedure performed by Leibniz before in 1695 and translated Newton's basic assumptions into the Leibnizian terminology. Du Châtelet assumed as basic concepts: the extension, the force of resistance and the active force.

> § 145. Tous les changemens qui arrivent dans les Corps peuvent s'expliquer par ces trois principes, *l'étendue, la force résistante,* & *la force active*; (…).[70]

Du Châtelet presented Newton's axioms completed by notions stemming from Leibniz's theory and, moreover, by those that Euler had introduced.

> § 229. La force active & la force passive des Corps, se modifie dans leur choq[71], selon de certaines Loix que l'on peut réduire à trois principales.[72]

In contrast to Euler who introduced a general and fundamental law for mechanics which is valid for the change of rest *and* motion of mass points[73] and the motion of

---

[65] Euler. 1736. *Mechanica sive motus scientia analytice exposita.*

[66] Newton. 1687. *Philosophiae naturalis principia mathematica.* Definitione.

[67] Leibniz. 1982b. *Specimen Dynamicum.* II (5).

[68] Newton. 1687. *Philosophiae naturalis principia mathematica.* Def. III.

[69] Leibniz. 1982b. *Specimen Dynamicum.* II.

[70] Inst1740 § 145.

[71] The active and the passive forces are *modified* due to the impact of bodies. Euler claimed that the forces are *generated* in the impact by the bodies.

[72] In the German translation, a part of the paragraph is not translated. "Die thätige und die leidende Kraft der Körper wird in ihrem Stoße (alternative translation: die im Stoße *modifiziert* werden) nach gewissen Gesetzen eingerichtet, welchen man hauptsächlich auf drey bringen kann." Naturlehre1743, § 229. Inst1740 § 229.

[73] Euler. 1736. *Mechanica sive motus scientia analytice exposita.* §§ 152 and 153.

solid and rigid bodies,[74] Du Châtelet treated Newton's Laws as "Lois générales du mouvement", i.e. preferentially for motion.[75] Moreover, Du Châtelet interpreted Newton's 2nd Law in terms of Leibniz's principle of sufficient reason.

> Premiere Loi: Une Corps persevére dans l'état où il se trouve, soit de repos, soit de mouvement, à moins que quelque cause ne le tire de son mouvement, ou de son repos.
>
> Seconde Loi: Le changement qui arrive dans le mouvement d'un Corps, est toujours proportionnel à la force motrice qui agit sur lui[76], & il ne peut arriver aucun changement dans la vîtesse, & la direction du Corps en mouvement que par une force extérieur; car sans cela ce changement se seroit sans raison suffisante.
>
> Troisième Loi: La reaction est toujours égale à l'action; car un Corps ne pourroit agir sur un autre Corps, si cet autre Corps ne lui resistoit; ainsi, l'action & la reaction sont toujours égales & opposées.[77]

In the original version, Newton described the change of motion due to an "impressed force" (1st Law and an "impressed moving force" (2nd Law).[78] The 1st Law determines the conditions for the preservation of state in general by the absence of *impressed* forces. Euler generalized Newton's proposition by substituting "impressed forces" with the true absence of "any external cause".[79]

> LEX I. Corpus omne perseverare in statu suo quiescendi vel movendi uniformiter in directum, nisi quatenus a viribus impressis cogitur statum illum mutare.[80]

In the translation, Du Châtelet replaced the word *force* with the word *cause* whose use in the *Institutions* may result from her reading of the *Mechanica*. Du Châtelet, however, used the words "quelque cause" instead of the Eulerian original version "external cause". This specification is of importance because it results from the principal distinction into *internal* and *external* principles. Moreover, it should be stressed that Euler, accentuating the relationship to statics, put the law for rest at the first place, the law for motion at the second one. Euler chose the same grammatical and logical structure emphasizing that the same *algorithm* is used for expressing the relation between the basic concepts.[81] Therefore, it is *one and the same law* that

---

[74] Euler. 1750a. Découverte d'un nouveau principe de Mécanique. § 20.

[75] Inst1740 § 229.

[76] Here, Du Châtelet omitted "impressed" because impressed refers to dead forces that are distinguished from moving forces.

[77] Inst1740 § 229.

[78] Newton. 1845. *The mathematical principles of natural philosophy*. Axioms.

[79] "56. Corpus absolute quiescens perpetua in quiete perseverare debet, nisi a causa externa ad motum sollicitetur." Euler. 1736. *Mechanica sive motus scientia analytice exposita*. § 56.

[80] Newton. 1687. *Philosophiae naturalis principia mathematica*. Axiomata.

[81] "Or on énonce communément ce principe par deux propositions, dont l'une porte, qu'un corps étant une fois en repos demeure éternellement en repos, à moins qu'il ne soit mis en mouvement par quelque cause externe ou étrangere. L'autre proposition porte qu'un corps étant une fois en mouvement, conservera toujours éternellement ce mouvement avec la même direction et la même vîtesse, ou bien sera porté d'un mouvement uniforme suivant une ligne droite, à moins qu'il ne soit troublé par quelque cause externe ou étrangere.

governs the *change of the state* of a body being either the state of rest or the state of uniform motion. The "external cause" for the change of the state is exclusively the presence of another body.[82] Du Châtelet renamed Newton's "force" and interpreted the principle of sufficient reason not mechanically, as Euler did, but metaphysically. According to Euler, the sufficient reason for the change of the state is the interaction with another body.[83]

In the translation of other parts of the *Principia* Du Châtelet maintained largely the previous interpretation presented in the *Institutions*. Here, only the Definition III and the 1st Law will be analyzed as far as the notion of inertia and preservation of state is concerned.

> Materia vis insita est potentis resistendi, qua corpus unumquodque, quantum in se est, perseverat in statu suo vel quiescendi vel movendi unformiter in directum.[84]

There are three interpretations: (i) Inertia is understood from "the inactivity of matter" or as a force that is to be activated by an impressed force (Newton), i.e. inertia is temporarily not a force, but operates like an "innate force of matter" (vis insita) to "preserve the present state" of the body,[85] (ii) the force of inertia is interpreted as a resisting force (Du Châtelet) and (iii) following Newton in the interpretation for the absence of an impressed force, but generalizing Newton's model for the presence of an external cause, inertia is not at all a force, but it is independent of any external cause and one and the same in non-interacting and interacting bodies, i.e. the inertia is preserved in the interaction of bodies (Euler).

C'est en ces deux propositions que consiste le fondement de toute la science du Mouvement, qu'on nomme la Méchanique." Euler. 1760–1762. *Lettres à une princesse d'Allemagne sur divers sujets de physique & de philosophie*. Lettre LXXIII.

The equivalence between the laws for rest and motion is demonstrated by (i) the equivalence of the *syntactic* and *logical* structure of these basic statements and (ii) by the same analytical form of the equation of motion (compare Sect. 1.3.). Usually, there is a difference between the *syntactic* structure and the *logical* structure of statements and statements composed of different (elementary) statements. Euler gave a formulation of the basic laws of mechanics where the requirements are in harmony.

[82] Euler. 1750b. Recherches sur l'origine des forces. Newton. 1988. *De gravitatione et aequipondio fluidorum*.

[83] Euler. 1746b. *Anleitung zur Naturlehre*. § 49.

[84] Newton. 1687. *Philosophiae naturalis principia mathematica*. Def. III.

[85] "*The vis insita, or innate force of matter, is a power of resisting, by which every body, as much as in it lies, endeavours to persevere in its present state, whether it be of rest, or of moving uniformly forward in a right line*. This force is ever proportional to the body whose force it is; and differs nothing from the inactivity of the mass, but in our manner of conceiving it. A body, from the inactivity of matter, is not without difficulty put out of its state of rest or motion. Upon which account, this *vis insita*, may, by a most significant name, be called *vis inertiæ*, or force of inactivity. But a body exerts this force only, when another force, impressed upon it, endeavours to change its condition; and the exercise of this force may be considered both as resistance and impulse; it is resistance, in so far as the body, for maintaining its present state, withstands the force impressed; it is impulse, in so far as the body, by not easily giving way to the impressed force of another, endeavours to change the state of that other." Newton. 1845. *The mathematical principles of natural philosophy*. Definition III.

According to Westfall, it took Newton 20 years to formulate the 1st Law.[86]

> Once he adopted the principle of inertia, the rest of his dynamics fell quickly into place. He had seized the essence of his second law twenty years before and had never altered it as he was wrestled with the first law.[87]

Not surprisingly, the contemporaries and followers had the same trouble to comprehend and to interpret it. Du Châtelet referred to Leibniz ("vis primitiva patiendi seu resistendi"[88] whereas Newton referred to Descartes ("quantum in se est").[89]

The 2nd Law determines the conditions for the change of the previously described states of rest and uniform motion.

Du Châtelet reinterpreted Newton's 3rd Law in terms of inherent and impressed forces ("car un Corps ne pourroit agir sur une autre Corps, si cet autre Corps ne lui resistoit"). Newton did not specify these forces into "active" and "resisting" forces, but stressed that they appeared in the mutual actions of two bodies.

> To every action there is always opposed an equal reaction; or the mutual actions of two bodies upon each other are always equal, and directed to contrary parts.[90]

Du Châtelet added that one body is acting and the other body is resisting. Thus, the force to act and the force to resist come into the play.[91] Du Châtelet combined notions of different origin and reinterpreted the basic laws in terms of *extension*, *active* and *passive forces*.

Euler prepared the theory for the next step. The relics of the old theory are removed, e.g. the "force of inertia", without any change in the validity of the theorems formulated before. This is the only difference between Euler and Du Châtelet: Euler had already constructed the new theory, but maintained the old names for calming the readers who were accustomed to the traditional interpretations. Du Châtelet paved the way for the splitting of Newton's and Descartes' legacy into those parts which are either compatible or incompatible with the representation of mechanics in the language of the calculus. The crucial point was to save Leibniz's measure of living forces. Euler demonstrated that the principles of mechanics and principles of the calculus are *compatible* and, moreover, the Leibnizian measure and the

---

[86] Currently, the decisive role of the 1st Law is underestimated since it is usually interpreted only as a special case of the 2nd Law. Mach claimed that the theorem on the inertia is already included in the 2nd Law. Mach. 1991. *Die Mechanik in ihrer Entwicklung*. 241. This interpretation, however, cannot be confirmed by Westfall's analysis of the genesis of the 1st Law reconstructed from Newton's writings. Westfall. 1993. *The Life of Isaac Newton*. 16. Westfall's results fit, on the contrary, much better for Du Châtelet's treatment of Newton's axioms.

[87] Westfall. 1993. *The Life of Isaac Newton*. 167.

[88] Leibniz. 1982b. *Specimen Dynamicum*. I (3).

[89] Descartes. 1998–99. *The Principles of Philosophy* II. § 37.

[90] Newton. 1845. *The mathematical principles of natural philosophy*, transl. by Andrew Motte. Axioms.

[91] Inst1740 § 141.

Cartesian measure do not only coexist, but are also simultaneously valid and necessarily valid.[92] Euler commented on the probability of hypothetical relations that had been proposed by Daniel Bernoulli for the basic laws of mechanics[93] and rejected Daniel Bernoulli's assumption that different representations of mechanical laws are of *equal probability*.

> 152. Apparet igitur non solum verum esse hoc theorema, sed etiam necessario verum, ita ut contradictionem involveret ponere $dc = p^2 dt$ vel $p^3 dt$ aliamve functionem loco p. Quae omnes cum Clar. Dan. Bernoullio in Comment. Tom. I. aeque probabiles videantur, de rigidis harum propositionum demonstrationibus maxime eram sollicitus.[94]

The decisive step Euler did and Du Châtelet did not do[95] was to treat rest and motion on an equal footing which had been only implicitly done by Newton[96] and never accepted by Leibniz.[97] Reading Du Châtelet, the comprehension of the theory and the argumentation of Euler may be essentially promoted. Reading Euler, the esteem of Du Châtelet's analysis may be considerably enhanced.

Euler analyzed carefully the problems to be solved for the transition from statics or rest to dynamics or motion[98] and presented the solution. In 1736, the origin of forces is not discussed, but the effect of any kind of forces is investigated.[99] The effect of a force on a moving body is calculated from the effect of the same force on a resting body.[100] For that purpose, Euler assumed that the *increment* of velocity dv is *independent* of the velocity.[101] Then, it follows that the *analytical form* of the equation of motion is *independent* of the state of the body.[102] Therefore, the increment of velocity dv is proportional to the force K, the time element dt and the inverse of the mass m, i.e. the body's resistance to the change of the state depends on its mass and is as more pronounced as greater the mass is. Supposing a

---

[92] Euler. 1736. *Mechanica sive motus scientia analytice exposita*. § 152.

[93] Not surprisingly, Euler acknowledged Du Châtelet's treatment of the same subject. Hypotheses, Du Châtelet said, are important for the progress of sciences since they are tools for the discovery of truth. It is necessary to estimate the degree of probability that represents the reliability and the certitude of hypotheses.

> 62. Les hipotheses n'étant faites que pour découvrir la vérité, on ne les doit point faire passer pour la vérité elle-même (…). Il est donc très-important pour le progrès des sciences (…) il faut estimer le degré de probabilité qui s'y trouve (…). 67. Les hipotheses ne sont donc qui des propositions probables qui ont un plus grand ou un moindre degré de certitude (…). Inst1740 §§ 62 and 67 (compare Sect. 2.).

[94] Euler. 1736. *Mechanica sive motus scientia analytice exposita*. § 152.
[95] Inst1740 §§ 224 and 225.
[96] Newton. 1687. *Philosophiae naturalis principia mathematica*. Lax. I.
[97] Leibniz. 1982b. *Specimen Dynamicum*.
[98] Euler. 1736. *Mechanica sive motus scientia analytice exposita*. § 101.
[99] Euler. 1736. *Mechanica sive motus scientia analytice exposita*. § 102.
[100] Euler. 1736. *Mechanica sive motus scientia analytice exposita*. § 118.
[101] Euler. 1736. *Mechanica sive motus scientia analytice exposita*. § 131.
[102] Euler. 1746b. *Anleitung zur Naturlehre*. Chapter 10.

body of the mass m, the change of its state from rest into motion and from uniform inot accelerated motion caused by the same force K and performed in the same time interval dt is described by the same relation between the increment of velocity dv and the force K Eq. 2.

Rest into motion	Uniform into accelerated motion	
$v = 0 \to v \neq 0$	$v = const \to v \neq const$	(1)
$m \cdot dv = K \cdot dt$	$m \cdot dv = K \cdot dt$	(2)

The analytical form and the form-invariance of the equation of motion play an essential role in the twentieth century. Without reference to Euler, Einstein generalized this principle in the theory of relativity whose origin, nevertheless, can be traced back to Euler.

Euler's Mechanica and Du Châtelet's Institutions

In the very short time window between 1735 and 1740, Du Châtelet is in a top position because of the simultaneous treatment of Descartes', Newton's and Leibniz's theories. The only equivalent in goal and spirit, but different in the method of presentation is Euler's *Mechanica* (1736) programmatically entitled *Mechanics or the science of motion analytically demonstrated* (*Mechanica sive motus scientia analytice exposita*).[103] Like Du Châtelet, Euler accentuated the educational goal as a main purpose next the construction of a scientific system of notions. The achieving of these two aims is guaranteed by the special form the suppositions are presented and the results are obtained and interpreted. For this reason, Euler developed *algorithms* for reckoning[104] that play the same role in mechanics which the mathematical

---

[103] Euler's program reads as follows: "Those laws of motion which a body observes when left to itself in continuing either rest or motion pertain properly to infinitely small bodies, which can be considered as points. (…) The diversity of bodies therefore will supply the primary division of our work. First indeed we shall consider infinitely small bodies (…). Then we shall attack bodies of finite magnitude which are rigid. (…) Thirdly, we shall treat of flexible bodies. Fourthly, of those which admit extension and contraction. Fifthly, we shall subject to examination the motions of several separated bodies, some of which hinder [each other] from executing their motions as they attempt them. Sixthly at last, the motion of fluids will have to be treated." Euler. 1736. *Mechanica sive motus scientia analytice exposita*. § 98.

[104] "That which is valid for all the writings which are composed without the application of analysis is especially true for the treatises on mechanics (*). The reader may be convinced of the truths of the presented theorems, but he did not attain a sufficient clarity and knowledge of them. This becomes obvious if the suppositions made by the authors are only slightly modified. Then, the reader will hardly be able to solve the problems by his own efforts if he did not take recourse to the analysis developing the same theorem using the analytical method." Euler. 1736. *Mechanica sive motus scientia analytice exposita*. Preface. (*) Newton's Principia (1687).

algorithms played in the *differential calculus*.[105] These algorithms are to be formulated *analytically*, i.e. by means of calculus,[106] observing the same rigour in the foundation known from geometry.[107] The work on this subjects had been continued in the following decades and the results had been summarized and axiomatically formulated in the *Anleitung* written around 1746[108] and in the *Theoria* published in 1765.[109]

Euler's *Mechanica* and Du Châtelet's *Institutions* are complementary representations of the state of art in natural sciences after Newton and Leibniz in the 1730s. Euler formulated the principles of mechanics and posted and solved a huge variety of sophisticated problems. Neither the *mathematics* nor the *methodology* behind this approach were presented in the same complete form, but only given in later writings.

Du Châtelet filled this gap as far as the *methodology* is concerned. The name of *methodology* covers those principles which had been earlier formulated by Newton and Leibniz like the *principle of contradiction*, the *principle of continuity* and the *principle of sufficient reason*.[110]

The esteem of the merits of Euler and Du Châtelet may be considerably enhanced by considering the fact that most of the literature, especially most of Newton's and Leibniz's writings which are available for us today were only published in the second half of the eighteenth century and still later in the nineteenth and twentieth centuries. The authors living in former times were faced with the challenge to reconstruct the missing parts like Newton's preliminary version of the *Principia* entitled *De gravitatione* and *De motu* by their own efforts. Newton's path to the *Principia* had been analyzed by Westfall.[111]

> In the Principia itself, he further eliminated the reference to inherent force from the statement of the first law, thus obliterating the principal record of the path by which he arrived at it.[112]

Euler and Du Châtelet developed programs in science and, moreover, in education to make scientific knowledge available for the public. Nevertheless, also Euler had

---

[105] Leibniz G. W. Nova methodus pro maximis et minimis, itemque tangentibus, quae nec fractas nec rationales quantitatis moratur, et singulae pro illis calculis genus.

[106] In this paper, the word "analytical" will be used in the meaning that was introduced by Euler in order to distinguish the new version from the former version in the representation of the science of motion. This denotation had been later maintained by Lagrange who, following Euler in goal and spirit, entitled his treatise on mechanics *Mécanique analytique*. Lagrange. 1788. *Mécanique analytique*. In the nineteenth century, the main contributions to analytical mechanics are due to Jacobi and Hamilton. Currently, this science is also known under the name of *Analytical Mechanics*.

[107] Euler. 1760–1762. *Lettres à une princesse d'Allemagne sur divers sujets de physique & de philosophie*. Lettre LXXI.

[108] Euler. 1746b. *Anleitung zur Naturlehre*.

[109] Euler. 1765. *Theoria motus corporum solidorum seu rigidorum*.

[110] Leibniz. 1982a. *Monadologie*. 31 to 36.

[111] Westfall. 1993. *The Life of Isaac Newton*. Chapter 8.

[112] Westfall. 1993. *The Life of Isaac Newton*. 167.

to be cautious in presenting new truths that were in contradiction or tension to the commonly accepted interpretations. Thus, in most cases Euler used to present the established truths embedded in the context of their refutation. In the Preface to *Mechanica*, Euler rejected the notion of the "force of inertia", but preserved this name in the bulk of the treatise.[113] Though he emphatically fought against the force of inertia in later works,[114] people maintained this name and the notion and, moreover, referred to him.[115] The *Anleitung* was never published in the eighteenth century.

Methods: Hypotheses, Models and the Calculus

> Il me paraît d'ailleurs qu'il seroit aussi injuste aux Cartésiens de refuser d'admettre l'attraction comme hipothese, qu'il est déraisonnable à quelques Newtoniennes de vouloir en faire une propriété primitive de la matiere.[116]

> Now, since in the law of absorption just assumed the hypothesis of quanta has yet found no room, it follows that it must be come into play in some way other in the emission of the oscillator, and this provided for by the introduction of the hypothesis of *emission of quanta*.[117]

In the Chap. III of the *Institutions*, Du Châtelet discussed the necessity of hypothesis as tools of promoting the progress in science.[118] Euler commented:

> Mais surtout le Chapitre sur les hypothèses m'a fait le plus grand plaisir, voyant, que Vous combattez, Madame, si fortement et si solidement quelques Philosophes Anglois, qui ont voulu bannir tout à fait le hypothèses de la Physique qui sont pourtant à mon avis le seul moyen de parvenir à une connoissance certaine des causes physiques.[119]

In the end of the unfinished letter to Du Châtelet, Euler started to discuss the basic principles of mechanics and to clear up the notion of forces, especially the difference between the force of inertia and other sorts of forces. This distinction had been already discussed in the Preface to the *Mechanica* in 1736, but only implicitly used or presented in an abridge version in the text.[120] Reading Du Châtelet's treatise, Euler may have realized that his cautious critique with regard to the notion of the force of inertia did not met the expected resonance since even such open minded

---

[113] Euler. 1736. *Mechanica sive motus scientia analytice exposita.* §§ 68, 74 to 76.

[114] Euler. 1750b. Recherches sur l'origine des forces. Euler. 1765. *Theoria motus corporum solidorum seu rigidorum.* §§ 18, 75, 76, 88, 92 to 96.

[115] Wundt. 1921. *Erlebtes und Erkanntes.* 228.

[116] Inst1740 Preface VII.

[117] Planck. 1913. *Vorlesungen über die Theorie der Wärmestrahlung.* § 147.

[118] Inst1740 §§ 53 to 71.

[119] Euler. 1963. Письма к ученым. Pis'ma k učenym (Letters to scholars). 278.

[120] Euler. 1736. *Mechanica sive motus scientia analytice exposita.* §§ 56, 57, 68, 74 to 76.

scholar like Du Châtelet were not aware of its importance for the foundation of mechanics. Therefore, Euler clarified:

> Mais j'espère qu'une bonne partie de ces gens changeront bientot leur sentiment après avoir lu Votre admirable dissertation sur les hypothèses: et je ne doute nullement, que Mr. Demairan ne soit entierement convaincu par les solides raisons que Vous avez opposées à ses idées si mal fondées sur la force des corps (…). Agréez donc, que je Vous présente mes pensées la dessus. Je commence par le premier principe de la Mecanique que tout corps par lui-même demeure dans son état ou de repos ou de mouvement. A cette propriété on peut bien donner le nom de force, quand on ne dit pas que toute force est une tendance de changer l'état, comme fait Mr. Wolf.[121]

Newton formulated theorems on the behaviour of *all* bodies, e.g. in the 1st Law, and specified in the 2nd Law this theorem to those bodies which are suffering from the action of a *moving force impressed* upon them. The connection between non-interacting bodies is guaranteed by their common relation to the absolute space whereas, in case of interacting bodies, the connection is governed by the 3rd Law on the equality of action and reaction. This method to describe the bodies and their motions was completely new in that time and the contemporaries underestimated the difficulties[122] to comprehend Newton's *Mathematical principles of natural science*.[123] Almost the same reactions are reported after Leibniz had published the principles of the calculus in 1684.[124] Both authors, moreover, intended to obscure and to obliterate the origin and the foundation of their new methods by several reasons. Newton was afraid that the postulate of a universal attraction between bodies might be misinterpreted by people.[125] The intentions in the post-Newtonian period had been considerably changed and the scholars of this time endeavoured to make their theories as clear as possible when they presented them to the public.[126]

Du Châtelet discussed the relations between hypotheses, the verification and falsification of theorems and theories.

> 64. Une expierence ne suffit pas pour admettre une hipothese, mais une seule suffit pour la rejetter lorsqu'elle lui est contraire.[127]

---

[121] Euler. 1963. Письма к ученым. Pis'ma k učenym (Letters to scholars). 279.

[122] "The young mathematician opened the book and deceived by its apparent simplicity persuaded himself that he was going to understand it without difficulty. But he was surprised to find it beyond the range of his knowledge and so to see himself obliged to admit that what he had taken for mathematics was merely the beginning of a long and difficult course that he had yet to undertake." Westfall. 1993. *The Life of Isaac Newton*. 192. The young mathematician was Abraham DeMoivre.

[123] Newton. 1687. *Philosophiae naturalis principia mathematica*.

[124] Leibniz. *Nova methodus pro maximis et minimis, itemque tangentibus, quae nec fractas nec rationales quantitatis moratur, et singulae pro illis calculis genus*.

[125] "Newton had good reason to be cautious. Weaned on the mechanical philosophy himself, he could not doubt how the concept of a universal attraction of all particles of matter for one another would be received. (…) [In the following, Westfall quoted Newton:] 'We said, *in a mathematical way*, to avoid all questions about the nature or quality of this force, which we would not be understood to determine by any hypothesis (…)'." Westfall. 1993. *The Life of Isaac Newton*. 188.

[126] Euler. 1736. *Mechanica sive motus scientia analytice exposita*. Euler. 1760–1762. *Lettres à une princesse d'Allemagne sur divers sujets de physique & de philosophie*. Inst1740.

[127] Inst1740 § 64.

Accentuating the decisive role of experience, Du Châtelet may be considered as a forerunner of Popper whose proposal for a methodology of science is currently known to be based upon the possibility of falsification.

Hypotheses and Models

Euler appreciated hypotheses as appropriate models for the discussion of the basic assumptions upon the mechanical systems under investigation comprising bodies and the vacuum or, as it becomes later a convention to speak of *models* of these objects. This approach to investigate the content of Newton's theory and to provide a basis for criticism had been also developed by Berkeley who rejected Newton's concept of absolute space and was in favour of relative motion.[128] Euler generalized the procedure and analyzed the notion of rest, motion, inertia, the origin of forces, the conservation of states and the change of states of bodies.

(a) The *one body model*: The simplest case is a world, where only one body exists. The world is composed of one body and the empty space being infinitely extended.

> 2. En effet si nous ne considérons qu'un seul corps, en supposant que tout le reste du monde soit anéanti, et que ce corps existe tout seul dans l'espace vuide et infini, la vérité de ce que je viens d'avancer sur la conservation de l'état, sautera d'abord aux yeux.[129]

Though the conservation of state is guaranteed, it is impossible to decide whether the state is the state of rest or the state of uniform motion. Therefore, essential elements of the theory remain to be not only indeterminate, but even indeterminable within the frame of this model. Nevertheless, one decade later in 1760, Euler referred to this model to demonstrate the basic principles in the theory of motion.

> Cette hypothese, quoique impossible, peut faire distinguer ce qui est opéré par la nature du corps même, de ce que d'autres corps peuvent opérer sur lui.[130]

Though this world made up of only one body is impossible, one can draw essential conclusions from that model. Here, Euler demonstrated that the reference to the *impossibility* of a world may be as important as the references to

---

[128] Berkeley. 1820. De Motu or The Principle and Nature of Motion and the Cause of the Communication of Motions. §§ 52 to 64.

[129] Euler. 1750b. Recherches sur l'origine des forces. § 2.

[130] Euler. 1760–1762. *Lettres à une princesse d'Allemagne sur divers sujets de physique & de philosophie*. Lettre LXXI.

the infinity of *possible* worlds known from Leibniz's theory.[131] Adding new constitutive parts to the model, here it is one additional body being included into the world, the previously indeterminable elements and notions can be now determinate. The status of world and the model of the world will be considerably changed by the inclusion of a second body since it becomes a *possible* mechanical model of a *possible* world. Following Archimedes and having discussed the properties of bodies in the empty space, the results are transferred into the real world[132] that is made up of coarse matter or bodies and subtle matter or ether.[133]

(b) Euler's *two body model*: There is another world, where only two bodies exist.[134] Since the state of a body can only be changed by an *external or foreign cause*, the external reason can be only provided by the *other body*. Thus, the general term "external cause" had been denoted without taking recourse to any idea of a force. Except this additional element of the enlarged system or the model of world, there are no other things that provide a sufficient reason to disturb the body and to modify its state. An interaction takes place, because a given space region *can only be occupied by one body*. It is impossible, that two bodies occupy the same space region.[135] This principle had been already formulated by Newton,[136] but Euler made systematically use of it to explain the *origin of forces*.[137] If two bodies interact, the *bodies change mutually their states*.[138]

(c) Leibniz's *two body model* for the conservation of living forces: Leibniz discussed the relations between bodies *before* and *after* the impact to demonstrate the reliability of his notions of bodies and forces. Before and after the impact, the bodies are not interacting, i.e. the theorem on the conservation of living forces is not only valid for these states, but also *during* the interaction.

Ex nostris quoque corporis viriumque notionibus id nascitur, *ut quod in substantia fit, sponte et ordinate fieri intelligi possit*.[139]

Having defined "living forces", the *two body model* is appropriate to demonstrate the conservation of living forces and the preservation of order.[140]

$$m_1 v_{1,before}^2 + m_2 v_{2,before}^2 = m_1 w_{1,after}^2 + m_2 w_{2,after}^2 = const \qquad (3)$$

---

[131] Leibniz. 1982a. *Monadologie*. §§ 53 and 54.

[132] Euler. 1736. *Mechanica sive motus scientia analytice exposita*. § 56.

[133] Euler. 1746b. *Anleitung zur Naturlehre*. Chap. 13.

[134] Euler. 1760–1762. *Lettres à une princesse d'Allemagne sur divers sujets de physique & de philosophie*. Lettre LXXI.

[135] Euler. 1746b. *Anleitung zur Naturlehre*. § 35.

[136] Newton. 1988. *De gravitatione et aequipondio fluidorum*.

[137] Euler. 1750b. *Recherches sur l'origine des forces*.

[138] Euler. 1765. *Theoria motus corporum solidorum seu rigidorum*. § 131.

[139] Leibniz. 1982b. *Specimen Dynamicum*. II (3).

[140] Leibniz. 1982b. *Specimen Dynamicum*. II (3).

Here, the masses of the bodies 1 and 2 and the velocities before and after the interaction are indicated by the subscripts. Leibniz did not discuss the intermediate state which is just described by Newton's 2nd Law where the change in motion of each of the bodies being involved in the interaction is described by dv_1 and dv_2 whose magnitudes are, however, indeterminate, i.e. the quantities $w_{1,after}$ and $w_{2,after}$ are unknown. Therefore, an additional equation is needed in order to calculate the velocities after the impact if the velocities before the impact are known. This relation is obtained by means of the Cartesian measure of motion or the conservation of momentum.[141]

$$m_1 v_{1,before} + m_2 v_{2,before} = m_1 w_{1,after} + m_2 w_{2,after} = const \qquad (4)$$

This relation had been already discussed by Leibniz in the first unpublished version of the *Specimen*.[142] Leibniz, however, commented erroneously that the interaction described by Eq. 4 is not compatible with the principle that the "full effect is equal to the whole cause"[143], but restricted the validity of Eq. 4 to the conservation of the centre of gravity.[144] The debate on the true measure of living forces was governed by the unsolved problem to represent *simultaneously* the mechanical laws in terms of the differential *and* the integral representation[145]. Nevertheless, referring to Galileo, Leibniz is under constraint to discuss the transition from an infinitesimal to a finite change of the velocity in terms of the relation between dead and living forces.[146]

---

[141] Euler. 1750b. Recherches sur l'origine des forces. Though Leibniz emphasized the "errors of Descartes" (Leibniz *Brevis demonstratio erroris memorabilis Cartesii et aliorum circa legem naturalem.*) concerning the definition of the measure of forces in 1686, he made use simultaneously of both equations (3) and (4) in a letter to Johann Bernoulli written in 1696: "His positis, ex lege virium absolutarum conservandarum fiet (1) Avv+Byy=Axx+Bzz; ex lege conservandae directionis, fiet (2) Av+By=Ax+Bz, quae regula, certo tantum casu, coincidit cum regula conservandae quantitas motus Cartesiana (...)." GM II, XXV, 260. The masses are labelled by A and B, the velocities before and after the impact are labelled by v, y and x, z, respectively. (See also Leibniz *Essay de Dynamique sur les loix du mouvement*. 215). Hence, as far as the analytical part or the integral representation was concerned, the affair was already over for Leibniz in 1696. Leibniz, however, was not ready to give the corresponding and mathematically equivalent analytical formulation in terms of the differential calculus. This part of the work had been done later by Euler. Euler. 1736. *Mechanica sive motus scientia analytice exposita* (compare Sect. 3.3.).

[142] Leibniz. 1982b. *Specimen Dynamicum*. UV (11).

[143] "(...) sed Effectum plenum esse causae integrae aequalem (...) falsam esse et cum his principiis pugnare; verissimam autem Hugenianam quae statuit Avv+Byy aequ. Axx+Bzz." Leibniz. 1982b. *Specimen Dynamicum*. UV (11).

[144] Leibniz. *Essay de Dynamique sur les loix du mouvement*. 227.

[145] The indispensability to make use of both representations for a "complete description" of the system is readily demonstrated: In Eq. 4 the real change of the states is just *interposed* between the "state before" and the "state after". Thus, it cannot be represented by Eq. 4 which describes *implicitly* the interaction, but only by an equation of another type like that one given by Eq. 2 which describes *explicitly* the change of the state.

[146] Leibniz. 1982b. *Specimen Dynamicum*. I (6).

Forces Interpreted as Magnitudes in the Frame of the Calculus

Leibniz developed a model where the "impressed or dead forces" generate "motion" or "living forces" by "infinitely many impressions of the dead force"[147]:

> (…) vis est viva, ex infinitis vis mortuae impressionibus continuitatis nata.[148]

Following Leibniz, Newton's "impressed *moving* force" being always of finite magnitude is resolved into a succession of "impressed *dead* forces". Du Châtelet who referred to this model accepted this interpretation,[149] but allowed for a *violation* of the 3rd Law:

> (…) de transferer de la force dans le corps pressé; aussi-tôt le corps céde, & ne renvoye plus le pressions de cette cause, mais il ce reçoit & les accumule dans lui (…). [Inst1740, § 561]
>
> (…) und schicket die Drückungen nicht mehr zurück.[150]

Here, Du Châtelet revealed unintentionally the *internal contradiction* of the Leibnizian construction, i.e. the generation of living forces is only possible by a *violation* of Newton's 3rd Law or the action-reaction principle.

Comparing Newton's model to Leibniz's model, the question has to be answered whether the forces are always of finite magnitude or may be sometimes also of infinitesimal magnitude.

Mathematically, Leibniz discussed the motion-motion and the force-force relations as the relations between "elementary quantities" and "finite quantities" where the *finite* quantity is generated by the accumulation of *infinitesimal* contributions which are consecutively excited. Later, Du Châtelet treated the generation of motion and the relation between dead and living force within the same frame.

Following Newton, the relation between "errors" and "forces" had to be given in terms of infinitesimal and finite quantities, respectively. The errors are the representation of the finite change in motion $\Delta(mv)$ in terms of *finite* spatial differences of second order $\Delta\Delta s$ or, alternatively, in terms of *infinitesimal* quantities by Leibnizian differentials $d(mv)$ and dds. Newton discussed these relations within a geometrical frame. Adding the analytic expressions we obtain:

> Cor. 1. (…) that the errors of bodies describing similar parts of similar figures proportional times, are nearly in the duplicate ratio of the times in which they are generated, if so be these errors are generated by any equal forces similarly applied to bodies, (…).
>
> Cor. 2. But the errors … are as the forces and the squares of the times conjunctly. ($s_{begin}^{errors} \sim K \cdot t_{begin}^2$) (which is equivalent to $dds \sim Kdt^2$).

---

[147] According to the rules of the calculus (Euler. 1755. *Institutiones calculi differentialis cum eius usu in analysi finitorum ac doctrina serierum*) the result is only described by a velocity of finite magnitude which is necessary for the definition of a "living froce" if the "elementary impressions" of the dead force are not of finite, but of infinitesimal magnitude. Euler. 1746b. *Gedancken von den Elementen der Cörper* II (60).

[148] Leibniz. 1982b. *Specimen Dynamicum*. I (6).

[149] Inst1740 § 559.

[150] Naturlehre1743 § 561.

Cor. 3. The same thing is to be understood of any space whatsoever described by bodies urged with different forces. All which, in the very beginning of motion, are as the forces and the squares of the times conjunctly.

Cor. 4. And therefore the forces are as the spaces described in the very beginning of the motion directly, and the squares of the time inversely. ($K \sim s_{begin}^{errors} / t_{begin}^{2}$)

Cor. 5. And the squares of the times are as the spaces described directly and the forces inversely. ($t_{begin}^{2} \sim s_{begin}^{errors} / K$)[151]

The errors s_{begin}^{errors} are of *infinitesimal* magnitude because the time element t_{begin} in the very beginning of motion is of *infinitesimal* magnitude ("de sorte qu'elle est toujours détruite dans une instant infiniment petit."[152]). According to the basic rules of the calculus,[153] the force K is necessarily of *finite* magnitude. In the Leibnizian calculus, the errors are called differentio-differentials or differences of second order[154] are functions of time and determinate in their magnitude by the square of the time-element (see above, Cor. 2). Following Leibniz,[155] Du Châtelet applied this theorem preferentially to the discussion of the relation between different types of forces.[156] Du Châtelet's model will be discussed in Sect. 3.4.

Newton obtained these relations from a *geometrical* model where the *uniform motion* is represented by a straight line and the errors by the deviation from this line. In case of rest there is nothing to be represented geometrically except a position, i.e. a certain point. Later, the parameter mass (labelled by m) had been strictly included[157] and the set of relations had been reduced to the equation of motion $m \cdot s_{begin}^{errors} = K \cdot t_{begin}^{2}$ or $m \cdot dds = K \cdot dt^{2}$. Obviously, the interpretation of the mass in terms of the force of inertia is impossible because the inertia is supposed to be independent of the magnitude of the impressed moving force K.

Bodies and Forces

In 1687, Newton formulated his program for mechanics whose main purpose was to investigate the relation between phenomena and forces.[158]

---

[151] Newton. 1845. *The mathematical principles of natural philosophy*, transl. by Andrew Motte. Book I, Section I, Lemma X.

[152] Inst1740 §§ 532, 535, 560, 565, 567.

[153] Newton. 1740. *La Méthode des Fluions et de suites infinies*; Leibniz G. W. *Nova methodus pro maximis et minimis, itemque tangentibus, quae nec fractas nec rationales quantitatis moratur, et singulae pro illis calculis genus.*

[154] [Wolff.1734. *Vollständiges Matheruabisches Lexicon*. 371 to 375.

[155] Leibniz. 1982b. *Specimen Dynamicum*. I (6).

[156] Inst1740 §§ 532, 535, 560, 565 and 567.

[157] Euler. 1736. *Mechanica sive motus scientia analytice exposita*. §§ 118, 130, 136, 138, 140, 151 to 158.

[158] "En effet toute la difficulté de la Philosophie paroit consister à trouver les forces qu'employe la nature, par les Phénomenes du mouvement que nous connoissons, & à démontrer ensuite, par là,

> Omnis enim Philosophiae difficultas in eo versari videtur, ut a Phaenomenis motuum investigemus vires Naturae, deinde ab his viribus demonstremus phaenomena reliqua.[159]

The program for mechanics formulated by Leibniz and presented in the treatise entitled *Specimen Dynamicum pro admirandis Naturae Legibus circa corporum vires et mutuas Actiones detegendis et ad suas causas revocandis*[160] which was published in 1695 looks very similar as far as the relation between phenomena and forces are concerned. The phenomena are explained by local motion:

> Vim ergo derivativam, qua scilicet corpora actu in se invicem agunt, aut a se invicem patiuntur, hoc loco non aliam intelligimus, quam quae motui (locali scilicet) chaeret, et vicissim ad motum localem porro producendum tendit. Nam per motum localem caetera phenomena materialia explicari posse agnoscimus. Motus est continua loci mutatio, itaque tempore indiget.[161]

Relative motion is not something real or absolute, "quasi reale quiddam esset motus et absolutum", hence, the phenomena are described with respect to a frame of reference which is not "given", but "chosen" (therefore, this relation is a "contingent" truth).

> Sic igitur habendum est, si corpora quotcunque sint in motu, ex phaenomenis non posse colligi in quo eorum sit motus absolutus determinatus vel quies, sed cuilibet ex iis assumto posse attribui quietem ut tamen eadem phaenomena prodeant.[162]

Leibniz accentuated that the phenomena did not depend on the choice of the frame of reference, e.g. the description by Ptolemy and Corpernicus are only two appropriate representations for different purposes. The motions of bodies belong to the class of respective phenomena.[163] The absolute element is to be expressed in terms of forces. Contrary to Leibniz, Du Châtelet accentuated the difference between Ptolemy and Copernicus.[164]

Time and Space

Euler treated always space and times together and as magnitudes "of the same sort".[165] Later, Lagrange claimed that mechanics can be regarded as a four-dimensional geometry.

---

les autres Phénomenes." PrincChat1756, Preface. The distinction between *primitive* and *derivative* forces can be related to the distinction between *internal* and *external* principles, respectively. The correspondence is almost complete since the primitive force ("quae in omni substantia corporea per se inest") will be confined by the derivative force that emerged in the impact of bodies.

[159] Newton. 1687. *Philosophiae naturalis principia mathematica*. Praefatio.

[160] Essay in Dynamics showing the wonderful Laws of Nature concerning the forces of bodies and the mutual Actions of bodies and tracing back them to their causes. Leibniz. 1982b. *Specimen Dynamicum*.

[161] Leibniz. 1982b. *Specimen Dynamicum*. I (4).

[162] Leibniz. 1982b. *Specimen Dynamicum*. II (2).

[163] Leibniz. 1982b. *Specimen Dynamicum*. II (2).

[164] Inst1740 § 57.

[165] Euler. 1748. Réflexions sur l'espace et le tems. § XX.

Contrary to Newton[166] and Leibniz,[167] but following Descartes and supposing extension as a basic notion, Du Châtelet discussed space and time in different Chapters of the *Institutions*. Not surprisingly, there is a pronounced preference of "space" over "time". The Leibniz-Clarke debate is reviewed: "Space is the order of things" (instead of the order related to the *states* of things as Leibniz claimed). There is no reason for the *limitation* of things[168] and it is indeterminate at which of the *positions* the body is to be placed. There is no sufficient reason to create the world at a certain position in the space.[169]

Reconsidering Leibniz's construction and comparing Newton to Leibniz and vice versa Leibniz to Newton Du Châtelet guessed that it is unlikely that Leibniz's theory of simple things can be confirmed once a day.[170] However, accepting Leibniz's union of "motion (velocity) and forces" Du Châtelet treated rest and motion *separately*[171] and did not acknowledge Euler's version of the relational theory of rest and motion already presented in the *Mechanica* in 1736. This separation of rest and motion is based on the separate treatment of space and time. Although Du Châtelet accepted the Leibnizian relational concept of time and space, she inverted the Newtonian and Leibnizian *order* and treated space before time is defined in the Chaps. 5 and 6, respectively.[172]

Du Châtelet decoupled the *internal* changes from the *external* changes, i.e. the motion of bodies.

> "Il y auroit un Tems, quand même il n'y auroit point de mouvement. Car certainement je pourrois ne jamais remuer de ma place & avoir des idées successives; (…). Ainsi, tant qu'il y aura des Etres dont l'éxistence se succedera, il y aura nécessairement un Tems, soit que les Etres soient en mouvement, soit qu'ils soient en repos."[173] "So lange demnach Dinge sind, von denen die Vorstellungen aufeinander folgen, so lange ist nothwendig eine Zeit; die Dinge mögen in Ruhe oder in Bewegung seyn."[174]

Here, Du Châtelet anticipated basic features of Kant's later theory of time and introduced a *universal common measure* of time, the instant which is related to "our mind".

> La seule qui soit universelle, c'est celle que l'on appelle un *instant*; car tous les hommes connoissent nécessairement cette portion de tems, qui s'écoule pendant qu'une seule idée reste dans notre esprit.[175]

The universal and common measure of time is represented by an *indivisible* time element. Indeed, this is a very interesting generalization of Leibniz's construction of

---

[166] Newton. 1687. *Philosophiae naturalis principia mathematica*. Definitiones.
[167] Leibniz. 1715. Initia rerum mathematicarum metaphysica. GM VII, III.
[168] Inst1740 § 73.
[169] Inst1740 § 74.
[170] Inst1740 § 136.
[171] Inst1740 Chap. XI.
[172] Inst1740 Chap. V and VI.
[173] Inst1740 § 110.
[174] Naturlehre1743 § 110.
[175] Inst1740 § 114.

simple things. Leibniz assigned these "time elements" as "fulgurations continuelles" to God's activity to preserve the world.[176] As a consequence, all *measures of time* are based on duration of our existence and the duration of the things coexisting with us.[177]

Leibniz distinguished (a) all bodies in the world (or all places in the world or the plenum) by their *internal* differences which are due to the monads[178] and (b) the monads by their different perspectives they take while perceiving the world, i.e. the other monads.[179] Du Châtelet assumed version (a) decoupled from version (b) whereas Euler assumed version (b) by the introduction of relative motion and observers.[180] Then, each of the observers perceived the system of bodies and his perception is distinguished from the perception of all other observers. Therefore, Du Châtelet interpreted Leibniz's principle of distinction only metaphysically, but not also mechanically, whereas Euler transformed the metaphysical Leibnizian principle into a mechanical model.

Place Defined Either as a Relation of Coexisting Things or Occupied by a Body

The basic difference in Euler and Du Châtelet that determines the subsequently discussed mechanical models of the bodies is found in Chap. V *De l'Espace*.[181] Following Leibniz, Du Châtelet excluded and rejected Newton's model of relation between *bodies and space*. Newton excluded the idea of *situation* from the definition of the *place*.[182] Obviously, Newton did not refer to Leibniz, but to Descartes. Supposing extension as the basic property of bodies[183] and defining *situation* in the Leibnizian version as the relation between *one* body and *all other* bodies,[184] the substitution of "all" bodies by "one other" body is not performed and, as a consequence, the relation between any *two* bodies remains to be partially indeterminate. This missing relation is not discussed by Du Châtelet, but only by Berkeley and Euler. Berkeley rejected the idea of absolute space[185] that is the basis for Newton's

---

[176] Leibniz.1982a. *Monadologie*. § 47.

[177] Inst1740 § 114.

[178] Leibniz. 1982a. *Monadologie*. § 8.

[179] Leibniz. 1982a. *Monadologie*. § 57.

[180] Euler. 1746b. *Anleitung zur Naturlehre*. Chap. IX.

[181] Inst1740 Chap. V.

[182] Newton. 1687. *Philosophiae naturalis principia mathematica*. Definitiones, Scholium.

[183] Inst1740 § 137.

[184] Inst1740 §§ 88 and 93.

[185] Berkeley. 1820. *De Motu or The Principle and Nature of Motion and the Cause of the Communication of Motions*. § 63.

definition of place. Euler replaced the absolute space with a frame of reference made up of bodies.[186]

Du Châtelet rejected Newton's definition of the place.[187] The consequence is that the body is not to be characterized by its *occupation* of places, but by *inherent* forces.[188] There are two kinds of matter called *coarse* matter from which the bodies are formed and *subtle* matter from which the ether is made up.[189] The subtle matter is supposed to be a fluid and plays the former role of the plenum whereas the bodies are the discrete things moving in the ether and interacting with each other. The gravitation is caused by the subtle matter.[190] Here, Du Châtelet is in conformity with Euler.

Du Châtelet discussed the notion of situation and place in the Chap. V entitled *De l'Espace*.[191] The notion of place has been already defined by Newton who excluded categorically any relationship to the Cartesian relativism.

> III. Place is a part of space which a body takes up, and is according to the space, either absolute or relative. I say, a part of space; not a situation nor the external surface of a body.
>
> IV. Absolute motion is the translation of a body from one absolute place into another; and relative motion, the translation from one relative place into another.[192]

The place is occupied by the body. In an earlier version, Newton has already emphasized that the space is occupied by bodies so that other bodies are excluded.[193] Contrary to Newton, Leibniz defined the place relationally by the situation.[194] Therefore, it is expected that Du Châtelet was inclined to follow rather Leibniz than Newton.

In Chap. 5 entitled *De l'Espace*, Du Châtelet developed the theory of space in the spirit of Leibniz's relational theory of things.[195] The remarkable result of the

---

[186] Euler. 1736. *Mechanica sive motus scientia analytice exposita.* §§ 6 to 12, 20.

[187] Inst1740 §§ 90 to 93.

[188] Inst1740 Chap. VIII and XXI.

[189] Inst1740 Chap. X. Euler. 1746b. *Anleitung zur Naturlehre.* Chap. 12 and 13.

[190] Euler. 1746b. *Anleitung zur Naturlehre.* Chap. 19. Inst1740 § 341.

[191] Inst1740 Chap. 5.

[192] Newton. 1845. *The mathematical principles of natural philosophy*, transl. by Andrew Motte. Definitions, Scholium.

[193] "Def. 1: Place is part of space which things fill evenly. Def. 2: Body is that which fills place. Def.3: Rest is remaining in the same place. Def. 4: Motion is change of place. NOTE: I have said that body fills place; that is that it so fills it that another thing of the same kind or some other body is completely excluded, forasmuch as it is impenetrable. Let it be, however, that place is called the part of space in which things inhere evenly, but so far as here bodies and not penetrable things are being examined, I have chosen to define (that to be) a part of space which something fills." Newton. 1988. *De gravitatione et aequipondio fluidorum.*

[194] *A Collection of Papers which passed between the late Learned Mr. Leibniz and Dr. Clarke, in the Years 1715 and 1716.* 5th letter to Clarke.

[195] Inst1740 §§ 72 to 87.

analysis is the postulate of a similarity between space and the real things and numbers and the counted things.

> Ainsi, l'Espace est aux Etres réels, comme les Nombres aux choses nombrées, (…) parce qu'on fait abstraction des déterminations internes de ces choses, (…).[196]

Du Châtelet's analysis of the relations between "real things" is based on Leibniz's concept of the coexistence of the "states of things".[197] In the further treatment, the "real thing" is necessarily to be specified as a thing called "body". Following Leibniz, Du Châtelet claimed that the place of the whole world is indeterminate in Newton's theory since there is no sufficient reason for the determination of its position in the space.[198]

Alternatively to Newton,[199] and following Leibniz, Du Châtelet defined the place as a "situation". There is no explicit statement that the place is *occupied* by the body since Du Châtelet did not distinguish between *occupied* and *non-occupied* regions of the vacuum.[200] The exclusion of a thing, e.g. of a book in a library, from its place is only due to the condition that only some parts of the book can coexist.[201]

Arguing in favour of Leibniz's relational theory of space and defining "place" as a "situation", Du Châtelet ignored completely the essential relation between the body and the spatial region the body is occupying, but substituted this relation by the assignment of forces to bodies.

> Il n'y a point de Matiere sans force, ni de force sans Matiere. (…) il y faut ajouter encore le pouvoir d'agir: ainsi la force qui est le principe de l'action (…) il ne sçauroit y avoir de Matiere sans force motrice, ni force motrice sans Matiere, comme quelques Anciens l'avoient fort bien reconnu.[202]

The definitions of rest and motion are given in Chap. 11 entitled *De Mouvement, & du Repos en général, & du Mouvement simple*[203] whose title is very similar to the title of Euler's *Consideratio de motu in genere*.[204] We will demonstrate that this similarity is not by chance, but can be observed also in essential features and arguments. Du Châtelet supposed the relational theory of motion, but treated motion

---

[196] Inst1740 § 87.

[197] Leibniz. 1715. Initia rerum mathematicarum metaphysica. GM VII, III. 18.

[198] Inst1740 § 74.

[199] It is remarkable that Newton rejected categorically the notion of "situs" which is nothing else than an entire refutation in advance of the later Leibnizian theory of *situs*. The reason is obvious. It is Newton's aversion against Descartes' relativism. "He may have capitulated to Descartes on motion. On the issue of relativism, which in Newton's view smacked of atheism, he continued to shout his defiance until his dying day." Westfall. 1993. *The life of Isaac Newton*. 166.

[200] Inst1740 §§ 90 to 93.

[201] Inst1740 § 92.

[202] Inst1740 § 141.

[203] Inst1740 .

[204] Euler. 1736. *Mechanica sive motus scientia analytice exposita*. Chapter 1.

always in relation to forces. Du Châtelet and Euler claimed that motion is the change of the place, but the place is defined differently:

> Le Mouvement est le passage d'un Corps du lieu qu'il occupe dans un autre lieu.[205]

> Motus est translatio corporis ex loco, quem occupabat, in alium. Quies vero est permansio corporis in eodem loco.[206]

> On appelle *le lieu* ou la *place* d'un Etre, sa maniere déterminée de coéxister avec les autres Etres (…). (…) & un autre Etre occupe la même place que cette table lorsqu'il obtient la même maniere de coexister qu'elle avoit avec tous les Etres.[207]

The statements agree word by word, the only difference is the omission of Euler's definition of rest. Therefore, one can guess that Du Châtelet read Euler's treatise whereas from Euler's correspondence it follows that he knew the *Institutions*.[208] Nevertheless, remembering the previous definition of the place, the word "occupation" may indicate Du Châtelet's reading of Newton and Euler, but she interprets "occupation" in terms of a situation.

Du Châtelet: Extension Is Independent of Forces. Euler Impenetrability Is Independent of Forces

Du Châtelet's supposed extension and only *two* kinds of forces, the *active* and the *passive* force or the force to resist, as basic properties of bodies. If there is an active force, there should be necessarily a passive force.[209] The active force is taken from Leibniz.[210] Contrary to Leibniz[211] Du Châtelet maintained the Cartesian notion of extension and supposed that extension is independent of forces. Following Du Châtelet, there are necessarily three principles: (i) extension, (ii) active force and (iii) passive force.

> § 141. L'étenduë qui résulte de la composition n'est donc pas le seule proprieté qui convient au Corps (…). § 145. Tous les changements qui arrivent dans les Corps peuvent s'expliquer par ces trois principes, *l'étendue, la force résistante*, & *la force active*; (…).[212]

The theory of an internal *resisting force* is an immediate consequence of assumption that the bodies are characterized by an internal *active* principle.

---

[205] Inst1740 § 211.

[206] Euler. 1736. *Mechanica sive motus scientia analytice exposita*. § 1.

[207] Inst1740 § 88.

[208] Euler. 1963. Письма к ученым. Pis'ma k učenym (Letters to scholars). 278.

[209] Inst1740 §§ 142 to 145.

[210] Leibniz. 1982b. *Specimen Dynamicum*. I (3).

[211] Leibniz. 1982b. *Specimen Dynamicum*. I (1).

[212] Inst1740 §§ 141 and 145.

Following Leibniz, Du Châtelet rejected the empty space,[213] but stressed the similarity between space and time.[214] Going beyond the Leibnizian frame where the order of successions is partially indeterminate because it is not quantified,[215] Du Châtelet related not only the space to the numbered things[216] (see above), but also the time which may be an essential step towards a synthesis of order and quantification.

> Cette comparaison du Tems & du Nombre peut servir à se former la véritable notion du Tems; & à comprendre que le Tems de même que l'espace, n'est rien d'absolu hors de choses.[217]

There are no parts of the time except those which are related to real things.[218] In contrast to Leibniz (and Newton) who only assumed that the line is the path of a point,[219] Du Châtelet specified the generation of a line as to be generated by the *uniform* motion of a point.[220] The time intervals are measured by uniform motion of things. Measurement is based on the comparison to a previously defined unit.[221]

The relation between different kinds of forces had been already tackled by Newton.

> The crux of Newton's dynamics lay in the relation of inherent force and impressed moving force, what he later called (as he struggled to clarify them) 'the inherent, innate and essential force of a body' and 'the force brought to bear or impressed on a body'. The continuing development of his dynamics hinged on the two concepts.[222]

Though Newton's results were finally available for his contemporaries, the reception of his work was considerably hampered because Newton used to give only very preliminary comments on the background he had have in mind.

> Once he adopted the principle of inertia, the rest of his dynamics fell quickly into place. He had seized the essence of his second law twenty years before and had never altered it as he was wrestled with the first law. (…) In the Principia itself, he further eliminated the reference to inherent force from the statement of the first law, thus obliterating the principal record of the path by which he arrived at it.[223]

Thus, the ambiguity in the status of the "force of inertia" was not removed, but survived and caused trouble in the eighteenth century mechanics. The most radical

---

[213] Inst1740 § 85.

[214] Inst1740 § 94.

[215] This criticism was due to Clarke. *A Collection of Papers which passed between the late Learned Mr. Leibniz and Dr. Clarke, in the Years 1715 and 1716*. 3rd letter to Leibniz.

[216] Inst1740 § 87.

[217] Inst1740 § 103.

[218] Inst1740 § 105.

[219] Leibniz. 1715. Initia rerum mathematicarum metaphysica. GM VII, III. 20.

[220] The exceptional role of uniform motion for the definition of temporal and spatial intervals had been stressed by Euler. Euler. 1748. Réflexions sur l'espace et le tems. §§ 20 and 21.

[221] Inst1740 § 106.

[222] Westfall. 1993. *The Life of Isaac Newton*. 166.

[223] Westfall. 1993. *The Life of Isaac Newton*. 167.

cut was due to Euler who substituted the Cartesian notion of extension with the notion of *impenetrability*[224] and argued that the *inertia* is independent of forces.

> Mit dem Worte Trägheit ist man auch gewohnt, eine Kraft zu verbinden und dem Körper die Kraft der Trägheit zuzuschreiben, wodurch große Verwirrungen veranlasset werden; denn da eine Kraft eigentlich dasjenige genannt wird, welches vermögend ist, den Zustand eines Körpers zu verändern, so kann dasjenige, worauf sich die Erhaltung eben desselben Zustandes gründet, unmöglich eine Kraft genannt werden.[225]

Though Euler warned of serious confusions, the temptation to made use of the idea of *inherent* or *innate* forces of matter was still alive through the end of the eighteenth century (compare Kant's *Metaphysical foundation of natural science* (1786)).

Du Châtelet on Dead and Living Forces

Du Châtelet claimed that the dead force and the element of living force have the *same measure*. However, after a grandiose beginning

> la force morte & l'élément de la vive ont une même mesure qui est la masse du corps multipliée par la vitesse infiniment petit que la pression lui communique[226]

making an essential step towards the discovery of the *common* force behind dead and living forces, Du Châtelet returned to the commonly accepted interpretation of the difference between impressed and living forces

> (...) mais j'aime cependant mieux les distinguer ici, parce qu'il y a une différence réelle entre elles; car dans le premier cas les degrés de force infiniment petits sont détruits à tout moment, au lieu que dans le second, ils s'accumulent dans le corps qui reçoit le mouvement.[227]

and refused to compare Kdt (à chaque *instant* infiniment petit) to Kds (à chaque *mouvement purement progressif* infiniment petit). Following Du Châtelet, the difference is only due to the difference between infinitesimal and finite quantities, i.e. dv and v, respectively, related to each other by the mathematical relation $\int dv = v$. However, it is not justified to interpret the result as a *mechanical* difference. Taking into account each of the momentary elementary processes of "destruction" and "accumulation", Du Châtelet assumed implicitly that the change of the dead force is finally proportional to the change of the living force $dK_{dead} \sim dK_{living}$ which had been later represented in terms of energy instead of forces, i.e. the difference (decrease/increase) in the potential energy is transformed into a difference (increase/decrease) of kinetic energy $dE_{pot} \sim dE_{kin}$. The total energy or the sum of kinetic and potential energies $E_{pot} + E_{kin} = const$ is preserved.

---

[224] Euler. 1746b. *Anleitung zur Naturlehre*. §§ 28, 31 and 35.
[225] Euler. 1746b. *Anleitung zur Naturlehre*. 225.
[226] Inst1740 § 561.
[227] Inst1740 § 561.

Du Châtelet argued rightfully (the analytical expressions are added):

§. 562. Lorsque la pression imprime au corps qui lui céde, le premier degré de force [dK], ou l'élement de la force vive [$dK_{vive} = d(mv^2)$], cet élément est proportionnel au petit espace [$d(mv^2) \sim ds$] que la pression fait parcourir au corps dans un petit tems donné [$d(mv) \sim dt$], ou à la vîtesse infiniment petite [dv] qu'elle lui communique dans ce petit tems [dt], & un pression (…) comme cette pression, qui produit dans le premier moment [dt] un élement de force vive lorsque l'obstacle céde infiniment peu, est la même qui produisoit une force morte [$dK_{vive} \sim dK_{morte}$], lorsque cet obstacle ne cedoit point du tout à son effort, on connoît la quantité de la pression qu'un obstacle invincible détruit, par rapport à une autre pression à laquelle l'obstacle céde infiniment peu [ds] dans un tems infiniment petit [dt], par l'espace [s], que cette pression, qui agit contre un obstacle invincible, seroit parcourir à cet obstacle dans un tems donné [t], si la force qu'elle communique au corps sur qui elle agit, devenoit vive de morte qu'elle étoit auparavant, comparé à l'espace, qui l'autre pression à laquelle l'obstacle céde infiniment peu, fait parcourir dans le même tems à un corps égal en masse au premier, en considérant toujours les effets dans un instant infiniment petit [dt].[228]

Euler's solution of the problem is based on the theorem that *masses* and *forces* in mechanics are always of *finite* magnitude whereas the *increments* of the measures of dead and living force, i.e. *mdv* and *mvdv* may be of infinitesimal magnitude. This is due to the infinitesimal increment of velocity dv whereas the mass m is finite. It is, therefore, impossible that the living force is generated by an infinity of infinitely little impressions of the dead force, as Leibniz supposed.[229] Replacing this relation between *different forces* by the postulate of *different effects* of *one and the same force* Euler investigated[230] the infinitesimal *increments*

$$(mdv)_{dead} \sim K_{Euler} \text{ and } (mvdv)_{living} \sim K_{Euler} \quad (5)$$

of those quantities which had been of formerly called "dead force" and "living force", respectively, which are proportional to one and the same force labelled by K_{Euler}. The difference between the formerly discussed "dead" and "living" forces is now traced back to the distinction between the purely time dependent and the purely path (coordinate) dependent changes of the states, expressed by the integrals

$$\int K_{Euler} dt \text{ and } \int K_{Euler} ds \quad (6)$$

These are finite quantities and the difference is traced back to the difference between space and time. In the simplest case, the force K_{Euler} is of one and same magnitude and constant, i.e. a time and coordinate independent quantity $K = const$. The complete set of equations of motions in terms of the *increments* of dead and living "forces" is obtained by completing the expressions in Eq. 5 with the corresponding infinitesimal temporal and spatial elements labelled by dt and ds, respectively.

$$(mdv)_{dead} = K_{Euler} dt \text{ and } (mvdv)_{living} = K_{Euler} ds \quad (7)$$

---

[228] Inst1740 § 562.

[229] Leibniz. 1982b. *Specimen Dynamicum*. I (6).

[230] Euler. 1736. *Mechanica sive motus scientia analytice exposita*. §§ 150 to 152.

The latter representation may be called Leibniz representation, the former one Descartes-Newton representation.[231] Euler demonstrated that these expressions can be transformed into each other substituting $dt = ds/v$ and $ds = vdt$ in the first and the second formula, respectively.[232]

Therefore, the debate on the true measure of living forces would be finished already in 1736. Unfortunately, but being characteristically for the delay in reception of innovations, neither Du Châtelet nor Kästner made use of Euler's representation. On the contrary, Euler's interpretation was questioned by d'Alembert.[233] Euler demonstrated that both equations (Eq. 7) are not only simultaneously valid, but, moreover, also necessary for the calculation of the final velocities after the impact of bodies from their initial velocities before the interaction. Furthermore, Euler adduced evidence that the expressions formerly assigned to "dead" and "ling forces" never deserve the name of a force. Forces originate from the *impenetrability* of bodies and are correlated with the mutual change of the states. Making use of these equations, the interaction of the bodies or the change of the state as well the conservation of states can be described by the same algorithm. Obviously, this result is only obtained by the application of the calculus.

Relative Motion in Euler and Du Châtelet

Euler did most of his work on relative motion only after 1740. Du Châtelet had only the chance to read the *Mechanica* and to be aware of the essential difference between the Leibnizian model of relative *transitions* and the Eulerian model of relative *motion*. The difference is mainly due to the different roles the observer played.[234] Additionally, the latter model is explicitly related to "us", "thus to say we are accustomed" to determine the position of body relatively to fixed boundaries.

> Sic dicere solemus corpus, quod respectu horum limitum situm eundem conservat, quiescere, id vero, quod situm eodem respectu mutat, moveri.[235]

These laws of motion being valid for extended bodies are transferred to the model of mass points.

> 98. Istae motus leges, quas corpus sibi relictum vel quietem vel motum continuando observat, spectant proprie ad corpora infinite parva, quae ut puncta possunt considerari.[236]

---

[231] "158. Propositia ista complectitur omnia principia hactenus tradita motus naturam definientia omnesque leges motus, si quidem potentia directio cum motus directione congruit." Euler 1736. *Mechanica sive motus scientia analytice exposita.* § 158. Euler preferred the Leibnizian representation. This interpretation (complectitur *omnia* principia) had been later strengthened. "(…) und den Begriff der lebendigen Kraft über die Größe der Bewegung weit erhebet." Euler 1746b. *Anleitung zur Naturlehre.* §§ 75 and 76.

[232] Euler. 1736. *Mechanica sive motus scientia analytice exposita.* §§ 127 to 160.

[233] d'Alembert. 1743. *Traité de Dynamique.* Chapter 1, Explanation 1.

[234] Euler. 1736. *Mechanica sive motus scientia analytice exposita.* § 97.

[235] Euler. 1736. *Mechanica sive motus scientia analytice exposita.* § 7.

[236] Euler. 1736. *Mechanica sive motus scientia analytice exposita.* § 98.

Here, Euler preserved the Cartesian assumption that bodies, as far as they are not disturbed, neither by an intervention of a spirit nor by an interaction with another body[237] preserve in their states, and transferred the principles being valid for bodies of finite size to bodies of infinitesimal magnitude. The observation of the relative motion is performed by an observer (Zuschauer) who is associated with the body.

> Spectator vero in corpore B relative quiescens ponitur et ipsum B ut quiescens considerans.[238]

The position of the observer is determined by the motion of the body. The perspective the observer obtained from looking around is unique, i.e. it is different from the perspectives of all other observers who are populating the world. Here, Euler transformed Leibniz's metaphysical model of monads[239] into the *methodology of observation*. The metaphysical Leibnizian pattern is transformed into a mechanical model. The perspectives of the observers and the comparison between different experimental findings become a part of the theory. The theory of relative motion is constructed as a theory of relative motion of bodies and, simultaneously, the relative motion of observers. The decisive step done by Euler is the reduction of the unlimited variety of Leibnizian observers just to two observers and two bodies. Obviously, these two elements are necessary and sufficient to model and discuss relative motion. The preliminary version had been already presented by Leibniz by the construction of the two body model for the conservation of living forces (compare Sect. 2.1.). Leibniz, however, was not ready to reduce the infinitely many monads just to two monads and the relation between them.[240]

---

[237] Euler. 1746b. *Anleitung zur Naturlehre*. § 49.

[238] Euler. 1736. *Mechanica sive motus scientia analytice exposita*. § 97.

[239] "57. And as the same town, looked at from various sides, appears quite different and becomes as it were numerous in aspects [perspectivement]; even so, as a result of the infinite number of simple substances, it is as if there were so many different universes, which, nevertheless are nothing but aspects [perspectives] of a single universe, according to the special point of view of each Monad." Leibniz. 1982a. *Monadologie* (Latta). § 57.

[240] Euler has a similar aversion to the model of the body made up of atoms of Epicurean type. Referring to Leibniz, Euler argued that there are insurmountable difficulties foreseen by Leibniz that prevented him to hypothesize such a model: "63. Nicht geringeren Schwierigkeiten ist aber das Lehr-Gebäude des Herrn von Wolff unterworfen, welcher behauptet, dass die Anzahl der einfachen Dinge, aus welchen ein Körper zusammengesetzt ist, würklich bestimmt und endlich sei. Und es ist kein Zweifel, dass nicht schon der Herr von Leibniz diese Meinung angenommen haben sollte, wenn er nicht nach seiner tiefen Einsicht dabei unüberwindliche Schwierigkeiten vorausgesehen hätte." Euler. 1746b. *Gedancken von den Elementen der Cörper*. II (63). Du Châtelet also rejected Epicurean atoms. Inst1740 Chapter VIII to IX "§ 121. Les atomes, ou partie insécables de la Matiere ne peuvent être les Etres simples; (…). (…) parce qu'il ne s'agit pas de sçavoir pourquoi l'étenduë existe, mais comment & pourquoi elle est possible." Inst1740 § 121. Here, Du Châtelet paid attention to the question how to investigate the reasons for the "possibility of extension" that is as important as the investigation of the "reality of extension" which had been later generalized by Kant who was not only asking for the reason for the "possibility of extension" (Kant. 1997. *Metaphysische Anfangsgründe der Naturwissenschaft*. Zweites Hauptstück), but also for the "possibility of experience".

Models of Relative Motion

The analysis of relative motion of bodies can be traced back to the ancient science. In Ptolemy's model of the world the onlooker is placed at the earth which is resting as an immobile body in the centre of the universe. Relative motions have been discussed by Zeno who introduced instructive models which had been reconsidered by his followers. Aristotle discussed Zeno's paradoxes on motion. Descartes introduced the relativity of motion beyond the frame where motion is considered as being paradoxical, but in the Cartesian version motion remained to be *indeterminate*. A body can take different motions and velocities or, it is possible to assign to a body different velocities in dependence on the choice of the frame of reference, i.e. with respect to the other bodies whose state in independent of the state of the chosen body. Descartes explained the relativity of motion using the model system of a moving ship where a passenger is observing his environment on the ship and the shore the ship is leaving.

The difficulty Du Châtelet and Kästner and even Leibniz were confronted with was caused by the implicit assumption they made that the observer occupied finally a position *outside* the system, e.g. the observer is never sitting at Zeno's flying arrow. Euler, however, described just such a situation modelling the observer's motion analytically by means of thought experiments performed with frames of reference that had been later called inertial systems.[241]

Du Châtelet. Motion as Illusion. Kästner's "Spitzfindigkeiten"

Du Châtelet and later the famous mathematician Kästner[242] analyzed relative motion in terms of the model known from the Leibniz-Clarke correspondence. Leibniz introduced an operational definition of relative *translation* in terms of *geometrical* relations which may be easily transferred to relative motion.

> 47. When it happens that one of those co-existent things changes its relation to a multitude of others, which do not change their relation among themselves; (…). And, to give a kind of a definition: *place* is that, which we say is the same to A and, to B, when the relation of the co-existence of B, with C, E, F, G etc. agrees perfectly with the relation of the co-existence, which A had with the same C, E, F, G, etc. (…) Lastly, *space* is that, which results from places taken together.[243]

---

[241] Euler. 1746b. *Anleitung zur Naturlehre*. Chapter 10.

[242] Kästner's comments on relative motion are basically equivalent to Du Châtelet's analysis. Though Kästner was familiar with the calculus, he did not construct an abstract model being exclusively composed of the relative uniform motion of bodies and observers, but discussed relative translations. Kästner's comments are included in order to demonstrate that it was impossible to surpass Leibniz without reference to Descartes or the recovery of Cartesian principles.

[243] *A Collection of Papers which passed between the late Learned Mr. Leibniz and Dr. Clarke, in the Years 1715 and 1716.* 5th letter to Clarke, § 47.

Descartes reduced the multitude of coexisting things to two bodies AB and CD and emphasized the *reciprocity* in the relations between the bodies, i.e. if body AB is translated relatively to a body CD, then body CD is also translated relatively to body AB.

> Ipsa enim translatio est reciproca, nec potest intelligi corpus AB transferri ex vicinia corporis CD, quin simul etiam intelligatur corpus CD transferri ex vicini corporis AB.[244]

The relations between AB and CD are defined *analytically* without reference to a further body. The only quantities are two distances, (i) the distance measured between AB and CD taken from body AB and (ii) the distance measured between CD and AB taken from body CD as point of reference. All the bodies and all the possible observers are *equivalent internal* parts of the system. Though Du Châtelet referred to Descartes as far as the *extension* is concerned, she preferred the Leibnizian model of *relative translations* whereas Euler recovered the *Cartesian reciprocity* between the moving of bodies. The velocity of body 1 relatively to body 2 is equal in magnitude to the velocity of body 2 relatively to body 1.[245]

In 1740, Du Châtelet discussed a model for relative motion preserving Leibniz's assumptions on the relation between rest and motion, especially that a motion can only be generated by a motion.[246] As a result, Du Châtelet *overestimates* finally the state of motion and *underestimates* the state of rest. Both the states are never treated on an equal footing. This will be demonstrated for the case of relative motion which is discussed in the famous Cartesian model of a ship travelling along a shore while a passenger, sitting or walking on the deck of the ship, is watching the shore and the desk positioned on the deck of the ship.

> Celui qui est dans le Vaisseau & qui croit que la pierre a marché d'Orient en Occident, attribue à la pierre le mouvement qui n'appartient qu'an Vaisseau; & il est trompé par ses de la même manière que nos sommes, quand nous croyons que le rivage que nous quittons s'ensuit, quoique ce soit le Vaisseau qui nous porte qui s'en éloigne, car nous jugeons les objets en repos, quand leurs images occupent toujours les mêmes points sur notre retine.[247]

Du Châtelet claimed that we are cheated by our senses if we believe that the shore is escaping from us. The *phenomenological* part of the problem is described rightfully as far as Du Châtelet considered an observer on the ship who is *resting*

---

[244] Descartes. 1998–99. *The Principles of Philosophy* II (XXIV).

[245] Not surprisingly, this proposition is also found in Leibniz, but expressed in terms of forces. "Sequitur etiam ex natura motus respectiva eandem esse corporum actionem in se invicem seu percussionem, modo eadem celeritate sibi appropinquent, id est manente eadem apparentia in phaenomenis datis, quaecunque demum sit vera hypothesis seu cuicunque demum vere ascribamus motum aut quietem, eundem prodire eventum in phaenomenis quaesitis seu resultantibus, etiam respectu actionis corporum inter se. Atque hoc est quod experimur, eundem nos dolorem sensuros sive in lapidem quiescentem es filo si placet suspensum incurrat manus nostra, sive eadem celeritate in manus quiescentem incurrat lapis." Leibniz. 1982b. *Specimen Dynamicum*. II (2).

[246] Leibniz. 1982b. *Specimen Dynamicum*. I (1).

[247] Inst1740 § 219.

relatively to the ship. Here, Du Châtelet assumed the Leibnizian relational model of spatial positions where a *privileged group* of bodies is assumed which is considered as a system of reference (the bodies C,F,E,G in the Leibnizian model), i.e. the observer is always attached to that group bodies. The other bodies in this world which are moving relatively to that privileged group are not equipped with observers. However, following Leibniz we have to take into account *all possible perspectives*.[248] The intuitive conviction, however, that the ship is to be moved by a force whereas the shore is at rest, destroys the Cartesian reciprocity by the assignment of forces to bodies.

In 1766, Kästner summarized the state of art as far as relative motion is concerned. Kästner referred also to the Leibnizian model that is still in power 50 years after its creation.

> Wenn wir einen Menschen in der Ferne auf dem Felde sehen, und nicht eigentlich erkennen können, ob er fortgeht oder stille steht, so werden acht geben, ob er seine Lage gegen unbewegliche Gegenstände, einen Baum, einen Hügel u.s.f. ändert oder nicht. (...) Wenn also der Mensch stille stünde, und der Baum oder der Hügel sich von ihm entfernte oder näherte, würde wohl dieses heissen der Mensch veränderte seinen Ort, und bewege sich also? Mit dieser Spitzfindigkeit hat man noch im vorigen Jahrhunderte die verwirrt, die die corpenicanische Weltordnung verketzerten. (...) Wer nicht Lust zu Zanken hat, wird leicht unterscheiden, ob ein Körper seinen Ort verändert, oder ob andere Körper ihren Ort um ihn verändern.[249]

Making use of the configuration analyzed by Du Châtelet for the motion of the ship relatively to the shore, Kästner discussed the spatial relation between a walker and some trees distributed in a landscape, and demonstrated unintentionally that the walker, the trees and the observers are *not* equivalent internal parts of the system. Kästner referred to the Newtonian definition of place that is different from a situation.

Euler's Early Relativistic Theory

Euler's decision in favour of relative motion together with the introduction of observes can be considered as a basic part of mechanics or the science of motion that was already introduced in the *Mechanica*.

> 80. Quia omnis idea, quam de motu habemus, est relative (§ 7), hae quoque leges non sufficiunt ad cognoscendum, qualis sit cuispiam corporis motus absolutus.[250]

---

[248] Leibniz. 1982a. *Monadologie*. § 57.

[249] "When we see a man standing at distance from us in the field and if we are not able to distinguish whether he walks or stands at his place, we will refer whether he will change his position with respect to resting objects, e.g. a tree, a hill etc. (...) If the man would remain at his place and the tree or the hill would approach him or went away, would this mean that the man would change his place, i.e. would he move? Such was the sophistry used in the last century to confuse those who intended to accuse the Copernican order of the world of heresy. (...) Those who are not in favour of disputing are able to distinguish without trouble whether a body is changing its place or the bodies of his environment do it." Kästner. 1766. *Anfangsgründe der höheren Mechanik*. I. Cap., § 1 and 2.

[250] Euler. 1736. *Mechanica sive motus scientia analytice exposita*. § 80.

The theory of relative motion was comprehensively elaborated in the *Anleitung* where the Chap. 10 is entitled *Von der scheinbaren Bewegung*.[251] Euler discussed the role of observers that are independent of the model of the closed cabin on a ship without windows (Galileo, Leibniz) or open cabin with windows (Descartes, Du Châtelet). In contrast to these models, Eulerian observers are watching each other and estimate their distances and motions and, additionally, are using the same general equation (see below on the validity and invariance of that equation)[252] of motion and exchange information about their measurements.

> 77. Die scheinbare Bewegung bezieht sich auf einen Zuschauer und wird durch zwei Stücke bestimmt, erstlich aus der Gegend, nach welcher dem Zuschauer der Körper erscheint, und hernach aus der Entfernung desselben vom Zuschauer (…). Von dieser [der scheinbaren] Bewegung ist um so viel nöthiger hier zu handeln, da wir uns in der Welt keinen anderen Begriff als von der scheinbaren Bewegung machen können; denn wir können die Oerter der Körper nicht anders als nach dem Orte unseres Aufenthaltes schätzen.[253]

If the observer is moving uniformly in along the same direction (Gegend) then all bodies which are either resting or also moving along a straight line appeared to him to persist in the same state. No force is necessary for the maintenance of such a motion of bodies. The observer can conclude that no force is acting upon the body. Then, all observers estimate the forces from the change or deviation in uniform motion.

> 81. Wenn der Zuschauer gleichgeschwind in einer geraden Linie fortrücket und die Gegenden richtig schätzet, (…). so werden zur Unterhaltung der scheinbaren Bewegung, (…) eben diejenigen Kräfte erfordert, als zur Unterhaltung der wahren Bewegung.[254]

In contrast to Leibniz, Euler described the body only by the concepts of motion and rest. It is impossible that there is a body which is neither moving nor resting.[255] Du Châtelet preserved essential Leibnizian assumptions on the relation between rest and motion, especially the assumption that a motion can only be generated by a motion.[256] As a result, Du Châtelet *overestimated* finally the state of motion and *underestimated* the state of rest.

> (…) car il n'y a de mouvement réel que celui qui s'opère par une force résidente dans le Corps qui se meut, & il n'y a de repos réel que la privation de cette force.[257]

---

[251] Euler. 1746b. *Anleitung zur Naturlehre*. Chapter 10.

[252] Euler. 1746b. *Anleitung zur Naturlehre*. Chapter 10 § 81.

[253] "77. The apparent motion is related to an observer and is determinate by two quantities, first the direction in what the body is appearing and, second, by the distance of the body from the observer (…). In any case and for principal reasons, we are in great need to consider apparent motion since we cannot get another idea as of apparent motion if we are in the world, since we are not able to estimate the positions of the body in another way except according to the place of our stay." Euler. 1746b. *Anleitung zur Naturlehre*. § 77.

[254] "If the observer is moving uniformly along a straight direction and the distances are rightfully estimated (…) the forces being necessary for the maintenance of the apparent motion are the same as for the maintenance of absolute motion." Euler. 1746b. *Anleitung zur Naturlehre*. § 81.

[255] Euler. 1736. *Mechanica sive motus scientia analytice exposita*. § 3.

[256] Leibniz. 1982b. *Specimen Dynamicum*. I (1).

[257] Inst1740 § 225.

Du Châtelet returned to the foundation of motion by forces.[258] This supposition accentuates the difference between rest and motion. The decisive role the conservation of the states of motion and rest played in mechanics is not explicitly formulated. The decoupling of the conservation of states from any kinds forces cannot be adequately formulated.

Euler introduced the following essential topics which make his mechanics different from the theories of his predecessors, (i) the rigorous statement on the priority of relative motion, combined with the introduction of an *observer*, called *spectator*.[259] In the following paragraph Euler presented his program for mechanics.[260] The theory was comprehensively elaborated in the *Anleitung* where the observer is called *Zuschauer*[261] and maintained in the *Theoria*,[262] (ii) the introduction of more than one observer who are comparing the results of their observations, which results in the confirmation of (iii) the invariance of the equation of motion, (iv) the explanation of the origin of forces and (v) the harmony between mathematics and mechanics resulting from Euler's procedure to coordinate his the progress mathematics with his progress in physics.

Summary

In 1740, Du Châtelet published her treatise *Institutions de physique* as a *methodological* and *historical* analysis of the controversial debates on the foundation of mechanics that completes advantageously Euler's *systematic* presentation of the principles of mechanics in the *Mechanica* in 1736. Euler translated Newton's *Principia* in the language of the Leibnizian calculus. In the 1740s, Du Châtelet translated Newton's *Principia* into French and Euler composed the *Anleitung zur Naturlehre* that corresponds in goal and spirit to Du Châtelet's *Institutions*.

In view of the educational attitude, both treatises fulfilled the same purpose to made science available to the common reader. This interpretation may also fit for Euler's *Anleitung* that was written in German. Reading Du Châtelet, the comprehension of the theory and the argumentation of Euler may be essentially promoted. Reading Euler, the esteem of Du Châtelet's analysis may be considerably enhanced.

Newton and Leibniz were exceptionally successful in the enrichment and completion of the Cartesian legacy by new concepts and methods. Though developing differently this rich legacy and surpassing Descartes in essential topics, their attempts were dominated by an anti-Cartesian attitude that hampered the construction of a satisfactory post-Cartesian physics. As a consequence, the Cartesian basic concepts survived only in two incomplete representations.

---

[258] Inst1740 § 225.
[259] Euler. 1736. *Mechanica sive motus scientia analytice exposita*. §§ 7, 80 and 97.
[260] Ibid., § 98.
[261] Euler. 1746b. *Anleitung zur Naturlehre*. §§ 77 to 83.
[262] Euler. 1765. *Theoria motus corporum solidorum seu rigidorum*. §§ 1 to 11.

Johann Bernoulli who was very influential in the education of the young Euler may have also stimulated Du Châtelet in her reconsideration of Newton and Leibniz intended to join the advantages and to remove the deficits of the Cartesian and the Newtonian celestial mechanics.[263] It is very likely that he draw the attention of the young Euler to Newton and Descartes.[264] Johann Bernoulli's paper was published just in that time Du Châtelet composed the *Institutions*.

Bernoulli's paper and Du Châtelet's *Institutions* are a characteristic examples of the investigations appearing in a *transitional period*, i.e. in a period of the development of science where a new theory comes into being, but its reception, interpretation and acceptance is still conditioned in content and shape by the former traditional frame. In such periods, the hypotheses became a powerful tool to bring forward the progress in science. Euler and Du Châtelet were right in emphasizing the important role of hypotheses for the development of science. In the beginning of the twentieth century, the most famous examples of hypotheses are Planck's *hypothesis* of the *discrete action* being related to the *hypothesis of quantum emission* and Einstein's subsequently introduced *hypothesis of light quanta*.

It is the merit of Du Châtelet that she did not highlight those parts of the seventeenth century legacy that were in opposition to each other, but those that are compatible and supplement one another.

Euler solved successfully the exceedingly difficult problem to create a new frame of reference for making available the complete legacy of the seventeenth century science. This was also Du Châtelet's attitude. Euler created a prototype for all further attempts of frontier crossings dealing with the relations between mathematics and physics and presented the suppositions, the method and the final solution whereas Du Châtelet explained to us the troublesome way to the solution people had to go. The extraordinary role Du Châtelet's *Institutions de physique* played in the eighteenth century is not only confirmed by the later translation of Newton's *Principia* into French, but due to the esteem her treatise met in the public that was probably surpassed only by Euler's *Lettres à une princesse d'Allemagne*.

References

A Collection of Papers which passed between the late Learned Mr. Leibniz and Dr. Clarke, in the Years 1715 and 1716, with an Appendix by Samuel Clarke. London MDCCXVII.
Berkeley, G. 1820. De Motu or The Principle and Nature of Motion and the Cause of the Communication of Motions. In *The works of George Berkeley*, Vol. II. London.

---

[263] "De cette maniere, j'ai tâché de concilier ensemble les deux systèmes par leur beau côté, pour en former un nouveau." Bernoulli. 1735. *La nouvelle Physique céleste*.

[264] Since 1720, Euler was in contact with Johann Bernoulli. "A. 1720 wurde ich bey der Universität zu den Lectionibus publicis promovirt: wo ich bald Gelegenheit fand, dem berühmten Professori Johanni Bernoulli bekannt zu werden (…)." Euler. 2007. Autobiography. In 1723, the young Euler received his master in philosophy with a dissertation that compared the Cartesian to the Newtonian system of natural philosophy.

Bernoulli, J. 1735. *La nouvelle Physique céleste*. Paris.
D'Alembert, J. 1743. *Traité de Dynamique*. Paris.
Descartes, R. 1998–99. *The Principles of Philosophy*, transl. by George MacDonald Ros.
Euler, L. 1736. *Mechanica sive motus scientia analytice exposita*. In *Opera Omnia* II, 1.
Euler, L. 1746. *Gedancken von den Elementen der Cörper*. In *Opera Omnia* II, 2.
Euler, L. 1748. *Réflexions sur l'espace et le tems*. In *Opera Omnia* III, 2.
Euler, L. 1750. *Découverte d'un nouveau principe de Mécanique*. In *Opera Omnia* II, 5.
Euler, L. 1750. *Recherches sur l'origine des forces*. In *Opera Omnia* III, 2.
Euler, L. 1755. *Institutiones calculi differentialis cum eius usu in analysi finitorum ac doctrina serierum*, vol. 1. In *Opera Omnia* I, 10.
Euler, L. 1765. *Réflexions sur l'espace et le tems*. In *Opera Omnia* II, 3 and 4.
Euler, L. 1760–1762. *Lettres à une princesse d'Allemagne sur divers sujets de physique & de philosophie*. In *Opera Omnia* III, 11 and 12.
Euler, L. 1746, published 1862. *Anleitung zur Naturlehre*. In *Opera Omnia* III, 1.
Euler, L. 1823. *Letters of Euler on different subjects in Natural Philosophy addressed to a German Princess*, ed. David Brewster. Edinburgh.
Euler, L. 1848. *Mechanik oder analytische Darstellung der Wissenschaft von der Bewegung*, transl. by Wolfers J. Ph. Greifswald.
Euler, L. 1963. Письма к ученым. Pis'ma k učenym (Letters to scholars), Moscow/Leningrad: Izdat. Akad. Nauk SSSR.
Euler, L. 2007. Autobiography. In *Leonhard Euler*, ed. E. A. Fellmann. Basel.
Euler, L. 2006. *Leonhard Euler. Life, Work and Legacy*. ed. Robert E. Bradley and C. Edward Sandifer. Elsevier.
Falckenberg, R. 1892. *Geschichte der neueren Philosophie*. Leipzig.
Gehler, J. S. T. 1787. *Physikalisches Wörterbuch oder Versuch einer Erklärung der vornehmsten Begriffe und Kunstwörter der Naturlehre mit kurzen Nachrichten von der Geschichte der Erfindungen und Beschreibungen der Werkzeuge begleitet in alphabetischer Ordnung*. Leipzig.
Heisenberg, W. 1925. Über quantentheoretische Umdeutung kinematischer und mechanischer Beziehungen. *Z Physik* 33: 879 – 893.
Kant, I. 1997. *Metaphysische Anfangsgründe der Naturwissenschaft*, ed. Konstantin Pollok. Hamburg.
Kästner, A. G. 1766. *Anfangsgründe der höheren Mechanik*. Göttingen.
Lagrange, P. S. de. 1788. *Mécanique analytique*. Paris.
Leibnizens mathematische Schriften, ed. C. I. Gerhardt.
Leibniz, G. W. *Nova methodus pro maximis et minimis, itemque tangentibus, quae nec fractas nec rationales quantitatis moratur, et singulae pro illis calculis genus*. In GM V, 220.
Leibniz, G. W. *Brevis demonstratio erroris memorabilis Cartesii et aliorum circa legem naturalem*. GM VI, IV, 117.
Leibniz, G. W. 1982. *Monadologie*, ed. Herbert Herring. Hamburg. [Leibniz, Monadology (Latta)] Translation by Latta.
Leibniz, G. W. *Essay de Dynamique sur les loix du mouvement* In GM VI, XII.
Leibniz, G. W. 1982. *Specimen Dynamicum*, ed. H. G. Dosch. Hamburg. In GM VI, XV, 234 and GM VI, XVI, 246.
Mach, E. 1991. *Die Mechanik in ihrer Entwicklung*. Darmstadt: Wissenschaftliche Buchgesellschaft.
Maupertuis, P. L. M. de. Lois du repos de corps. In *Œuvres de Maupertuis*, vol. 4, 45–64. Lyon: Bruyset.
Maupertuis, P. L. M. de. Les lois du mouvement et du repos déduites d'un principe métaphysique. In *Œuvres de Maupertuis*, vol. 4,31–42. Lyon: Bruyset.
Nick, K. L. 2001. *Kontinentale Gegenmodelle zu Newtons Gravitationstheorie*. Frankfurt a. M.
Newton, I. 1740. *La Méthode des Fluxions et de suites infinies*. Paris.
Newton, I. 1988. *De gravitatione et aequipondio fluidorum* Klostermann Texte Philosophie and The Newton Project. http://www.newtonproject.sussex.ac.uk.
Newton, I. 1687. *Philosophiae naturalis principia mathematica*. London.

Newton, I. 1845. *The mathematical principles of natural philosophy*. New York.
Planck, M. 1913. *Vorlesungen über die Theorie der Wärmestrahlung*. Leipzig.
Schrödinger, E. 1926. Quantisierung als Eigenwertproblem. *Ann. Physik* 79: 361–376, *Ann. Physik* 79: 489–527, *Ann. Physik* 80: 437–490, *Ann. Physik* 80: 109–139.
Schrödinger, E. 1926. Über das Verhältnis der Heisenberg-Born-Jordanschen Quantenmechanik zu der meinen. *Ann. Physik* 79: 734–756.
Vorländer, K. *Geschichte der Philosophie*, Leipzig 1908.
Westfall, R. S. 1983. *Never at Rest*. Cambridge.
Westfall, R. S. 1993. *The life of Isaac Newton*. Cambridge.
Windelband, W. 1892. *Geschichte der Philosophie*. Freiburg i. B.
Wolff, Ch. 1734. *Vollständiges Mathematisches Lexicon*. Leipzig.
Wundt, W. 1921. *Erlebtes und Erkanntes*. Stuttgart.

Leibniz's Quantity of Force: A 'Heresy'? Emilie du Châtelet's *Institutions* in the Context of the *Vis Viva* Controversy

Andrea Reichenberger

The following article deals with the foundational programme of mechanics developed by Emilie du Châtelet in her work *Institutions* (1742) considering its historical background.

Today it is commonplace to differentiate between classical and modern physics to highlight the innovatory features of physics after 1900: the abandonment of the concepts of absolute space and time in Einstein's theory of relativity, and of causality and determinism in quantum mechanics.

These terms are not, however, very satisfactory as historical categories. Especially, it is not correct to identify 'classical' with 'Newtonian' physics. Newton's laws of motion were not perceived as straightforward and unproblematic by eighteenth century physicists. There was a diversity of opinion over the status and meaning of these mechanical laws. The critiques of the Newtonian concepts of space, time, matter and force should not be viewed as examples of a philosophical opposition to the emerging orthodoxy of classical mechanics, or as prefiguring non-classical developments, but as indicative of the diversity of eighteenth century systems of natural philosophy.

Many conceptual innovations of eighteenth century physics cannot meaningfully be described as Newtonian. For example, conservation laws did not play a central role in Newtonian thinking, at least not in that of Newton himself. However, the incorporation of conservation principles in mechanics played an important role in many physical treatises after Newton. This motivation also underlay the *vis viva* controversy, a dispute about the question whether the Cartesian quantity mv or the Leibnizian quantity mv^2 is the true measure of force.

From a modern point of view, the quarrel seems easy to explain. The Cartesian quantity of motion refers to what is now called momentum, the Leibnizian quantity of force to the kinetic energy of a moving body. Both quantities are conserved.

A. Reichenberger (✉)
Ruhr-University of Bochum, Department of Philosophy I, Universitätsstraße 150,
44801 Bochum, Germany
e-mail: Andrea.Reichenberger@ruhr-uni-bochum.de

In fact, it was not that simple. Far from being a 'logomachia', the *vis viva* controversy involved the confrontation with specific ontological presuppositions underlying the physico-philosophical theories of that time. In this context Du Châtelet came into play. She proposed a reformation of metaphysics as science offering a framework for the integration of Leibniz's *vis viva* theory into Newtonian mechanics.

The *Vis Viva* Controversy

The term *vis viva*, the Latin word for *living force*, was positioned in by Gottfried Wilhelm Leibniz, in opposition to the theory of the conservation of motion developed earlier by René Descartes.[1] Descartes had argued that the total quantity of motion (lat. *quantitas motus*) always remains constant. Further, he defined this quantity as the product of mass and velocity/speed (mv). Speed here is taken to be independent of direction; it is taken as skalar, not as a vector quantity. Leibniz rejected Descartes' quantity of motion as the true measure of motion. Using results obtained by Galileo Galilei, Christiaan Huygens and others, Leibniz claimed that the true measure of force was the product of mass and the square of the speed (mv^2), which is conserved when bodies fall, and which he called *vis viva*.

Reviewing the controversy from the standpoint of the late nineteenth century, Ernst Mach commented:

> We now know that both the Cartesian and the Leibnizian measure of the effectiveness of a body in motion have, each in a different sense, their justification. [...] The investigations of Newton really proved that for free material systems not acted on by external forces the Cartesian sum $\sum mv$ is a constant, and the investigations of Huygens showed that also the sum $\sum mv^2$ is a constant, provided work performed by forces does not alter it. The dispute raised by Leibniz rested, therefore, on various misunderstandings. It lasted 57 years, till the appearance of D'Alembert's Traité de dynamique, in 1743.[2]

Over hundred years after Mach John Roche comes to the same result:

> D'Alembert and Euler undermined the whole debate by showing that, during an arresting impact (sometimes modelled by compressing a spring), a contact force F is exerted for a certain distance ds, but *also* for a certain time dt. Mathematically, this was represented by the simultaneous equations $mv^2 = Fdx$ and $mv = Fdt$. This meant that both parties were wrong (with respect to a persisting impetus) and right (with respect to the mathematics of impact).[3]

In fact, the *vis viva* controversy was not a simple case. Neither Jean-Baptiste le Rond d'Alembert nor Leonhard Euler resolved the controversy, although both portrayed the struggle as a dispute about words. It was a widely-used manner of

---

[1] There are a lot of excellent articles and books about the *vis viva* controversy which I cannot discuss here. For further reading, see Diogenes. 1984. From *Vis Viva* to Primary Force in Matter; Hankins. 1965. Eighteenth-Century Attempts to Resolve the *Vis Viva* Controversy; Papineau. 1977. The *Vis Viva* Controversy; Terrall. 2004. *Vis Viva* Revisited.

[2] Mach. 1983. *The Science of Mechanics*. 366.

[3] Roche. 2006. What is Momentum. 1024.

speaking of that time to characterize the *vis viva* controversy as a semantic dispute. William s'Gravesande, Roger Boscovich, Joseph Saurin and Pierre Louis Moreau de Maupertuis spoke about a dispute about words as well. Daniel Bernoulli, to quote another example, supposed: 'An vero ista vis viva aëri inhaereat proprie externo an interno, logomachia est.'[4]

But if the discussions were merely skirmishes in a war of words, why took it so long to resolve the fundamental issues at stake? Although science policy, personal rivalries, national prejudices, alliances and careers played an important role, it is important to notice that the *vis viva* controversy was not just a pointless quibble over semantics and fuzzy definitions, but rather an examination of the ontological presuppositions underlying the metaphysics of the three dominant physico-philosophical theories of that time, the Cartesian, Leibnizian and Newtonian mechanics.

René Descartes' physics based on the central doctrine that the essence of matter (lat. *res extensa*) is three-dimensional spatial extension. Descartes assumed, e.g. in his *Principles of Philosophy* (1644), that the material realm contains nothing but matter in motion, that all action is by contact and that all of physics can be described geometrically.

Newton agreed with Leibniz that Descartes' explanation of the universe as extended matter in motion is insufficient, because extension cannot itself account for the activity found throughout nature. Further, this activity cannot be derived from corporeal motion, since motion either is impressed upon a body by another body, or originates from activity pre-existing in the body itself. Therefore, force must be a basic concept of mechanics, not a derivative phenomenal effect of the action of speed and size.[5]

There were profound differences in how scientists like Descartes, Newton and Leibniz viewed the structure of the world, differences which made it difficult for those who followed them to reach an agreement on the problems associated with contact between moving bodies:

> In retrospect, we can see that it [the dispute] involved a disagreement over the nature of matter if the ultimate particles of which matter is composed are completely inelastic hard atoms, *vis viva cannot be conserved* – but also depended on the fact that, despite Newton's work, clear distinctions among and definitions of various mechanical quantities such as impulse, momentum, work, power, and, above all, force, had not been established.[6]

The controversy came to a head when Newtonians intervened in the first decades of the eighteenth century. For example, Wilhelm Jacob's Gravesande, a famous Newtonian scholarship in Holland, did experiments in which he found out that if dents in clay were used to measure the force of a body in motion, the height from which a ball falls is as the square of the velocity acquired in falling and with which the body strikes the clay. Thus, force would have to be understood as *vis viva*. The

---

[4] Bernoulli. 1738. Hydrodynamica, sive de viribus et motibus fluidorum commentarii. 349.

[5] For Newton's criticism of Descartes see Newton's *De Gravitatione et Aequipondio*. The text is published in: Hall. 1962. *Unpublished Scientific Papers of Isaac Newton*. 89–156.

[6] Home. 2003. Mechanics and Experimental Physics. 361.

Italian physicist Giovanni Poleni conducted similar experiments. He dropped balls onto tallow to compare impacts. He demonstrated as well that the force of motion is proportional to the square of the velocity.

Counterarguments were presented by the Newtonians Henry Pemberton, John T. Desaguliers, John Eames, and Samuel Clarke in explaining the experiments from within the Newtonian framework attempted to restore the system to its established state by demonstrating the successful handling of mechanical problems using momentum considerations.[7]

A deeper look at the controversy shows that the common categorization between the Cartesian, Leibnizian and Newtonian school is oversimplified. Especially in France the situation was much more complicated. It is significant that those philosophers and scientists who defended mv^2 as the true measure of force were called «forceviviers».

In 1724, the Paris Academy of Science offered a prize for the best proposal discussing the laws of impact.[8] Johann I Bernoulli, a well known Swiss mathematician, submitted a paper in which he defended *vis viva*.[9] Bernoulli's essay «Discours sur les loix de la communication du mouvement» based on the hypothesis of the elasticity of matter. He maintained that absolutely hard bodies did not exist and went on to model the force of collision by analogy with the compression and release of springs.

Bernoulli's contribution was declined by the Jury because it had missed the point. The Academy's question about hard-body collisions were not answered, but rejected. Colin Maclaurin won the prize. In his contribution «Démonstration des loix du choc des corps» he argued for a mechanics based on Newton's laws of motion, with the force of bodies defined as mv.[10] In the following years the question of the measure of force became a controversial subject at the Paris Academy:

> The most animated debate took place in 1728 and 1729, when it turned out that Bernoulli had a significant number of allies within the institution who defended his model of elastic bodies and who promoted his distinctly Leibnizian mathematical methods for doing analytical mechanics. Meetings were filled with presentations going back and forth on the question, explicitly attacking and responding to each other.[11]

In April and May of 1728, Jean-Jacques Dortous de Mairan gave a lecture at the Academy in which he rebuilt a Cartesian position by reducing acceleration to uniform inertial motion. He hoped he had spoken the last word at the Academy and

---

[7] For further reading, see Boudri. 2002. *What was Mechanical about Mechanics*. 108; Hankins. 1965. Eighteenth-Century Attempts to Resolve the *Vis Viva* Controversy; Hankins. 1985. *Science and the Enlightenment*. 31; Hankins. 1990. *Jean D'Alembert*. 206; Schaffer. 1995. The Show that Never Ends. 180.

[8] The prize question called: «Quelles sont les loix suivants lesquelles un corps parfaitement dur, mis en mouvement, en meut un autre de même nature, soit en repos, soit en mouvement, qu'il rencontre, soit dans la vuide, soit dans la plein».

[9] Bernoulli's essay was first published in 1727. It also appeared in the *Receuil des pièces qui ont remporté les prix de l'Académie* (1732).

[10] The English translation 'Demonstration of the Laws of the Collision of Bodies' is published in: Tweddle. 2007. *MacLaurin's Physical Dissertations*.

[11] Terrall. 2004. Vis Viva Revisited. 195.

the question would be no longer at issue. But he was mistaken. There were several advocates of Bernuolli's position in the Academy. Mairan was well aware of that fact. For example, the eminent Academician Pierre-Louis Moreau de Maupertuis, a celebrated as well as controversial supporter of Newton's theory in France, had learnt of Leibniz's views on mechanics from his teacher Johann I Bernoulli. Interestingly, he did not attack Mairan, but remained silent. It was Du Châtelet who hazarded to criticize Mairan in public.

Emilie du Châtelet's Programme

In her work *Institutions* (1742) Du Châtelet reads, shortly speaking, Newton with Leibniz's glasses, and Leibniz with Wolff's glasses.[12] However, Du Châtelet's programme was not only an outcome of her life long struggle with the theories of Newton, Leibniz and others searching for a way to combine them. She developed an architecture of mechanics that was pursuing two aims:

(i) firstly, to guarantee the secure foundation of mechanics on the basis of the principle of contradiction and the principle of sufficient reason. The principle of contradiction states, one cannot say of something that it is and that it is not in the same respect and at the same time. The principle of sufficient reason states that nothing happens without a reason or a cause why it should be so rather than otherwise.
(ii) secondly, to offer a methodological framework for the construction of theories via hypotheses.

Du Châtelet compared the first task to the foundations of a building, the second task, i.e. setting up hypotheses, to the scaffolding of a building: «il est vrai que lorsque le Bâtiment est achevé, les Echafauts deviennent inutiles, mais on n'auroit pu l'élever sans leur socours».[13]

According to Du Châtelet, the building metaphor illustrates how science is organized in its theory and practise: first principles regulate every theory construction. Mathematics is founded on the principle of contradiction. At least, this single principle would be sufficient to demonstrate every part of arithmetic. But propositions about the contingent world of phenomena, including everything physics deals with, depend not only upon the principle of contradiction, but also upon the principle of sufficient reason from which other rational principles derive, such as the principle

---

[12] The book was first published anonymously in Paris in 1740. A second edition appeared in Amsterdam two years later (1742), with a modified title and a clear statement of authorship: *Institutions physiques de Madame la marquise du Châtelet adressés à son fils*. One year later, in 1743, the book was translated into Italian and German: *Istituzioni di Fisica di Madama la Marchesa du Chastelet indiritte a suo figlio*, respectively: *Der Frau Marquisinn von Du Chastelet Naturlehre an Ihren Sohn*. See also Hutton. 2004. Emilie du Châtelet's *Institutions de physique*; Iltis. 1977. Madame du Châtelet's Metaphysics and Mechanics; Locqueneux. 1995. Les *Institutions de physique* de Madame du Châtelet ou d'un traité de paix entre Descartes, Leibniz et Newton.

[13] 1742Rep1988. Chapter VIII.

of the identity of indiscernibles, the principle of continuity and the principle of the equivalence of cause and effect. To quote Du Châtelet:

> Dans la Géométrie où toutes les vérités sont nécessaires, on ne se sert que du principe de contradiction. [...] Vouz avez vu ci-dessus, que tout ce qui n'implique point contradiciton est possible; mais il n'est pas actuel. Il est possible, par exemple, que cette table qui est quarrée devienne ronde, cependant cela n'arrivera peut-être jamais; ainsi tout ce qui éxiste étant nécessairement possible, on peut conclure de l'éxistence à la possibilité, mais non pas de la posssibilité à l'éxistence.[14]

On the basis of the principle of sufficient reason Du Châtelet interpreted and modified Newton's axioms. This can be shown by a comparison between Newton's *Axiomata, sive leges motus*, formulated in the third edition of the *Philosophiae Naturalis Principia Mathematica* (1726), and Du Châtelet's reading of the three laws of motion in the *Institutions*:

> LEX I: Corpus omne perseverare in statu suo quiescendi vel movendi uniformiter in directum, nisi quatenus a viribus impressis cogitur statum illum mutare.
>
> LEX II: Mutationem motus proportionalem esse vi motrici impressae, et fieri secundum lineam rectam qua vis illa imprimitur.
>
> Lex III: Actioni contrariam semper et æqualem esse reactionem: sive corporum duorum actiones in se mutuo semper esse æquales et in partes contrarias dirigi.[15]

Du Châtelet formulates the three laws of motion as follows:[16]

> La force active & la force passive des Corps, se modifient dans leur choc, selon de certaines Loix que l'on peut réduire à trois principales.
>
> PREMIÈRE LOI: Un Corps persévère dans l'état où il se trouve, soit de repos, soit de mouvement, à moins que quelque cause ne le tire de son mouvement, ou de son repos.
>
> SECONDE LOI: Les changement qui arrive dans le mouvement d'un Corps, est toujours proportionel à la force mortice qui agit sur lui; car sans cela ce changement se seroit sans raison suffisante.
>
> TROISIEME LOI: La réaction est toujours égale à l'action; car un Corps ne pourroit agir sur un autre Corps, si cet autre Corps ne lui resistoit: ainsi, l'action & la réaction sont toujours égales & opposées.

Comparing Newton's Latin original with Du Châtelet's interpretation, the following differences are conspicuous:[17]

(i) Newton took his *Axiomata sive leges motus* as fundamental principles based on experimental philosophy in opposition to the metaphysical arguments of the Cartesians and Leibnizians. Within Du Châtelet's theory principles are part of metaphysical knowledge providing a foundation for the phenomenal laws of nature.

---

[14] 1742Rep1988, § 8f.

[15] Quoted from the critical edition of *Isaac Newton's Philosophiae naturalis principia mathematica*, edited by I. Bernard Cohen and Alexandre Koyré in 1972.

[16] 1742Rep1988. § 229.

[17] See also Suisky. 2009. *Euler as Physicist*. 14–16.

(ii) Before presenting the three laws of motion, Du Châtelet distinguishes between active and passive forces. Indeed, Newton himself identified the inertial force with innate force or a passive principle by which bodies persist in their motion or rest. Inertia only tends to preserve existing states of motion, and only uniform motion in a straight line. For every change in motion, Newton argued, an active principle is necessary, i.e. impressed force.

Active and passive forces are also important in Leibniz's dynamics. Like Newton, Leibniz defined the quantity of force in terms of the effect, or the potential for effect, a moving body possesses. However, Leibniz's concepts have a completely different meaning embedded in his metaphysics of monads. Active and passive forces are twofold: either primitive or derivative whereas derivative forces are modifications of the dominant primitive forces of a substance. By neglecting the differences between Newton's ontology of forces and Leibniz' metaphysics of monads, Du Châtelet focuses on the common ground of both theories, namely the distinction between active and passive principles.

(iii) In the first law, Du Châtelet replaces the word *force* by the word *cause*. She does not write that every body perseveres in its state of being at rest or of moving uniformly straight forward, except insofar as it is compelled to change its state by force impressed. She says that every body perseveres in its state of being at rest or of moving uniformly straight forward, except insofar as it is compelled to change its state by cause «à moins que quelque cause ne le tire de son mouvement, ou de son repos».

(iv) In the second law, Du Châtelet omits the word *impressed*. She does not maintain that the alteration of motion is ever proportional to the motive force impressed. She only speaks about motive force (frz. *force motrice*). Further, Du Châtelet adds the note that otherwise the change of motion of a body would happen without sufficient reason «car sans cela ce changement se seroit sans raison suffisante».

(v) Finally, Du Châtelet translates the third law without noticeable modifications. But it is worth to notice that she states an analogy to Leibniz's principle *causa aequat effectui* to consolidate the principle of conservation.

Du Châtelet's reinterpretation of Newton's axioms aims at solving a difficult problem of Newton's mechanics: the apparent lost of force. From the description of the motion of globes on the balance, Newton concluded that force, is lost. In his *Opticks* Newton reintroduced divine intervention in the core of physics. He argued that force was lost to the Universe and thus necessitated a Deity who would replenish that force periodically.

From a mechanical and rationalistic point of view any such suggestion of a need for divine intervention would cast doubt on the adequacy of the laws of motion by themselves to maintain the world machine in perfect working order. God can be the initial source of natural laws but never intervenes within the sequence of physical causes.

Du Châtelet believed that the problem of the dissipation of force could be solved insofar as the proportionality of force to the quantity of motion might be explicable in terms of Leibniz's *vis viva* theory. She tried to make clear that Leibniz's formula is in no case a physical heresy:

> Vous avez vu au Chap. XIII, qu'il est démontré par la théorie de Galilée que les espaces que la gravité fait parcourir aux corps qui tombent vers la terre, font comme les quarrés des vitesses: donc les force vives que les corps aquièrent en tombant, font aussi comme les quarrés de leurs vitesses, puisque ces forces font comme les espaces. Cette assertion parut d'abord une espèce d'Hérésie Physique. D'où viendroit ce quarré, disoit-on?[18]

In the last chapter of the *Institutions* Du Châtelet tried to defend the Leibnizian concept of living force as follows:[19]

> Toutes les expriénces ont confirmé cette découverte, dont on a l'obligation à Mr. De Leibnits; & elles on sait voir que dans tous les cas, la force des corps qui sont dans un mouvement actuel & fini, est proportionnelle au quarré de leur vitesse multiplé dans leur masse, & cette estimation des forces est devenue un des principes les plus séconds de la Méchanique.

Du Châtelet offers at least four arguments for Leibniz's *vis viva:*

(i) The Newtonians thesis that force is lost by friction contradicts the principle of conservation. The universe is not in need of an intervention by God preventing the world from running down:

> M. Newton conclut de cette conidération, & de cette de l'inertie de la matière que le mouvement va sans cesse en diminuant dans l'Univers; & que'ensin notre Systême aura besoin quelque jour d'être reformé par son Auteur.[20]

(ii) Leibniz's *vis viva* perfectly fulfils the principle of sufficient reason. Therefore, it is well founded on metaphysical ground: «Or, cette conservation des forces seroit une raison Métaphysique très forte».[21]

(iii) Thirdly, the concept of *vis viva* can mathematically expressed as an infinitesimal element of action, based on Leibniz's idea to apply the differential calculus to the analysis of physical phenomena.[22]

(iv) Last but ot least, Leibniz' *vis viva* can be empirically confirmed and correctly measured. Here, Du Châtelet refers to the experiments of Jacob Hermann and Jacob William s' Gravesande.[23]

Du Châtelet knew about the bygone and current debates very well. See e.g. Kawashima 1990. La participation de madame Du Châtelet à la querelle sur les forces vives; Walters 2001. La querelle des forces vives et le rôle de Mme du Châtelet. She had studied the main treatises about the *vis viva* and had discussed them with Maupertuis and others. Especially, the correspondence with Maupertuis motivated her

---

[18] See 1742Rep1988 § 567.

[19] 1742Rep1988 § 568.

[20] 1742Rep1988 § 586.

[21] 1742Rep1988 § 588.

[22] 1742Rep1988 § 561.

[23] 1742Rep1988 § 583f.

to go public with her objections to the Cartesian measure of motion. The result was a flak of the positions held by Jean Jacques d'Ortous de Mairan and James Jurin.

Du Châtelet argued that Mairan made two elementary mistakes in his *Dissertation sur l'estimation et la mesure des forces motrices des corps* (1728). First of all, he did not differentiate between the quantity of motion and the quantity of force. Secondly, he confused accelerated motion with uniform inertial motion.[24]

Du Châtelet's critique provoked Mairan. He responded with an open letter to Du Châtelet, and she answered back. Both letters were reprinted in the second edition of her book in 1742.[25] The dispute about *vis viva* burned up again within the academic circle and moved out into the *Republic of Letters*.

Mairan feared the expansion of the controversy to his disadvantage. By this time he was secretary of the Academy, and his standing in the scientific community was never been higher. In a letter of 14 April 1741 he wrote to Gabriel Cramer, the most prolific of Mairan's Geneva correspondents and protagonist of the *vis viva* conception, he had heard that Cramer intended to support Du Châtelet.[26]

After all, neither Cramer nor other colleagues, e.g. Maupertuis or Johann II Bernoulli, stood up for Du Châtelet in public.[27] However, in private correspondence they all took Du Châtelet's part. For instance, Cramer payed tribute to the Marquise in a letter to Alexis-Claude Clairaut:

> Comme l'ouvrage de Mad. du Chatelet ne me parviendra apparemment qu'un peu tard [...] faites moi la grace de me dire en un mot ce que c'est. Je m'imagine que ce sera sa dispute avec Mr de Mairan sur les forces vives. Mais peut-être y aura-t-il quelques additions. Cette illustre dame m'a fait l'honneur de m'envoyer par le P. Jacquier une copie de la rêponse qu'elle a faite à M[onsieu]r de Jurin sur le même sujet. Il me semble qu'elle a raison sur tous les points, et qu'elle le prouve avec beaucoup de force, de netteté et d'élégance.[28]

Besides Mairan the Newtonian James Jurin became a target of Du Châtelet's criticism. Jurin kept well informed of the ongoing debate. Samuel Clarke's paper contesting the views of Leibniz and Bernoulli was read at the Royal Society meeting of 29 February 1728, and Jurin later discussed it with John Byrom. Cramer wrote Jurin an extensive letter in January 1729, once again arguing for the case of *vis viva*. Cramer, who resided in London in the late 1720s and became acquainted with Jurin at that time, had visited 's Gravesande in Leyden on his return to Geneva. Jurin corresponded with Mclaurin, Poleni, Mairan, Voltaire, and Du Châtelet as well.[29]

---

[24] 1742Rep1988 § 574.

[25] Both letters were also published in the *Journal de Trávoux* (*Mémoires pour l'Histoire des Sciences & des Beaux-Arts*) (1741) and at the end of Du Châtelet's *Dissertation sur la nature et la propagation du feu* (1739).

[26] See McNiven Hine. 1996. *Jean-Jacques Dortous de Mairan*. 103.

[27] Only two persons took Mairan's side: Abbé Dedier in his work *Nouvelle réfutation de l'hypothèse des forces vives* (1741) and Voltaire, who was well known for his aversion against Leibnizian thought.

[28] Speziali. 1955. Une correspondance inédite entre Clairaut et Cramer. 220.

[29] See Rusnock. 1996. *The Correspondence of James Jurin* (1684–1750). 41.

Like Mairan, Jurin argued that the right measure of force of a moving body is defined as the product of its mass and velocity. But unlike Mairan, he did not refer to Descartes, but based his thesis on Newtonian mechanics. He claimed he could furnish an *experimentum crucis* to refute the 'Leibnizian doctrine' once and for all.

Suppose a plane moving in a straight line with a velocity of 1, e.g. a boat on the water. On the plane is a body of mass 1 fastened to a bend spring. The body acquiring its velocity from the moving plane has the force of 1. Being released the spring pushed the body in the same direction as the plane. Consequently, the total force of the body as well as its velocity is 2 because the force of a body is proportional to the mass multiplied by the simple velocity. Thus, the Leibnizian thesis, its moving force must be 4 must be false.[30] Quod erat demonstrandum?

Du Châtelet expressed strong objection to Jurin's demonstration. In her opinion its implicit assumption was wrong, namely that the elastic spring communicates to a body transported on a movable plane the same force as to a body located on an resting plane. Jurin neglected the recoil effect supplied to the moving plane or ship.[31]

From a modern point of view, Du Châtelet recognized that one must consider the energy exchange between the moving object and the moving plane, i.e. the transformation of energy from the form of potential energy to kinetic energy and vice versa.

But at Du Châtelet's time a clear distinction between the concepts of force, energy and momentum was not available. An 'objective' evaluation leading to the legitimacy of both interpretations mv and mv^2 was not open to her because there were deep disagreements about the nature of matter and force.

The question was: What exactly is conserved during an impact? Why different conclusions are drawn from same experiments? Du Châtelet was convinced that the disagreements could be resolved referring to undoubted metaphysical principles upon which all reasoning is founded. But she did not realized that the *vis viva* theory she argued for was undertermined by the principle of sufficient reason.

Suppose the principle of conservation is a true metaphysical principle and no physical theory can be true which neglects this principle. Even if this is correct, the principle of conservation says nothing about the question whether the Cartesian quantity mv or the Leibnizian quantity mv^2 is the 'true' measure of force.

Du Châtelet argued like other *forceviviers* that time has nothing to do with the measure of force. The force destroyed is always equal to the effect that it produces, regardless of the time in which it was produced. Indeed, from a mathematical point of view, squaring the velocity removed concern about direction of motion and made Leibniz's quantity of force dependent on distance traveled, not time. However, the measure of force acting through time is mathematically derivable, too.

A further problem was that Leibniz's *vis viva* did not hold for inelastic collisions, but only for perfectly elastic collisions or pendulums. Du Châtelet was aware of this fact. Interestingly, she did not conclude, that therefore the conservation of *vis viva* is falsified. Her argument was that although the experimental results for inelastic

---

[30] See Jurin. 1745. An Inquiry into the Measure of the Force of Bodies in Motion. Jurin's *experimentum crucis* is also discussed by Iltis 1977. Madame du Châtelet's Metaphysics and Mechanics.

[31] See 1742Rep1988, § 585.

collisions appear to violate the conservation of *vis viva*, the experimental results could be false.[32] But it was an open question if there were false. Here, the problem of underdetermination of a theory appears again.

From the *Vis Viva* Controversy to the Principle of Least Action

Searching for a unified theory of mechanics is a stimulus for physical research until today. But perhaps no other principle of physics has to a larger extent nourished exalted hopes into a universal theory, and has ignited metaphysical controversies about causality and teleology than did the principle of least action. For further reading, see Pulte 1987. *Das Prinzip der kleinsten Wirkung und die Kraftkonzeptionen der rationalen Mechanik*; Stöltzner and Weingartner 2005. *Formale Teleologie und Kausalität in der Physik*.

The eighteenth century history of the principle of least action is deeply connected with the names of Pierre-Louis Moreau de Maupertuis and Leonhard Euler. Maupertuis formulated the principle of least action as a metaphysical principle from which all mechanical laws can be derived. It states that in all natural phenomena, a quantity called 'action' – for Maupertuis, the product of mass, distance travelled and velocity – tends to be minimized.[33] Maupertuis even combined his economic law of nature with an attempted proof of the existence of God.

Independently of Maupertuis, Euler developed a formulation of the principle of least action in the same year which is applied here for the first time in perfect form as a variational principle.[34] Euler did not share Maupertuis' metaphysical enthusiasm which blinded him to important difficulties.

Firstly, it remained to be shown how to express Maupertuis' «principe de la moindre action» in the form of a general equation. Secondly, Maupertuis did not see the problem that not only minima, but other types of extremal points (i.e. saddle points and maxima) need to be admitted as well.

Both, Maupertuis and Euler, were dissatisfied with the standard options within the *vis viva* controversy because neither the Cartesian *quantitas motus* nor the Leibnizian *vis viva* were able to applied in every relevant circumstances. Maupertuis and Euler were convinced that they were able to present an acceptable solution both to the representatives of the momentum view, i.e. the Cartesians, and to the supporters of living forces, i.e. the Leibnizians. Both quantities, *quantitas motus* and *vis viva*, can from a more fundamental principle, the principle of least action.[35]

---

[32] See 1742 Rep1988, § 589f.

[33] See Maupertuis. 1744. Accord de différentes lois de la nature qui avaient jusqu'ici paru incompatibles.

[34] See Euler. 1744. Methodus inveniendi lineas curvas maximi minimive proprietat gaudentes.

[35] Ironically, this view was the reason that Maupertuis rejected Newton's second law as a principle or axiom as well as the forces underlying it, and that Euler did not realize that Newton's second law of motion can be derived from the principle of least action via the conservation principle of *vis viva* See Schramm. 2005. The Creation of the Principle of Least Action.

It is not my aim to go into details here, nor to deal with the bitter controversy about the priority for the principle of least action at the Prussian Academy of Sciences initiated by Samuel König in 1751 who claimed that Maupertuis' principle had been invented by Leibniz. For our purpose it is worth to add a footnote to the history of the principle of least action referring to Du Châtelet.

The correspondence between Euler and Maupertuis started in 1738. In the same year Du Châtelet began to discuss the puzzling case of *vis viva* with Maupertuis. She considered Bernoulli's contribution to the prize question of the Paris Academy in 1724 as a cogent argumentation for *vis viva*, and did not accept that Mairan regarded the issue as settled by his own paper of 1728. Asking Maupertuis for a clear statement she was not confident with his answer the *vis viva* controversy would be a useless dispute about words. At Du Châtelet's urging Maupertuis finally agreed that Bernoulli's arguments in his *Discours* were correct. Up to this time, Maupertuis had never expressed an opinion about living forces. It was Du Châtelet's interest in the question which forced Maupertuis to deliberate it too.

At the same time, Euler heard about Du Châtelet. In 1737, both had participated in the prize question of the Paris Academy about the nature and properties of fire. Du Châtelet's *Dissertation sur la nature et la propagation du feu* was published in 1739 together with Euler's contribution *Dissertation sur le feu, sur sa nature et ses propriétés* which had won the prize.[36]

Euler read Du Châtelet's essay about fire with great interest, as he told Maupertuis in 1740. Further, Euler knew both editions of Du Châtelet's *Institutions* very well. In a letter to Du Châtelet (eventually written on 30 May 1744) Euler commented on her book at length. He praised her hypotheses chapter which would show that scientific hypotheses are necessary and fruitful for further research. It would be misleading to ban hypotheses from science as propagated by the «Philosophes Anglois». Further, Euler agreed with Du Châtelet in her critique of Mairan's ill-founded defence of *mv*. Euler added he regrets the aggressive tone with which the *vis viva* controversy would held up. Then he says:

> Je commence par le premier principe de la Mecanique que tout corps par lui même demeure dans son état ou de repos ou de mouvement. A cette propriété on peut bien donner le nom de force, quand on ne dit pas que toute force est une tendance de changer l'état, comme fait Mr. Wolf. Tout corps est donc pourvu […].[37]

Unfortunately, the letter ends at this point. Unfortunately, too little is known about Du Châtelet's correspondence with Maupertuis and Euler. Maupertuis' letters do not survive, so his voices must be extrapolated from Du Châtelet's comments and questions.

Regarding to the correspondence between Du Châtelet and Euler, only two letters from Euler to Du Châtelet are published in *Письма к ученым*. *Pis'ma*

---

[36] For further reading, see Hentschel. 2005. Die Pariser Preisschriften; Joly. 2001. Les théories du feu.

[37] Euler. 1963. Письма к ученым. Pis'ma k učeny. 279.

k učeny (1963), as far as I know. Are there further letters at Saint Petersburg Academy of Sciences? Did Du Châtelet discuss the principle of least action in her *Essai sur l'optique* of which allegedly only *a fragment* is preserved?[38]

Conclusion

Du Châtelet made an important contribution to bring the problem of quantity of motion and force to the attention of a wide variety of philosophers and physicists. Her intended reformation of «metaphysics as science»[39] throw new light on the role of principles in mechanics.

Du Châtelet's programme of mechanics can be characterized as an architectonic system following the idea that all of physics is founded on the principle of contradiction and the principle of sufficient reason. It was influenced by Christian Wolff's methodological considerations who was said to have methodized the thoughts of his great predecessor Gottfried Wilhelm Leibniz.[40]

References

Bernoulli, D. 1738. Hydrodynamica, sive de viribus et motibus fluidorum commentarii. In *Die Werke von Daniel Bernoulli*, vol. 5, ed. David Speiser and Patricia Radelet-de Grave, 93–424. Basel: Birkhäuser 2002.

---

[38] The fragment is published in Wade 1947. *Studies on Voltaire*. 188–208. Until today, Theodore Besterman's edition *Les Lettres de la marquise du Châtelet* (1958) is the critical standard edition. Ulla Kölving, editor of the book *Émilie du Châtelet. Éclairages et documents nouveaux* (2008), will publish a new edition of Du Châtelet's correspondence. Furthermore, Du Châtelet's letters and manuscripts in the Voltaire Collection, St. Petersburg, National Library of Russia, are being prepared by Ruth Hagengruber with assistence of Ana Rodrigues et al. to be published. Last, but not least, Du Châtelet's *Essai sur l'optique* which Fritz Nagel found in the Bernoulli-Archive of the University of Basel library in 2006, will be published by Nagel together with Sulamith Gehr. We shall not forget to mention: Zinsser and Hayes 2006. *Emilie Du Châtelet: Rewriting Enlightenment Philosophy and Science*; Zinsser and Bour 2009. *Emilie Du Châtelet. Selected Philosophical and Scientific Writings*. I am sure these and further results requires further research on Du Châtelet's contributions to philosophy and science.

[39] See Gardiner 1982. Searching for the Metaphysics of Science.

[40] For further reading, see Böttcher 2008. La réception des *Institutions de physique* en Allemagne; Gireau-Geneaux 2001. Mme du Châtelet entre Leibniz et Newton; Rey 2008. La figure du leibnizianisme dans les Institutions de physique.] Nevertheless, Du Châtelet's *Institutions* were more than a translation of Wollfian ideas for a French audience. Du Châtelet gave Newtonian mechanics a rationalistic interpretation and foundation, a re-definition of basic concepts (space, time, matter, force) and a fundamental modification of the ontology underlying Newton's *philosophia naturlis*. Insofar, her work was a catalyst for the integration of the principles of conservation of momentum and energy into classical mechanics.

Bernoulli, J. 1732. Discours sur les loix de la communication du mouvement. *Recueil des pièces qui ont remporté le prix de l'Académie Royale des Sciences*, vol. 2, 1–108.

Böttcher, F. 2008. La réception des *Institutions de physique* en Allemagne. In *Émilie du Châtelet: Éclairages & documents nouveaux*, ed. Ulla Kölving and Olivier Courcelle, 243–254. Ferney-Voltaire: Centre International d'Étude du XVIIIe Siècle.

Bourdi, C. J. 2002. *What Was Mechanical About Mechanics? The Concept of Force between Metaphysics and Mechanics from Newton to Lagrange*. Dordrecht: Kluwer.

Deidier, l'abbé. 1741. *Nouvelle réfutation de l'hypothèse des forces vives*. Paris: Ch. A. Jombert.

Descartes, R. 1644. *Principles of Philosophy*, trans., V. R. Miller and R. P. Miller. Dordrecht: Kluwer Academic Publishers 1983.

Diogenes, A. 1984. From *Vis Viva* to Primary Force in Matter. In *Leibniz' Dynamica. Symposion der Leibniz-Gesellschaft in der Evangelischen Akademie LOCCUM*, ed. Albert Heinekamp, 55–61. Franz Steiner: Stuttgart.

Euler, L. 1744. Methodus inveniendi lineas curvas maximi minimive proprietate gaudentes sive solutio problematis isoperimetrici latissimo sens accepti, Additamentum II. In *Leonhardi Euleri Opera Omnia sub auspiciis Societatis Scientarium Naturalium Herlveticae*, Ser. 1, Opera mathematica, Bd. 24, 298–308. Basel: Birkhäuser.

Euler, L. 1963. Письма к ученым. *Pis'ma k učeny (Briefe an Gelehrte)*. Moskau/Leningrad: Izdat. Akad. Nauk SSSR.

Gardiner, L. J. 1982. Searching for the Metaphysics of Science: The Structure and Composition of Madame du Châtelet's *Institutions de physique*, 1737–40. *SVEC* 201: 85–113.

Girau-Geneaux, A. 2001. Mme du Châtelet entre Leibniz et Newton. In *Cirey dans la vie intellectuelle: la réception de Newton en France*, ed. François de Gandt, 173–86. Oxford: Voltaire Foundation.

Hall, A.R., and M. B. Hall. 1962. *Unpublished Scientific Papers of Isaac Newton*. Cambridge: Cambridge University Press.

Hankins, T. 1965. Eighteenth-Century Attempts to Resolve the *Vis Viva* Controversy. *ISIS* 56: 281–297.

Hankins, T. L. 1985. *Science and the Enlightenment*. Cambridge: Cambridge University Press.

Hankins, T. L. 1990. *Jean D'Alembert: Science and the Enlightenment*. London: Taylor & Francis.

Hentschel, K. 2005. Die Pariser Preisschriften Voltaires und der Marquise du Châtelet von 1738 über die Natur und Ausbreitung des Feuers. In: *Physica et historia. Festschrift für Andreas Kleinert zum 65. Geburtstag*, ed. Susan Splinter, Sybille Gerstengarbe, and Horst Remane, 175–186, Halle: Leopoldina.

Home, R. W. 2003. Mechanics and Experimental Physics. In *The Cambridge History of Science: Eighteenth-century science*, ed. Roy Porter, 354–374. Cambridge: Cambridge University Press.

Hutton, S. 2004. Emilie du Châtelet's *Institutions de physique* as a Document in the History of French Newtonianism. *Studies in history and philosophy of science* 35 A, no. 3: 515–31.

Iltis, C. 1977. Madame du Châtelet's Metaphysics and Mechanics. *Studies in History and Philosophy of Science* 8, no. 1: 29–48.

Joly, B. 2001. Les théories du feu de Voltaire et madame du Châtelet. In *Cirey dans la vie intellectuelle: la réception de Newton en France*, ed. François de Gandt, 212–238. Oxford: Voltaire Foundation.

Jurin, J. 1745. An Inquiry into the Measure of the Force of Bodies in Motion: With a Proposal of an Experimentum Crucis, to Decide the Controversy about It. *Philosophical Transactions* 43: 423–440.

Kawashima, K. 1990. La participation de madame du Châtelet à la querelle sur les forces vives. *Historia scientiarum. International Journal of the History of Science Society of Japan* 40: 9–28.

Kölving, U., and O. Courcelle. 2008. *Émilie du Châtelet: Éclairages & documents nouveaux*. [... du Colloque du Tricentenaire de la Naissance d'Emilie du Châtelet qui s'est tenu du 1er au 3 juin 2006 à la Bibliothèque Nationale de France et à l'ancienne Mairie de Sceaux]. Ferney-Voltaire: Centre International d'Étude du XVIIIe Siècle.

Locqueneux, R. 1995. Les *Institutions de physique* de Madame du Châtelet ou d'un traité de paix entre Descartes, Leibniz et Newton. *Revue du Nord* 77, no. 312: 859–92.

Mach, E. 1883. *The Science of Mechanics: a Critical and Historical Account of its Development [= Die Mechanik in ihrer Entwicklung. Historisch-kritisch dargestellt*, dt. Orig.], translated by Thomas Joseph McCormack. Chicago: The Open court publishing.

Mairan, J. J. d'O. de. 1728. Dissertation sur l' estimation et la mesure des forces motrices des corps. *Mémoires de l'Académie Royale des Sciences:* 1–50.

Maupertuis, P. L. M. de. 1744. Accord de différentes lois de la Nature qui avaient jusqu'ici paru incompatibles. *Histore Acad. Sci. Paris:* 417–426.

Mclaurin, C. 2007. Demonstration of the Laws of the Collision of Bodies. In *MacLaurin's Physical Dissertations*, ed. Ian Tweddle, 55–68. London: Springer.

McNiven Hine, E. 1996. *Jean-Jacques Dortous de Mairan and the Geneva Connection*. Voltaire Foundation: Oxford.

Newton, I. 1972. *Isaac Newton's Philosophiae naturalis principia mathematica*. 2 vols., ed. I. Bernard Cohen and Alexander Koyré, Cambridge: Cambridge University Press.

Papineau, D. 1977. The *Vis Viva* Controversy: Do Meanings Matter? *Studies in History and Philosophy of Science* 8, no. 1: 111–142.

Pulte, H. 1987. *Das Prinzip der kleinsten Wirkung und die Kraftkonzeptionen der rationalen Mechanik: eine Untersuchung zur Grundlagensproblematik bei Leonhard Euler, Pierre Louis Moreau de Maupertuis und Joseph Louis Lagrange*. Stuttgart: Steiner.

Rey, A.-L. 2008. La figure du leibnizianisme dans les *Institutions de physique*. In *Émilie Du Châtelet: Éclairages & documents nouveaux*, ed. Ulla Kölving and Olivier Courcelle, 231–242. Ferney-Voltaire: Centre International d'Étude du XVIIIe Siècle.

Roche, J. 2006. What is Momentum? *European Journal of Physics* 27: 1019–1036.

Rusnock, A. A. 1996. *The Correspondence of James Jurin (1684–1750): Physician and Secretary to the Royal Society*. Rodopoi: Amsterdam.

Schaffer, S. 1995. The Show that Never Ends: Perpetual Motion in the Early Eighteenth Century. In *British Journal for the History of Science* 28: 157–189.

Schramm, M. 2005. The Creation of the Principle of Least Action. In *Formale Teleologie und Kausalität in der Physik*, ed. Michael Stöltzner and Paul Weingartner, 99–114. Paderborn: Mentis.

Speziali, P. 1955. Une correspondance inédite entre Clairaut et Cramer. *Revue d'histoire des sciences et de leurs applications* 8, no 3: 193–237.

Stöltzner, M., and P. Weingartner. 2005. *Formale Teleologie und Kausalität in der Physik*. Paderborn: Mentis.

Suisky, D. 2009. *Euler as Physicist*. Berlin et al.: Springer.

Terrall, M. 2004. Vis Viva Revisited. *History of Science* 42, no. 2: 189–209.

Wade, I. O. ed. 1947. *Studies on Voltaire. With Some Unpublished Papers of Mme du Châtelet*. Princeton: Princeton University Press.

Walters, R. L. 2001. La querelle des forces vives et le rôle de Mme du Châtelet. In *Cirey dans la vie intellectuelle: la réception de Newton en France*, ed. François de Gandt, 198–211. SVEC 11. Oxford: Voltaire Foundation.

Zinsser, J. P., and J. C. Hayes, eds. 2006. *Emilie Du Châtelet: rewriting Enlightenment philosophy and science*. Oxford: Voltaire Foundation.

Zinsser, J. P., and I. Bour, eds. 2009. *Emilie Du Châtelet. Selected Philosophical and Scientific Writings*. Chicago: Chicago University Press.

From Translation to Philosophical Discourse – Emilie du Châtelet's Commentaries on Newton and Leibniz*

Ursula Winter

> It would hardly be an exaggeration to call 28 April 1686 – the day Newton submitted his Principia to the Royal Society in London – as one of the great days in the history of mankind. [...] Arguably the Third Book entitled "The System of the World" which contained the universal law of gravity made the most profound impression.[1]

It is with this claim regarding the fundamental and paradigmatic influence of Newton's physics that Nobel prize winner Ilya Prigogine introduces his *Dialog mit der Natur* in 1980, a study which has the subtitle *Neue Wege naturwissenschaftlichen Denkens* ('*New paths in scientific thought*') and is dedicated to the most defining paradigms in the history of the natural sciences. Newton's discoveries, he claims, seemed to have single-handedly solved the "code of nature" by means of one most influential and and creative act.[2] It was not until Einstein's discoveries in the early twentieth century, Heisenberg similarly argues, that Newton's theories came to lose some of their groundbreaking influence in the field of physics.[3]

---

*To my son Sascha Alexander Winter with thanks for supporting me in my research work on Du Châtelet.

[1] Prigogine and Stengers. 1983. *Dialog mit der Natur. Neue Wege naturwissenschaftlichen Denkens*. 9. (my emphasis – U.W.). "Es ist kaum übertrieben, wenn man den 28. April 1686, an dem Newton seine *Principia* der Royal Society in London vorlegte, als einen der größten Tage in der Geschichte der Menschheit bezeichnet. [...] Den größten Eindruck machte wohl das Buch III der *Principia*, "The System of the World", welches das universelle Gesetz der Gravitation enthielt." For a more in-depth discussion of Newton's thoughts see Koyré. 1957. *From the Closed World to the Infinite Universe,* and Freudenthal. 1986. *Atom and Individual in the Age of Newton: On the Genesis of the Mechanistic World View*. 88.

[2] Prigogine and Stengers. 1983. *Dialog mit der Natur*. 36–37.

[3] See Heisenberg. 1959. *Physik und Philosophie*. 72.

U. Winter (✉)
TU Berlin, Institut für Philosophie, Literatur-Wissenschafts-und Technikgeschichte,
Straße des 17. Juni 135, 10623 Berlin. Privat: Dannenwalder Weg 92,
13439 Berlin, Germany
e-mail: ursulawinterdr@aol.com

Two Theories of Equal Value

In their paradigmatic explanation of cosmic structure, Newton's *Principia* and Leibniz's ideas of an energetic and dynamic universe, however, prove to be of equal importance to the scientific and philosophical discourses of the Enlightenment. For a long time, Emilie du Châtelet's French translation and her commentary on Newton's *Philosophiae naturalis principia mathematica* provided the only access to Newton's principles of natural philosophy in French[4] and, together with her commentary on Leibniz's philosophy in the *Institutions de physique*, it thus greatly influenced scientific discourse in the eighteenth century and can be said to have furthered both transnational and European scientific collaboration. In fact, Du Châtelet realized at an early stage the ground breaking potential of both Newton's and Leibniz's ideas on natural philosophy. Yet she also succeeded in presenting and commenting on their innovative approach in a most convincing manner, regardless of the fact that their ideas partly contradicted the official opinion of the "République des Savants". Today, a great number of Leibniz's ideas have proved to be of visionary nature even when seen in the light of more recent scientific developments, and renowned scholars have stated an affinity of his theories on space and time with Einstein's space and time theorem.[5] Leibniz's concepts of least action and the conservation of momentum remain valid in both mathematics and physics,[6] and his concept of a binary numeral system has proved to be essential for disciplines such as computer science[7] and contemporary communication systems. However, it was not until recently and in view of the latest developments in current mathematics that the full range of Leibniz's

---

[4] Du Châtelet's translation of Newton's work was republished in two volumes by Jacques Gabay in Sceaux in 1990. Also see Zinsser. 2001. Translating Newton's Principia: The Marquise du Châtelet's Revisions and Additions for a French Audience. 227–245, and Biarnais. 1982. *Les Principia de Newton. Genèse et structure des chapitres fondamentaux avec traduction nouvelle*; also compare Winter. 2007. Übersetzungsdiskurse der französischen Aufklärung: Die Newton-Übersetzung Emilie du Châtelets, as part of "Aufklärung und Moderne", t.12. 19–36; see also Guicciardini. 1999. *Reading the Principia: The debate on Newton's mathematical methods for natural philosophy from 1687 to 1736*.

[5] See Poser. 1989. Leibnizens Theorie der Relativität von Raum und Zeit. 123–138. Also see Mainzer. 1996. Der Krieg der Philosophen. Zum Verhältnis von Physik, Philosophie und Religion bei Leibniz bis zur Aufklärung. 72–95; compare also Freudenthal. 1986. *Atom and Individual in the Age of Newton: On the Genesis of the Mechanistic World View*. 7.

[6] See Mainzer. 1988. *Symmetrien der Natur. Ein Handbuch zur Natur- und Wissenschaftsphilosophie*. Mainzer considers Noether's theory to be rooted in Leibniz's thoughts. Noether's theory, in turn, seems to present a "kategoriale Rahmenbedingung der Physik" in Kant's sense of the Principle of the Permanence of Substance, which here has to be understood as a starting point for all kinds of theoretical research in physics. See Falkenburg. 1989. Kants Einwände gegen Symmetrieargumente bei Leibniz. 148–180.

[7] Cf. Zacher. 1973. *Die Hauptschriften zur Dyadik von G. W. Leibniz. Ein Beitrag zur Geschichte des binären Zahlensystems*; also see Mainzer. 2003. *KI. Künstliche Intelligenz. Grundlagen intelligenter Systeme*.

mathematical innovations have come to be acknowledged fully.[8] The term of "force vive", which was defended by Emilie du Châtelet against several well-respected scholars[9] in a notorious dispute towards the end of the eighteenth century, has been in use as an operative term in kinetic physical theories since the nineteenth century. According to Du Châtelet, both Leibniz's particular notion of force and Newton's concept of gravity are equally important for an in-depth understanding of nature and the secrets of its creation. In a letter to the physicist Maupertuis, she consequently states: "Enfin il les a découvertes [les forces vives, U.W.], et c'est avoir *deviné un des secrets du créateur.*"[10]

Both Du Châtelet's translations and her comments on Newton's and Leibniz's work greatly contributed towards one of the most influential paradigmatic changes within the European *Republique des Savants* during the eighteenth century, a paradigmatic change which furthermore transcended national boundaries and has remained highly influential to this day. Following Kuhn, paradigm is here understood as an elaborate concept which provides a valid referential framework for subsequent scientific theories and experimental research.[11]

Contrary to the predominant academic opinion of the Enlightenment, Du Châtelet believed that Newton's and Leibniz's theories were not to be seen as two mutually exclusive concepts. Instead, as she could show in her *Institutions de physique*, she argued that Newton's theories on gravity and Leibniz's ideas on energetic structures of the universe could be seen as complementary. Du Châtelet's censor Pitot, who had to approve the *Institutions de physique* in 1738, accordingly emphasizes her balanced presentation of the two concepts.[12] And indeed, even from a contemporary perspective, Newton's and Leibniz's philosophical concepts successfully introduced "two theories of equal value" for a new paradigm in natural philosophy at the time of the Enlightenment.

> The possibility of proposing different theories of the same subject on the basis of empirical research is grounded in the fact that a scientific theory does not consist merely in ascertaining supposedly theory-independent "facts of experience". […] the investigation of the same phenomena could support *the formulation of different theories of equal value.*[13]

In his analysis of both Leibniz's and Newton's approach, Freudenthal points out that "two theories of equal value" may indeed be formulated when founded on

---

[8] See Eberhard Knobloch and Siegmund Probst. 2003. Vorwort. series 7: *Mathematische Schriften*, vol. 3, XXII.

[9] See MaiCh1741; compare Terrall. 2004. Vis viva. 189–209. See also Walters. 2001. La querelle des forces vives et le rôle de Mme du Châtelet. 198–211.

[10] LetChBI 217, to Maupertuis, 10 February 1738 (my emphasis - U.W.).

[11] Kuhn. 1993. *Die Struktur wissenschaftlicher Revolutionen.*

[12] "Approbation". Inst1740.

[13] Freudenthal. 1986. *Atom and Individual in the Age of Newton: On the Genesis of the Mechanistic World View*. 2–3. (my emphasis – U.W.). Here, Freudenthal explicitly refers to Newton's and Leibniz's thoughts.

experimental facts then available at the time of their occurrence. Freudenthal, however, turns out to be rather more positive towards the methodological value of Leibniz's theorem.[14] This coincides with Emilie du Châtelet's view, as will be shown in the following.

For the context of this article, it is necessary to point out that the traditional view of Du Châtelet's high regard for Leibniz and of her later dismissal of his ideas in favour of Newton's work towards the end of her life is somewhat inaccurate. Contrary to common scientific opinion, Du Châtelet did not abandon Leibniz's ideas nor did she consider them premature. Rather, it was Voltaire[15] who seems to have been responsible for such a misleading interpretation. In his popular historical preface to Du Châtelet's posthumously published translation of Newton's *Principia* in 1759, he claimed:

> Apres avoir rendu les imaginations de Leibnitz intelligibles, son esprit qui avoit acquis encore de la force & de la maturité par ce travail même, comprit que cette métaphysique si hardie, mais si peu fondée, ne méritoit pas ses recherches. [...] Ainsi, après avoir eu le courage d'embellir Léibnitz, elle eut celui de l'abandonner: courage bien rare dans quiconque a embrassé une opinion, [...] ennemie des parties & des systèmes, qu'elle se donna toute entière à Newton.[16]

Besterman's collection of Du Châtelet's correspondence documents that Voltaire's assumption regarding Du Châtelet's progression from Leibniz to Newton and the later dismissal of the theories of the former was by no means accurate. On the contrary, Du Châtelet's correspondence rather proves that she was equally interested in both Leibniz's and Newton's ideas and texts from as early as 1738 (as far as they were available to her). In 1738 she told Bernoulli that she had come across and already examined Leibniz's writings for the *Acta Eruditorum*: "J'ai lu depuis que je vous ai écrit ce que mr de Leibnits a donné dans les acta Eruditorum sur les forces vives, et j'y ai vu qu'il distinguait entre la quantité du mouvement, et la quantité des forces."[17] In another letter dated 30th March 1739, she also expressed her interest in further and unpublished instalments of Leibniz's work and – to that end – asked Bernoulli to send her his father's and Leibniz's correspondence, which at that point had not even been published.

> Je vous prie monsieur de faire un peu ma cour à mr votre père. Je voudrais obtenir une grace de lui, ce serait la communication du *Commercium Epistolicum* qu'il avait prêté à mr Koenig. Il pourrait être sûr de la plus grande fidélité, et de la plus grande exactitude.[18]

---

[14] Ibid., 4: "The following chapter will show that the theory of Leibniz (which did not prevail) is not merely of equal value but also has some advantages."

[15] Voltaire's ironical polemics against Leibniz's thoughts and Du Châtelet's philosophical disagreements with Voltaire regarding Leibniz's philosophy are of course well known. See Belaval. 1993. Quand Voltaire rencontre Leibniz. 242–246.

[16] PrincChat1990 vol I, VI.

[17] LetChBI 220, to Maupertuis, 30 April 1738.

[18] LetChBI 352, to Bernoulli, 30 March 1739. Besterman points out that this is the notorious correspondence named *Virorum celeberr. Got. Gul. Leibnitii et Johan, Bernoullii commercium philosophicum et mathematicum,* which was published in Lausanne in 1745.

In April 1739 Emilie du Châtelet confirmed that she had received the named correspondence and informed Bernoulli of her intention to work through the letters as quickly as possible and to return them right away: "à vous remercier du *Comercium Epistolicum* de mr votre père et de mr de Montmort." At the same time she told Prault of her continuing research in relation to Newton's work: "J'ai l'optique de Neuton",[19] and in 1739 – one year prior to her own publication of the *Institutions de physique* – she also informed Bernoulli that she had finished writing a separate study on optics based on Newton's work.[20] As her correspondence with Jacquier shows, she had also planned to publish a further study on astronomy based on Newton's ideas. As with the correspondence between Leibniz and Bernoulli, Du Châtelet was again provided with hitherto unpublished scientific material.[21]

> Les *Eléments d'algèbre* de m. Cleraut vont paraître. [...] On imprime aussi la traduction de Keills de Monier. C'est son traité d'astronomie. *Cet ouvrage m'a fait suspendre celui que vous savez que je méditais sur cette matière.* Je le lis actuellement, il me l'a prêté en feuilles, quoi qu'il ne paraisse pas encore.[22]

At the time, Du Châtelet was deeply engaged with both Newton and Leibniz and, in addition to the *Institutions de physique*, she also made some written drafts of her own research into optics and astronomy.

Du Châtelet's correspondence definitely falsifies the claim Voltaire made in his introduction to the *Principia*. As opposed to Voltaire's remarks regarding her later abandonment of Leibniz's ideas, Du Châtelet's correspondence proves that she supported Leibniz's philosophy right to the end of her life – including his controversial theory on monads. Shortly before her death in 1746, she mentioned to Bernoulli that she had originally planned to present a paper on Leibniz's monads at the Berlin Academy, which was supposed to defend his thoughts. She regretted that she had to abandon her plans due to the immense work load she was facing in translating Newton's *Principia*. "Je serais bien tentée de défendre les monades",[23] she wrote on 6 July. It would be wrong, therefore, to assume that Du Châtelet abandoned Leibniz's ideas. On the contrary, in 1746 she specifically asked Bernoulli to present a paper on condition that his paper was in favour of the monad theory. Her request specifically stated "*[q]ue ce fût pour les* [the monads; U.W.] *défendre*",[24] which again

---

[19] LetChBI 329, to Prault, 16 Feb 1739.

[20] LetChBI 375, to Bernoulli, 3 August 1739.

[21] This, of course, bears witness to the fact that she treated these manuscripts with great care and also that she possessed a good reputation amongst the scholars of the "République des Savants", who repeatedly used to provide her with valuable manuscripts for her work. Thus she thoughtfully asks Bernoulli to confirm the receipt of his valued *Commercium epistolicum:* "Je sais comment ce manuscrit vous est cher, et je connais combien il vous le doit être, ainsi je ne serais point tranquille si je ne sache que vous l'avez à présent." LetChBI 375, to Bernoulli, 3 August 1739.

[22] LetChBII 143, to Jacquier, 12 November 1745 (my emphasis – U.W.). After reading Le Monnier's translation of the "traité d'astronomie", Emilie du Châtelet had abandoned her own work towards a publication on astronomy.

[23] LetChBII 152, to Bernoulli, 6 July 1746.

[24] LetChBII 152 (my emphasis – U. W.).

provides ample proof that Du Châtelet was always convinced that some of both Leibniz's theorems and Newton's theory of gravity were essential paradigms in the fields of natural science.[25]

Newton's System of the World

As was the case with many scientific books of the time, as well as physical theory and calculations, Newton's *Principia* also presented an entirely new view of the world.

> With these words Newton formulates the fundamental step forward in the history of science represented by the *Principia*: it is a systematic presentation of dynamics, and from its laws he derives the "system of the world."[26]

As early as 1728, Pemberton, who was Newton's publisher at the time, saw Newton's *Principia* as a contribution to both natural science and "philosophy",[27] and in England Newton's discoveries were even hailed as a new "cosmic law". Out of the spatial infinity of the universe, planets were made visible to the human eye via telescope and their orbits were suddenly open to calculations and the application of human reason.[28] This unbelievable triumph, of course, turned Newton into a scientific icon in his homeland. In his prologue to the *Principia*, Edmund Halley accordingly heralds Newton as a man who is closer to the gods than any other human being, since "no mortal being had ever been allowed to gain this much insight."[29] In his epitaph, Alexander Pope, author of the bestselling "*Essay on man*", even praises Newton as the creator of light in the world ("God said: *Let Newton be*! and all was *Light*"[30]) because he had succeeded in subjecting the infinite universe to human reason and had thus made it accessible to science.

Amongst the French scholars of the *Académie des Sciences*, and for Fontenelle in particular, however, Newton's new insights into physics and his most influential theories such as the principle of gravity and his mathematical calculations remained most controversial until the middle of the century.[31] Newton's force of gravity was

---

[25] Her contemporaries accordingly named her "Mylady Newton et Madame la baronne de Leibnitz" respectively. See also Winter. 2006. Vom Salon zur Akademie. Émilie Du Châtelet und der Transfer naturwissenschaftlicher und philosophischer Paradigmen innerhalb der europäischen Gelehrtenrepublik des 18. Jahrhunderts. 285–306.

[26] Freudenthal. 1986. *Atom and Individual in the Age of Newton: On the Genesis of the Mechanistic World View*. 14.

[27] Pemberton. 1728. *A View of Sir Isaac Newton's Philosophy*.

[28] Regarding the influence of Newton's *Principia* on the philosophical discourses of the Enlightenment see Winter. 1972. *Der Materialismus bei Diderot, Droz und Minard*. 90–103 and Winter. 2001. Naturphilosophie und Naturwissenschaften. 173–208.

[29] Halley. 1988. *Preface to Isaaci Newtoni Opera quae Extant Omnia*, ed. Samuel Horsley, 1779–1785. 7–8.

[30] Epitaph intended for Sir Isaac Newton. 1954. In: Pope, Alexander. Minor Poems, ed. Norman Ault, 317. (emphasis in the original).

[31] Badinter and Muzerelle. 2006. *Madame du Châtelet. La femme des Lumières*. 90–103.

considered an "occult" quality of matter, since it did not seem possible to explain how bodies could be attracted by other bodies from such a vast distance. It was not until Maupertuis's[32] Lapland expedition that Newton's theories could be validated by precise measurements. Regardless of Maupertuis's results, however, many scholars preferred to wait for yet another confirmation from La Condamine's later Peru expedition.

Contrary to the dominant opinion of the "République des Savants", Emilie du Châtelet was well aware of the paradigmatic change in physics Newton's ideas would cause.[33] Together with Voltaire[34] and Maupertuis, Du Châtelet rallied great support for Newton's theorems in France. As early as 1738, she published a peer-review of Voltaire's *Eléments de la philosophie de Newton* for the *Journal des Savants* in which she explicitly called Newton's ideas a ground-breaking turn in the natural science, as "une entière révolution dans la physique".[35]

It was not until the middle of the eighteenth century that Newton's theories on the planetary system and on gravity were generally acknowledged throughout Europe's academic world. It is remarkable in this context that the subtitle of Immanuel Kant's *Allgemeine Naturgeschichte und Theorie des Himmels* claims to be written in accordance with Newton's recent discoveries (*nach Newtonischen Grundsätzen abgehandelt*) and thus illustrates the fact that Newton's ideas caused "cosmological enthusiasm" everywhere. Newton's discoveries seemed to allow human reason to grasp and open a formerly infinite universe to mathematical calculation. "Reason", Kant consequently claims, was now allowed access to "the infinity of time and spaces" and was able to cover the "entire limitless extent of infinite spaces […] with numberless worlds without end".[36] At the same time, Rousseau pointed out how human reason, like the sun, had now been given the power to measure the infinity of

---

[32] Regarding Maupertuis' relationship with Emilie du Châtelet see Hagengruber. 1999. Eine Metaphysik in Briefen. E. du Châtelet an P. L. M. de Maupertuis. 189–211. Maupertuis would later become president of the Prussian Academy of Science.

[33] For a re-evaluation of Du Châtelet's scientific and metaphysical thought, compare Zinsser. 2006. *La Dame d'Esprit*. See also Kölving and Courcelle. 2008. *Émilie Du Châtelet: Éclairages et documents nouveaux*. On Du Châtelet's mathematical standing please refer to Klens. 1994. Mathematikerinnen im 18. Jahrhundert. 177–258. See also Zinsser and Hayes. 2006. *Emilie Du Châtelet: rewriting Enlightenment philosophy and science*.

[34] Du Châtelet's relationship with Voltaire is discussed in greater detail in Wehinger. 2003. Experimente auf der Suche nach dem Glück. Gabrielle Emilie du Châtelet und Voltaire. 89–110.

[35] LetSav1738 534–541. Also see de la Harpe. 1941. *Le Journal des Savants et l'Angleterre, 1702–1789*. 359. According to current research, the term of a scientific "revolution" in this context firstly came into use at the beginning of the eighteenth century.

[36] Kant. 1996. Allgemeine Naturgeschichte und Theorie des Himmels oder Versuch von der Verfassung und dem mechanischen Ursprunge des ganzen Weltgebäudes nach Newtonischen Grundsätzen abgehandelt. 335. The original reads: "durch alle Unendlichkeiten der Zeiten und Räume hindurch […] die ganze grenzenlose Weite der unendlichen Räume mit Welten ohne Zahl und ohne Ende".

the universe in that it was able to "parcourir à pas de géant, ainsi que le soleil, la vaste étendue de l'univers."[37] And in his well-known *Discours préliminaire* for the *Encyclopédie*, d'Alembert, who was an influential mathematician and philosopher at the time, praises Newton as the doorman to a new philosophical understanding of the world. "Newton [...]", he claims, "parut enfin, et donna à la philosophie une forme qu'elle semble devoir conserver".[38]

Des Maizeaux, in his widely read *Recueil de diverses pièces sur la philosophie*, which consisted of a collection of philosophical texts including Newton's and Leibniz's correspondence, also mentions the god-like appraisal Newton received. The *Principia*, he writes, were an "ouvrage, qui au jugement de M. Le marquis de l'Hôpital semblait être la production d'un Génie ou d'une Intelligence céleste, plutôt que celle d'un homme".[39] Until the publication of Kant's *Critique of Reason*, it can be said that Newton's and Leibniz's controversial theories on space and time formed one of the central philosophical discourses in the eighteenth century. The conflict between Leibniz's and Newton's actual theories of cosmic structure, their respective theories of space and time and their concepts of movement and force, were well laid out in Leibniz's and Clarke's[40] famous correspondence, in which Newton himself formulated Clarke's comments.

In the first chapter of his work, which appropriately bears the title "philosophia naturalis", Newton predefines essential philosophical terms such as space, time and motion, thus providing a solid base for his paradigm of nature and – at the same time – positing some of the most influential philosophical thoughts of the Enlightenment. A number of prominent sections in his *Principia*, such as the introduction and the concluding paragraphs of his "System of the World", here form a paradigmatic frame for Newton's most important definitions in natural philosophy and furthermore situate his mathematical calculations and arguments within the wider structure of his study.[41]

Newton's definition of absolute time precedes his initial axioms. "Le temps absolu" he understands as "vrai et mathématique, sans relation à rien d'extérieur, coule uniformément, et s'appelle durée."[42] In accordance with his theory of absolute time, he also introduces his theory of absolute space. Matter is here defined in such a way that even the smallest particles of all bodies are extended, solid, impenetrable, movable and endowed with the inertia. Newton's idea of the matter-implanted force only refers to inertia, which is invariable.

---

[37] Rousseau. 1971. *Discours sur les Sciences et les Arts*. 27.

[38] D'Alembert. 1904. *Discours préliminaire de l'Encyclopédie*. 100.

[39] Des Maizeaux. 1720. *Recueil de Diverses Pièces sur la Philosophie, la Religion naturelle, l'Histoire, les Mathématiques, etc. par Mrs. Leibniz, Clarke, Newton et autres Autheurs célèbres*. XII.

[40] This correspondence entitled *A Collection of Papers, Which passed between the late learned Mr. Leibnits and Dr. Clarke, In the Years 1715 and 1716, relating to the Principles of Natural Philosophy and Religion*, was first published in London in 1717 and was later added to Desmaizeaux's *Recueil* in 1720. The Recueil was reprinted in 1740.

[41] The title of the *Principia*, of course, already suggests a philosophical understanding of nature which was thoroughly grounded in mathematical calculations.

[42] PrincChat1990 vol I, 8.

Towards the end of the last part of the *Principia*, in his "System of the World", Newton further seeks to answer some of the most vital questions in the fields of metaphysics and religion, the extent to which God as creator and sustainer intervenes in a system of the world with universal gravity and its respective effects on the planets and comets. Newton affirms these two functions of God in the system of nature and discusses in some detail God's importance within the planetary system,[43] his influence on the cosmos, which is substantial and not just virtual, and on the universal laws governing its course.

> Cet admirable arrangement du Soleil, des planètes & des comètes, ne peut être que l'ouvrage d'un être tout-puissant & intelligent. [...] De plus, on voit que celui qui a arrangé cet Univers, a mis les étoiles fixes à une distance immense les unes des autres, de peur que ces globes ne tombassent les uns sur les autres par la force de leur gravité. [...] Il est présent partout, non seulement *virtuellement,* mais *substantiellement;* car on ne peut agir où l'on n'est pas; [...] Nous le connoissons seulement par ses propriétés & ses attributs, par la structure très-sage & très-excellente des choses, & par leurs causes finales.[44]

In Freudenthal's words, "Newton coupled the theory of absolute space to metaphysical-religious notions."[45] For Newton, the notion of God is therefore still inextricably bound to natural philosophy, since God's wisdom created the world in accordance with a systematic vision. These "causes finales" are evident in all the natural sciences, which shows that Newton's physical and mathematical theories were based on an elaborate paradigm inclusive of God, nature and the cosmos.

From the "Philosophia Naturalis" to Astronomical Theory – Du Châtelet's Commentary on Newton's Principia

In her commentary on the *Principia*, Du Châtelet on the other hand shifts her focus from Newton's theories of natural philosophy to his astronomical theories. Du Châtelet in fact excludes Newton's natural philosophy in favour of an in-depth discussion of his astronomy. In her "Exposition abrégéé du Système du monde", Newton's theories on absolute time and space, his notion of God and God's influence on cosmic structure and his concept of the "causae finales" are neither mentioned

---

[43] In his correspondence Newton calls God an antive, causal force behind the planetary system and refers to him as the sole explanation for the phenomenon of gravitation. For Du Châtelet's comment, see Zinsser. 2006. *La dame d'esprit.* 189: "In Newton's world, the Creator had to replenish the force periodically and in perpetuity. From Du Châtelet's perspective, accepting this image of the Supreme Being and his "continual miracles" undermined any claim to certain knowledge of the workings of nature's laws. [...] In contrast, in Leibniz's world of *forces vives* there was no need for God to intervene, for the German *philosophe* believed that this force was conserved in the universe."

[44] PrincChat1990, vol II, 177–178 (Newton's emphasis).

[45] Freudenthal. 1986. *Atom and Individual in the Age of Newton: On the Genesis of the Mechanistic World View.* 7.

nor explicitly commented on. Newton's concepts of mass and inertia, which form the foundation of all of his definitions of matter, are also only mentioned briefly.

> Le Livre des Principes commence par huit Définitions; [...] il définit dans la troisième la *force d'inertie* ou force résistante dont toute matière est douée; il fait voir dans la quatrième ce qu'on doit entendre par *force active*; il définit dans la cinquième *la force centripète*; et il donne dans la sixième, septième et huitième, la manière de mesurer sa *quantité absolue*, sa *quantité motrice*, et sa *quantité accélératrice*. Ensuite il établit les trois Lois de mouvement suivantes.[46]

This very brief list of Newton's basic definitions in Du Châtelet's "Exposition" is immediately followed by Newton's laws of motion. Newton's "scholias", which he inserted in between his definitions and these laws of motion, are missing from the commentary.

This limitation of Du Châtelet's commentary with regard to questions such as the "causae finales" and the concept of force is even more remarkable when considering that Du Châtelet intervened in the spectacular dispute over the "forces vives" with her own publications on the subject.[47] Her omissions could therefore by no means have been aimed at a strict division of physics and metaphysics, since Du Châtelet in both her *Institutions de physique* and her correspondence explicitly subscribed to the view that metaphysics, physics and mathematics were closely related.[48] It seemed essential to her that physical science was based on metaphysical prerequisites.[49]

However, Du Châtelet's commentary on the *Principia* surprisingly only deals with Newton's physical laws and their respective mathematical calculations, to which she now adds a number of her own calculations.[50] This shift in focus is aptly supported by her subtitle *"Explication des principaux phénomènes astronomiques. Tirée des principes de m. Newton"*, which shifts the emphasis from Newton's extensive paradigm of nature to descriptions and calculations of physical laws.

What this shows is that Du Châtelet understands her commentary on Newton to be a discourse in its own right; her work seeks to uphold an objective view of Newton's work, while at the same time introducing her own definitions and terms. Consequently, Du Châtelet also dedicates the beginning of her work to her own definition of gravitation:

> Au reste, *je déclare ici,* comme M. Newton a fait lui-même, qu'en me servant du mot d'*attraction*, je n'entend que la force qui fait tendre les corps vers un centre, sans prétendre assigner la cause de cette tendance.[51]

---

[46] Du Châtelet. 1990. PrincChat vol II, 8 (emphasis in the original).
[47] MaiCh1741.
[48] Inst1742Am. Reprint see Inst1742Rep1988.
[49] LetChBII 12, to Frédéric, prince royal de Prusse, 25 April 1740.
[50] Clairaut himself verified her calculations and considered them to be accurate. See also Gandt. 2001. *Cirey dans la vie intellectuelle. La réception de Newton en France*.
[51] PrincChat1990, vol II, 10 (my emphasis – U.W.).

Contrary to the quasi god-like adoration of Newton, Du Châtelet positions Sir Isaac along the lines of such great thinkers as Ptolemy, Pythagoras, Copernicus, Kepler, Bernoulli, Clairaut and other influential mathematicians. Instead of seeing Newton as an "intelligence céleste", therefore, she relates Newton's ideas to particular theoretical and historical developments while underlining the influence previous approaches generally have on any development of new scientific theory, as can be shown by her account of this approach in the *Institutions*. According to Du Châtelet, Newton's results were therefore mostly based on Kepler's and Huyghens' research. "C'est en profitant des travaux de Kepler, & en faisant usage des Théoremes de Huyghens, que Mr. Newton à découvert cette Loi universelle répandue dans toute la Nature".[52] Du Châtelet asserts that an objective view is necessary for any dealing with scientific theory: "Je crois encore plus nécessaire de vous recommander de ne point porter le respect pour les grandes hommes jusqu'à l'Idolatrie".[53] And although she positions Newton among a great line of extraordinary mathematicians and physicists, she also works with and examines Newton's theories from a critical distance.

> Mais ce sentiment de M. Newton étoit fondé sur ce qu'il pensoit que dans toute sphéroide dans l'équilibre, la pesanteur doit être toujours en raison renversée de la distance au centre, proportion qui n'est vraie que si le sphéroide est homogène.[...] La conclusion à laquelle conduit le calcul ci-dessus, rend la théorie précédente assez difficile à concilier avec les observations qui concernent la figure de la terre.[54]

Du Châtelet here quotes Clairaut's hypotheses in an attempt to present a possible amendment to Newton's theories by stating that "[d]ans les calculs qu'a employé M. Clairaut, il a supposé à la vérité la forme elliptique aux couches extérieures [...] l'hypothèse des couches elliptiques a une grande raison à être préférée aux autres."[55]

In accordance with her call for an objective view even in relation to Newton's theories, Du Châtelet challenges her contemporaries not to blindly follow and argue in line with established authorities. Not unlike Kant's notorious claim "sapere aude", she also demands that one have recourse to one's own reason for scientific validation.

> Lorsqu'on a l'usage de la raison, il ne faut en croire personne sur sa parole, mais qu'il faut toujours examiner par soi-même, en mettant à part la considération qu'un nom fameux emporte toujours avec lui.[56]

There is no place for Newton's theories on space, time and divine influence on cosmic structure in Du Châtelet's exposition, which is dominated by her account of

---

[52] Inst1742Am, 6. Du Châtelet elaborates further on the matter: "Nous nous élevons à la connaissance de la vérité [...] en montant sur les épaules les uns des autres. Ce sont Descartes & Galilée qui ont formé les Huygens, & les Leibnits."

[53] Inst1742Am, 11.

[54] PrincChat1990, 258.

[55] PrincChat1990 vol II, 259. In contrast to other members of the "République des Savants", whose eulogies frequently displayed a rather uncritical euphoria, Emilie du Châtelet here proves her scientific ability by implying that *any* scientific theory needs to be verified via empirical observation.

[56] Inst1740, 11.

astronomical principles and their respective calculations. In her commentary, Du Châtelet presents Newton as an astronomer and physicist rather than someone presenting a new "philosophia naturalis". On her account, Newton's theories are limited to the physical and astronomical world and are not understood as definite findings, as the cosmic law *per se*, as was the case with some of the English commentators. In presenting Newton's ideas that way, Du Châtelet understands his theories as a considerable step towards scientific progress while at the same time reminding her readers to keep an open mind regarding their subsequent modifications and corrections.

Du Châtelet's Commentary in the Eyes of the "République des Savants"

While recent research has focussed very much on Du Châtelet's translation and less on her commentary on Newton's *Principia*, her commentaries were well received as philosophical studies in their own right during the time of the Enlightenment. The simple fact that a renowned scientist such as Dortous de Mairan published his own study of the "forces vives" as a response to Du Châtelet's critique of his approach in her *Institutions de physique*[57] points out Du Châtelet's exceptional scientific standing among the scholars of her time. This high regard for Du Châtelet's *Institutions* is even further underlined by a very early translation of her work into German. The renowned Adolph von Steinwehr, whose name is mentioned inclusive of all his numerous titles in the *Naturlehre,* translated Du Châtelet's *Institutions* into German in as early as 1743, shortly after its original publication.[58] The "Journal des Savants", which only discussed research of the highest scientific value, also commented on the *Institutions* in two extensive articles.

Immanuel Kant himself commented extensively on Du Châtelet's theses during the notorious dispute on the "forces vives" at the time. In fact, he not only refers to Du Châtelet's thesis in his early work *Thoughts on the True Estimation of Vital Forces,* but he also links her to leading mathematicians of the time, among them illustrious names like Bernoulli, Wolff and Huygens.[59] Kant was quite familiar with the 1743 German translation of Du Châtelet's *Institutions de physique.* He repeatedly refers to her work in his current argument and admits that Du Châtelet provided him

---

[57] The following quotes "*Institutions de physique*" rather than "*Institutions physiques*", since the Enlightenment also preferred this title over the one from the 1742 edition (as the article "Hypothèse" in the *Encyclopédie* clearly shows).

[58] Naturlehre1743.

[59] Kant. 1996. Gedanken von der wahren Schätzung der lebendigen Kräfte und Beurtheilung der Beweise, derer sich Herr von Leibnitz und andere Mechaniker in dieser Streitsache bedienet haben nebst einigen vorhergehenden Betrachtungen, welche die Kraft der Körper überhaupt betreffen. 15–218.

with an answer to a philosophical question.[60] Kant's early work further shows that he valued Du Châtelet's comments on Leibniz (in the *Naturlehre*) and also that he had studied them thoroughly. He even refers to one of Du Châtelet's footnotes, which he considered to be of great importance: "I have found an answer to this objection in Mme du Châtelet's *Lessons in Physics*, which, as it appears, goes back to no other than the famous Mr Bernoulli".[61]

Even some of the most fundamentally important contributions to the *Encyclopédie* have Du Châtelet's commentary on Leibniz as a reference for both scientific and philosophical questions.[62] The *Encyclopédie*, which might well be considered a 'bestseller' in the modern sense of the word,[63] was heralded by its publishers as an inventory of the entire knowledge of its time, especially in the fields of mathematics, physics and natural philosophy. It is quite remarkable in this context that d'Alembert in it praises Du Châtelet's translation of the *Principia* as one of the most important publications on Newton's theories[64] and furthermore positions it among the most influential studies on Newton available at the time. Voltaire's publication *Eléments de la philosophie de Newton*, on the other hand, is not even mentioned.

Both d'Alembert's scientific and philosophical contributions to the *Encyclopédie* repeatedly mention Du Châtelet's commentaries. In fact, it is her *Institutions de physique* which is referred to in a number of articles on matters both scientific and philosophic. In his article on "Feu", d'Alembert furthermore goes as far as to refer to Du Châtelet's *Dissertation sur la nature et la propagation du feu* and points out that – in keeping with the original call for papers – Du Châtelet's work demonstrates a thorough familiarity with a great number of physical studies, that she presents her own theories well and also that she possesses great knowledge in the fields of empirical sciences; "remplies de vûes & de faits très-bien exposés."[65] Diderot, who was a philosopher, editor of the *Encyclopédie* and a friend of d'Alembert's, agreed with Du Châtelet's authority on Newton's ideas. In a letter to Voltaire, Diderot asks him to forward some of his remarks and calculations relating to Newton's theories directly to Mme du Châtelet, for she alone was in a position to proofread them accurately.[66]

It is remarkable that d'Alembert's contributions to the *Encyclopédie* prominently refer to Du Châtelet's commentary as a work in its own right and that he seems to

---

[60] Compare Winter, Ursula. 2006. Vom Salon zur Akademie. Emilie du Châtelet und der Transfer naturwissenschaftlicher und philosophischer Paradigmen innerhalb der europäischen Gelehrtenrepublik des 18. Jahrhunderts. 296.

[61] Kant. 1996d. 71. The original reads: "Ich finde in der Frau von Chastelets *Naturlehre* eine Antwort auf diesen Einwurf, die, wie ich aus der Anführung ersehe, den berühmten Herrn Bernoulli zum Urheber hat."

[62] See also Carboncini. 1988. Vorwort. In: Inst1742Rep.

[63] For a discussion of the *Encyclopédie* as a 'bestseller' and its significantly high number of sold copies which were distributed all over Europe see Darnton. 1979. *The Business of Enlightenment*.

[64] "Newtonianisme". In *Encyclopédie ou Dictionnaire raisonnée des sciences, des arts et des métiers, par une société de gens de lettres*. vol. XI. 122–123.

[65] "Feu", in *Encyclopédie ou Dictionnaire raisonné des sciences, des arts et des métiers*, vol. VI. 603.

[66] Diderot. 1955–1973. *Correspondance*. vol. I. 79–80. To Voltaire, 11 June 1749.

prefer her commentary over her highly influential translation. He repeatedly mentions "le commentaire que madame la marquise de Chatelet nous a laissé sur les principes de Newton avec une traduction de ce même ouvrage".[67]

During the time of the Enlightenment and within the "République des Savants", Du Châtelet's commentaries – both her *Institutions de physique* and her commentary on Newton – were considered to provide a fundamentally important contribution to scientific discourse.

From Translation to Philosophical Discourse

In keeping with the idea of a "progrès de l'esprit humain", the work of the translator has to be seen as fundamentally contributing to the European reception of scientific and philosophical discourse and progress during the time of the Enlightenment. The great reputation translators such as Pierre Coste, the translator of John Locke's work, had in the realm of scientific and philosophical discourses of the time may serve to illustrate the argument here. Du Châtelet demands that translation should be considered and accepted as a form of "interpretation" rather than a literal translation of the original. "Je n'ay point pour mon autheur le respect idolatre de tous les traducteurs", Emilie du Châtelet writes in her reflections on translating Mandeville's *Fable of the Bees* and puts forth a concept of critical translation practice. "J'ay pris aussi la liberté d'y ajouter mes propres réflexions."[68] In addition to the actual translation, reflections of her own constitute a critical dialogue with the author. Du Châtelet's publisher consequently calls her translation of Newton's *Principia* a 'work in its own right' and refers to Châtelet as an "interpreter" rather than a translator of scientific discourse.

> L'illustre Interprète, plus jalouse de saisir l'esprit de l'Auteur, que ses paroles, n'a pas craint en quelques endroits d'ajouter ou de transposer quelques idées pour donner au sens plus de clarté. En conséquence, on trouvera souvent *Newton* plus intelligible dans cette traduction que dans l'original.[69]

During the time of the Enlightenment, basic cultural transfers were considered to be almost as important as philosophical theorems.[70] Even famous philosophers such as Diderot and Buffon, names which considerably contributed to the paradigm of nature during the second half of the century, translate introductory and influential philosophical works into French. Buffon himself, whose work on the *Histoire Naturelle générale et particulière* and the *Epoques de la Nature* made him stand out

---

[67] "Newtonianisme". In Encyclopédie ou Dictionnaire raisonnée des sciences, des arts et des métiers, par une société de gens de lettres. 122–123.

[68] See Wade. 1967. *Studies on Voltaire, with some unpublished papers of Mme du Châtelet*. 135. ("Préface du traducteur" regarding Du Châtelet's 1735 draft of Bernard de Mandeville's *The Fable of the Bees or Private Vices, Publick Benefits* from 1729). Compare Wehinger and Brown. 2008. Übersetzungskultur im 18. Jahrhundert. Übersetzerinnen in Deutschland, Frankreich und der Schweiz.

[69] "Avertissement de l'éditeur". PrincChat1990 vol I, I.

[70] Compare Winter. 2007. Die Newton-Übersetzung Emilie du Châtelets. 28.

as an important "naturaliste" of the eighteenth century, translated one of Newton's studies, his *Calcul des Fluxions*. Buffon's translation of Newton and Diderot's translation of Shaftesbury's *Inquiry Concerning Virtue and Merit* were both reviewed and extensively quoted from in the influential *Journal des Savants*, which proves that translations were considered to be as much a scientific "event" as original scientific studies.[71] In other words, during the time of the Enlightenment, translations of and studies on philosophical discourse were closely linked.

"L'amour de l'étude"

Du Châtelet's scientific and philosophical development presents itself as a continuous process from critical translations of a predefined discourse to independent commentaries, research, calculations and theoretical reflections.[72] In Besterman's words, her correspondence shows "un esprit scientifique qui a bien exactement conçu les principes fondamentaux de la physique et de la méthodologie scientifique".[73] What her correspondence shows, then, is that her philosophical and scientific comments were based on a thorough examination of and a familiarisation with all the seminal studies on philosophy and natural science available to her. A letter Du Châtelet wrote to her bookseller in February 1739 may provide ample proof of her extensive scientific and philosophical research here. As her letter shows, her library at the time already comprises the following:

> J'ai l'optique de Newton, Rohaut commenté par Clark, Vhiston, la figure de la terre, figure des astres,[74] Musembrok phisique,[75] s'Gravesende phisique,[76] Receuil de Lettres de Leibnits et de Clark[77], les entretiens phisiques du père Renaut, pour ce qu'ils valent, Euclide, Pardies, Malesieux, l'application de l'algèbre à la géométrie de Guinée, les sections coniques de Mr de Lhopital, les mathématiques universelles, et les oeuvres[78] de des Cartes.[79]

At the time of this letter, Du Châtelet had already studied all of the above mentioned introductory works on philosophy and science, some of which had been published in Latin exclusively. In order to further her "recherches scientifiques", Du

---

[71] See de la Harpe. 1941. *Le Journal des Savants et l'Angleterre*. 358.

[72] Compare Ira O. Wade's introduction to the *Unpublished papers,* where Wade mentions the extensive scope of Du Châtelet's translation work.

[73] Besterman, Théodore. Notes préliminaires. In: LetChBI, 16.

[74] Besterman's footnote here refers to Maupertuis, LetChBI, 330.

[75] Here Besterman refers to Pieter van Musschenbroeks *Epitome elementorum physico-mathematicorum* from 1725, ibid. 330.

[76] This footnote refers to Willem Jacob 'S-Gravesande's *Physicae elementa mathematica experimentii confirmata* from 1720, ibid. 330.

[77] The "Lettres" Besterman here quotes refer to Kortholt's edition of Leibniz's correspondence. See *Viri illustri Godefridi Guil. Leibnitii Epistolae ad diversos* (1734–1741); ibid, 330.

[78] His footnote refers to Descartes's *Opera posthuma physica et mathematica.*, which were published in Amsterdam in 1701 and also to his *Opera philosophica omnia,* published in 1697; ibid, 331.

[79] LetChBI, 328–331, to Laurent François Prault, 16 Feb. 1739.

Châtelet now requests further titles from Prault. Apart from the "*Principia mathematica* de Mr Neuton d'une belle edition", she orders the complete protocols of the *Académie des Sciences*, which Prault had just received at that time, the *Philosophical Transactions* of London's Royal Society, the *Nouvelles de la Republique des Lettres* up to Bayle's death and any other publication on the natural sciences available.

> Je veux encore les transactions philosophiques, la république des lettres jusqu'à la mort de Bayle, et tous les livres de physique que vous trouverez dans votre chemin. [...] Je vous prie de me chercher les *Principia mathematica* de Mr Neuton d'une belle édition.[...] Je relis votre lettre et je trouve que le recueil de l'Académie va à 500 livres. Il n'est plus question que du temps du paiement.[80]

It was due to her continuous and extensive scientific studies (of which the scholars of the *République des Savants* were, of course, well aware) that one of Voltaire's students dedicated to her an *Epitre sur l'amour de l'étude, à Madame la Marquise du Chastelet, par un élève de Voltaire*.[81] Du Châtelet's extensive research in preparation for her commentaries on Newton and Leibniz are well recorded in her correspondence and furthermore illustrate her transition from translating predefined scientific discourse to writing her own commentaries on the subject matter and to setting down her own scientific theory.

It is worth noticing the date of Du Châtelet's above mentioned order. 1739, the date of her letter to Prault, falls right into a period where Du Châtelet was completing her first manuscript and preparing for her first print of the *Institutions de physique* in 1740. Emilie du Châtelet had withheld the publication of her *Institutions de physique* for 2 years, although it had already been approved by her censors. As her letter to Prault aptly illustrates, this delay in publication was certainly due to Du Châtelet's extensive and thoroughly conducted research between 1738 and 1740; a period which she primarily used for an update and verification of her intensive research into Leibniz's work.

As Pitot, Du Châtelet's censor, points out, the 1738 version of the *Institutions de physique* already contained an introduction to both Leibniz's and Newton's philosophy:

> J'ai lu, par ordre de Monseigneur le Chancelier, un Manuscrit qui a pour titre: Institutions physiques; cet Ouvrage *dans lequel on a exposé les Principes de la Philosophie de Mr. Leibnits & ceux de Mr. Newton,* est écrit avec beaucoup de clarté, & je n'y ai rien trouvé qui puisse en empêcher l'impression. A Paris, ce 18 Septembre 1738 signé Pitot.[82]

Little research has been done into this particular aspect, and scholars so far have primarily focussed on Du Châtelet's mathematical acquaintance with König and how she was apparently introduced to Leibniz's thoughts via their discussion of

---

[80] LetChBI, 239.

[81] Epitre sur l'amour de l'étude, à Madame la Marquise du Chastelet, par un élève de Voltaire. 1815 (attributed to Helvétius). In his correspondence, Voltaire mentions the "matériaux immenses" Du Châtelet had compiled for her philosophical studies and ascribes to her a "patience et une sagacité qui m'effraient". See Besterman. Notes preliminaires. LetChBI 17.

[82] "Approbation", in: Inst1740. (my emphasis – U.W.).

Wolff's *Ontologia* in 1739.⁸³ Pitot's account of Du Châtelet's 1738 version of the *Institutions de physique* and the fact that she had already discussed Leibniz's theories in this earlier version, however, clearly invalidate such an explanation.⁸⁴ Current research has also failed to take into account Du Châtelet's own remarks on the completion of her manuscript in 1738, where she clearly states – in accordance with Pitot's observations – that the *Institutions* had already been completed in 1738 and that she subsequently only had to make some minor changes to the original version. There is further evidence which negates the above mentioned assumption that Du Châtelet had been introduced to Leibniz's thoughts by König.

There are numerous references to Leibniz's and Wolff's work in Du Châtelet's correspondence from earlier years, as for example Voltaire's private contact to Clarke and his relation to Des Maizeaux, the publisher of *Recueil de Diverses Pièces sur la Philosophie, la Religion naturelle, l'Histoire, les Mathématiques, etc. par Mrs. Leibniz, Clarke, Newton et autres Autheurs célèbres*, which was published in 1720, clearly shows. In 1737, Du Châtelet had already met Baron von Keyserlingk, Wolff's translator, on a private visit he paid to Cirey.⁸⁵ In addition to the above, Voltaire himself had pointed out that Emilie du Châtelet was well acquainted with Locke's, Clarke's, Leibniz's and Wolff's thoughts in a letter fragment dating from 1737.⁸⁶

One of the reasons for Du Châtelet's secrecy regarding the publication of her *Institutions de physique* was certainly that it dealt with Leibniz's theories and that Voltaire had repeatedly expressed his utter dislike of Leibniz's thoughts.⁸⁷ Du Châtelet's first edition of the *Institutions de physique* was published anonymously and was written *"in complete secrecy"*,⁸⁸ as Barber points out. In her correspondence with Frederick II, Emilie du Châtelet mentions the potential conflicts her work could provoke in relation to Voltaire's polemics against Leibniz and how

---

⁸³ Compare Barber. 1967. Madame du Châtelet and Leibnizianism: the genesis of the Institutions de physique. 200–222; see p. 209: "There is no real evidence *that this process began before König's arrival at Cirey in March 1739*." (my emphasis – U.W.).

⁸⁴ Pitot, Du Châtelet's censor, was impressed with the conceptual quality of her commentary to Newton's and Leibniz's philosophical principles, as he mentions in his correspondence with Voltaire. He received the manuscript in 1738, and there is no evidence that either Pitot or Voltaire had already been informed of the author's name at the time.

⁸⁵ Crown Prince Frederick had Keyserlingk work on several translations of Wolff's texts, which he then forwarded to Voltaire. Barber's comment reads (Madame du Châtelet and Leibnizianism: the genesis of the Institutions de physique. 205): "Madame du Châtelet can scarcely have failed to acquire some awareness of Leibnizian ideas from this, but there is no evidence that they attracted her at that date." However, his verdict seems hardly convincing, especially when considering that Pitot confirms how principles of Leibniz's philosophy were presented convincingly and ready to print as soon as only 1 year later in 1738.

⁸⁶ See Barber. 1967. Madame du Châtelet and Leibnizianism: the genesis of the *Institutions de physique*. 205.

⁸⁷ See Belaval. 1993. Quand Voltaire rencontre Leibniz. 242. He identifies a certain "théâtralisation de l'ironie" regarding Voltaire's comments on Leibniz's philosophy.

⁸⁸ See Barber. 1967. Madame du Châtelet and Leibnizianism: the genesis of the *Institutions de physique*. 205 (my emphasis – U. W.).

this could possibly affect their relationship. In fact, she points out the differences in opinion between her and Voltaire as regards Leibniz's theories but also affirms her right to make independent philosophical decisions: "[l]a liberté de philosopher est aussi nécessaire que la liberté de conscience."[89]

The Early Drafts of the Institution de Physique

When Du Châtelet had finished her early draft of the *Institutions* in 1738, Leibniz's work was not yet fully available. In fact, it was not until 1768 that Dutens first published a more extensive corpus of Leibniz's work in the six volumes of his *Opera omnia*, and Leibniz's influential *Nouveaux Essais sur l'entendement humain* were available for the first time in 1765. Regarding the availability of Leibniz's texts in 1738, some of Leibniz's theories were accessible to Du Châtelet only via Wolff's *Ontologia*, which she received in May 1739 in a translation Frederick II had ordered on her behalf.[90] Du Châtelet was also in the fortunate position of receiving several of Leibniz's texts which were available as first drafts only.

The above mentioned request for authentic Leibniz texts in 1739 and her interest in the as yet unpublished correspondence between Leibniz and Bernoulli, together with her extensive research efforts as illustrated by her letters to Prault, all provide ample reason to assume that Du Châtelet spent most of her time between 1738 and 1740 considerably reworking her first version of the *Institutions*.

Barber's assumption that Du Châtelet had only been able to write her extensive introduction to Leibniz's thoughts as it was laid out in chapters I – VIII of the *Institutions* after her acquaintance with König and his summaries in 1739[91] does certainly not hold here, given that Pitot's 1738 report on the *Institutions* clearly negates such a reading. In addition, it is also clear that Du Châtelet's 'minor changes' to Leibnizian metaphysics as pointed out by the *Avertissement du Libraire* could not have been referring to an addition of an 180 pages strong corpus: "l'Auteur ayant voulu y faire quelques changemens". As the following will show, the structure of the opening chapters also provides enough evidence to falsify this thesis.

When considering the content of Chaps. V and VI, for instance, it is clear that their account of space and time undeniably refers to Leibniz's correspondence with Clarke, which Du Châtelet was already familiar with at that time. The references in the margins of her chapter on Leibniz's theories of space and time explicitly indicate the *Commercium epistolicum* and also the "Dispute de Mr de Leibnits, & du Docteur Clarke sur l'Espace"[92] as her sources. Nor does Chap. IV, which is titled "Des Hipotheses" refer to Wolff's *Ontologia* as an influence. There is in fact no

---

[89] LetChBII, 13–14. To Frédéric, prince royal de Prusse. She writes: "Vous serez peut-être étonné que nous soyons d'avis si différents."

[90] LetChBII, 18.

[91] Barber. 1967. Madame du Châtelet and Leibnizianism: the genesis of the Institutions de physique. 217.

[92] Inst1742Am, 97 (identical with Inst1740).

indication that Wolff influenced these chapters at all. It rather shows that Chap. IV and the subsequent chapters had already been drafted in 1738 because they make specific use of those texts Du Châtelet had already been long acquainted with.

Contrary to these chapters, however, Chaps. II and III, entitled "De l' Essence des Attributs & des Modes" respectively and also parts of Chap. I "Des Principes de nos Connoissances" show considerable similarities to Wolff's line of thought. It seems reasonable, therefore, to assume that only Chaps. I, II and III were added in part or fully after Du Châtelet's acquaintance with Wolff's work in 1740 – "à la tête de l'ouvrage", as Du Châtelet points out herself.[93]

This theory is further supported by the structure of the *Institutions*.

In her decision to introduce the 1738 version of the *Institutions* – manifestly a treatise on the principles of physics – by some methodological prolegomena on the cognitive value of hypotheses, together with definitions of space and time and of the structure of bodies, Du Châtelet closely follows the structure of Newton's *Principia*. Newton had also introduced his third book, the "System of the World," via an exposition of methodological concepts and he introduced definitions of terms like space, time and the characteristics of solid matter, before treating the principle of universal gravitation and calculus. Apart from the first three chapters, which can therefore be assumed to have been added later, Châtelet's 1738 version of the *Institutions* shows itself to be closely following the structure of Newton's *Principia*.

The fact that 1742 saw a new edition of the *Institutions* published in Amsterdam and that this new edition bore the subtitle "Nouvelle édition, corrigée et augmentée considérablement" also hints at further in-depth studies Du Châtelet must have conducted between 1740 and 1742. Compared to the first edition from 1740, this second edition of her work shows some changes to the "Avant-Propos" section and to Chaps. IX and X, which dealt with the structure of matter.[94] Similar to the changes Du Châtelet made regarding the initial title of Chap. X due to the results of her research, she also altered the title of Chap. IX ("De la Divisibilité & Subtilité de la Matière").[95] In a letter she wrote to Frederick II, Du Châtelet claims that Descartes's term of matter's "subtilité" was no longer appropriate as a scientific concept and had to be discarded after the latest observations English scholars had published in the *Philosophical Transactions*.

> La privation de l'air ne causa aucune altération au mouvement de cette montre, ce qui est une belle preuve contre l'explication que les cartésiens donnaient du ressort; car, si la Matière subtile en était la cause, l'air, qui est une Matière très subtile, devrait y contribuer. Il y a d'ailleurs d'autres raisonnements qui prouvent, [...] *que cette matière subtile n'existe pas*. L'expérience [...] a été faite à Londres par mr Derham, & v. a. r. peut en voir le détail & le succès dans les *Transactions philosophiques*, nro 194.[96]

---

[93] This corresponds to Châtelet's decision "de mettre *à la tête de l'ouvrage* quelques idées de mr. de Leibnits sur la métaphysique.", LetChBII 18, to Bernoulli, 30 June 1740 (my emphasis – U.W.).

[94] Apart from Chaps. IX and X, all chapters follow the titles and the order of the 1740 edition.

[95] Similar comments have to be made regarding the changes in Chap. X. Its title "De la Figure & de la Porosité des Corps" was extended considerably for the 1742 edition and now reads "De la figure, de la porosité, & de la solidité des Corps, et des causes de la cohésion, de la dureté, de la fluidité, & de la mollesse." The new titles of chap. IX and X of the 1742 edition now refer to precise physical terms supported by thorough scientific research.

[96] LetChBI, 340. To Frédéric, prince royal de Prusse (my emphasis – U.W.).

In 1742 Du Châtelet also adds some interesting changes to the introduction. In 1740, she had pointed out that some important principles of Leibniz's theorems had been taken from Wolff.[97] Du Châtelet, however, was well aware of the differences between the two, although many of her contemporaries considered her philosophy to be nearly identical. "Il [Wolff, U.W.] explique avec tant de clarté & d'éloquence le systeme de Mr. de Leibnits, qui a pris entre ses mains une forme toute nouvelle."[98] However, in her private correspondance, Du Châtelet spoke rather deprecatingly of Wolff after having read the three volumes of his Physics. "Je connais Mr Wolff pour un grand bavard en métaphysique. Il est plus concis dans les 3 tomes de sa physique mais il ne me paraît pas avoir fait de découvertes ni dans l'une ni dans l'autre."[99]

Still, regardless of Du Châtelets criticism in her private correspondence, Wolff is mentioned in her 1740 introduction to the *Institutions*. One of Châtelet's footnotes even prominently displays Wolff's name and refers to several chapters of "l'ontologie de WOLFF"[100] as a source for her own work. In the edition of 1742, however, the aforementioned reference to Wolff is missing.

The "Advance Copy" of 1740

In the following section I would like to point out some important differences between the above versions of the *Institutions* and another version, which I have discovered in the *Staatsbibliothek Preußischer Kulturbesitz*[101] and had the chance to examine in context with my research on Emilie du Châtelet. It is very likely that the version I have found is the "advance copy" Du Châtelet had sent to Frederick II of Prussia prior to the publication of the *Institutions des physique* in 1740.[102] The advance copy bears the note "Ex Biblioth. Regia Berolinensi." A handwritten comment reads "par Mad La Marquise Duchatelet, 1ere Edition". Although this edition from the royal library is exactly the same in both layout and formatting, it is missing the title page and therefore does not provide any information on the publisher and the date and place of publication. Apart from some minor orthographic changes, the text itself, however, is identical to the aforementioned edition of 1740. My assumption that this particular version of the *Institutions* is in fact an advance copy of the published study is further supported by the fact that the beginning of each chapter shows a large blank space probably left intentionally for the illustrations included in the first

---

[97] Inst1740, 12–13, par. XII. She adds: "Un de ses disciples … m'en faisoit quelquefois des extraits," even though she does not refer to König by name.

[98] Inst1740, 131.

[99] LetChBI, 264, à Maupertuis. 29 september 1738.

[100] The footnote of the 1740 edition shows WOLFF's name in capitals, see Inst1740, 12–13.

[101] The remaining chapters of the "advance copy" follow the edition from 1740 apart from some minor orthographic variants in 178 pages.

[102] Compare Barber. 1967. Madame du Châtelet and Leibnizianism: the genesis of the Institutions de physique. 213.

edition. I therefore argue that this find is indeed the "advance copy" Emilie du Châtelet sent to Frederick II.

It is remarkable in this context that the "advance copy" comprises two handwritten pages in the "Avant-propos" section, viz. paragraph XII on pages 13 and 14 of the 1740 edition. It is on these pages that Du Châtelet draws on the importance of Leibnizian metaphysics and also provides the aforementioned reference to Wolff's work.

The prominent footnote, however, is missing in this "advance copy" of the *Institutions*.[103] This of course poses the question of whether Du Châtelet subsequently added the footnote herself or, if she didn't, how and to what purpose it was added to the "advance copy" just before the printing of the first edition of the *Institutions* in 1740.

This omission of Du Châtelet's reference to Wolff's work is important for our understanding of Wolff's and Leibniz's influence on Du Châtelet's thought, even more so because the footnote with reference to Wolff was again omitted in the revised second edition in 1742.

It shows that König's influence and the significance of Wolff's *Ontologia* in favour of Du Châtelet's original studies of Leibniz's philosophy have been considerably overrated. Du Châtelet's extensive research into Leibniz and her work on his authentic material between 1737 and 1739 on the contrary prove that her reference to Wolff's work appeared no longer adequate for the second edition of her *Institutions* in 1742. The question remains, however, of why Du Châtelet omitted the footnote in her "advance copy", only to add it shortly before printing in 1740 and then cut it from the revised and second edition 2 years later. In view of Du Châtelet's severe criticism of Wolff's work in september 1738 in her letter to Maupertuis, I argue that the footnote could be inserted at the request of Frederick II whose appreciation for Wolff's philosophy is well known.

There are considerable changes to be found in parts of paragraph XII at the end of her introduction to the 1742 edition, which now explicitly states the importance of metaphysics and its close ties with the disciplines of physics and mathematics. "Plusieurs vérités de Physique, de Métaphysique & de Géométrie sont évidemment liées entre elles. La mètaphysique est le faîte de l'édifice."[104]

Towards the end of her introduction for the 1742 edition, Du Châtelet points out that metaphysics may not offer many clear-cut truths, yet the few it is in a position to offer are of great importance. According to Du Châtelet, it is vital to explain these "vérités métaphysiques" before introducing physics-related principles.

> Mais ce peu de vérités métaphysiques, que nous pouvons connaître a une si grande influence sur toutes celles qu'on peut découvrir dans les autres parties de la Philosophie, que je crois indispensable de vous en expliquer.[105]

---

[103] If I am right in assuming that this is the lost advance copy sent to Frederick II, then the question arises of whether Du Châtelet herself added the footnote and its reference to Wolff's influence, for according to my research, the first edition shows none but typographical additions by the publisher, when compared to the "advance-copy". In her letter to Frédéric, prince royal de Prusse, in april 1740 she announces her "essai de métaphysique" and explicitly states that Leibniz's philosophy is "la seule qui m'est satisfaite". See LetChBII, 13.

[104] These two page long passages were added to the advance copy by hand and were kept for the first edition in 1740 (apart from the above mentioned footnote).

[105] Inst1742Am, 15. This passage is particularly significant, for it was added in 1742.

She then proceeds to discuss and comment on Leibniz's philosophical principles. In Du Châtelet's 1740 edition of the *Institutions*, Leibniz's philosophy had been introduced as fundamentally important to the purposes of her study. In 1742, however, some of Leibniz's theorems were already being considered as philosophical truth and were consequently defined as *"vérités* métaphysiques".

A Change of Paradigms in an Energetic Universe – Du Châtelet's Institutions de Physique

In addition to her commentaries on Newton's and Leibniz's work, Du Châtelet's *Institutions* also provided her with an outlet for her own theories and can thus indeed be said to constitute a philosophical discourse in its own right. § XI illustrates Châtelet's view of her contributions to Leibniz's philosophy and her own theories as follows:

> La Physique est un batiment immense, qui surpasse les forces d'un seul homme. Les uns y mettent une Pierre, tandis que d'autres batissent des ailes entières. Mais tous doivent travailler sur les fondements solides qu'on a donnés à cet edifice dans ce dernier siècle, par le moyen de la géométrie & des observations; *il y en a d'autres qui levent le plan du Batiment; et je suis du nombre de ces derniers.*[106]

This shows that Emilie du Châtelet is not lacking confidence when she positions herself among those thinkers who are in a position to gain insight into the construction plan of the universe, "qui levent le plan du Batiment".

Along with mathematics and natural science, Du Châtelet here again ascribes great explanatory value to the discipline of metaphysics. Kant's posthumously published essays seem to agree with Du Châtelet's thoughts here, when he states that "it would be erroneous [...] to think that pre-definitions and mathematics, even when added to a good deal of observation and experiment, were enough to construct a physical science of its own; *it calls for metaphysics to ground it all in a general concept.*"[107]

Based on her assumption that mathematics, physics and metaphysics are inextricably linked, Du Châtelet was convinced that any attempt at physics was in fact strongly influenced by basic metaphysical concepts, since "la physique ne peut se passer de la métaphysique, sur laquelle elle est fondée."[108] Current research has considered this concept to be somewhat 'uncritical' in its approach to philosophy and natural science.[109]

---

[106] Inst1740, 12.

[107] See Kant. 1996c. Übergang von den metaphysischen Anfangsgründen der Naturwissenschaft zur Physik. 30–31. The German original reads: "Daß mit metaphysischen Vorbegriffen und Mathematik, ja auch mit einem reichen Vorrat von Beobachtung und Versuchen man sich schon zur Zimmerung einer Physik als einem System anschicken könne, ist nicht weniger irrig, wenn *keine Metaphysik den Plan zum Ganzen* [bietet]". (my emphasis – U.W.).

[108] LetChBII, 12–13, to Frédéric, prince royal de Prusse, 25 April 1740.

[109] See Barber. 1967. Madame du Châtelet and Leibnizianism: the genesis of the Institutions de physique. 222.

In contrast to this critique, even Kant in his basic *Metaphysical Foundations of Natural Science* emphasizes that every legitimate approach to natural science has to be grounded in a foundation of metaphysics.[110] Hegel also believed that any scientific research is to be based on metaphysical decisions, since every scientific decision as such obviously evokes its metaphysical implications.[111]

As Einstein points out in his *The Evolution of Physics*, the terms and definitions on which physics as a natural science is based, are in fact "pure inventions of the human mind that cannot be legitimised *a priori*, neither by the nature of human reason nor in any other way."[112] "Philosophical concepts", as Einstein goes on to say, "as far as they are based on scientific findings, influence the further development of scientific spirit".[113]

In her *Institutions de physique*, Du Châtelet seeks to form a clearly structured system out of various elements of both Leibniz's and Newton's theoretical systems, elements each of which she considered capable of analysing important aspects of philosophy, physics, astronomy and mathematics respectively. In accordance with her view that philosophy and natural science are closely related, Du Châtelet propounds the metaphysical principles of the structure of the universe (following Leibniz's concept) in Part I of the *Institutions*, and partly its physical principles in Part II.

Moreover, in the section of physics, she adopts Newton's principles of gravity and the structure of the planetary system as laid out in the *Principia*. Her debt to Leibniz centres on his concept of "force vive", his concept of time and space and some of his basic philosophical principles which introduce her study as "vérités métaphysiques".

This high value Du Châtelet ascribes to Leibniz's principles was by no means shared by the "Republique des Savants", and her introductory comments provide ample proof of the unfavourable situation, the dismissive attitudes and the wide-spread ignorance regarding Leibniz's ideas among the "Hommes de lettres" of her time.

> Les idées de M. de Leibnits sur la Métaphysique, sont encore peu connues en France, mais elles méritent assurément de l'être. Malgré les découvertes de ce grand homme, il y a sans doute encore bien des choses obscures dans la métaphysique.[114]

---

[110] Kant. 1996b. Metaphysische Anfangsgründe der Naturwissenschaft. In: Werkausgabe, Schriften zur Naturphilosophie. 13. Kant's title already indicates the close relationship between metaphysics and natural science. Not unlike Du Châtelet, Kant refers to the principles of metaphysics as "Anfangsgründe" for natural science.

[111] Hegel. 1959. *Einleitung in die Geschichte der Philosophie*. 41–42.

[112] Einstein. 1993. Zur Methodik der theoretischen Physik. 135. The original reads: "freie Erfindungen des menschlichen Geistes, die sich weder durch die Natur des menschlichen Geistes noch sonst in irgendeiner Weise *a priori* rechtfertigen lassen."

[113] Einstein and Infeld. 1950. *Die Evolution der Physik*. (The Evolution of Physics, dtsch.). 66: "Philosophische Konzepte, sofern sie auf wissenschaftlichen Forschungsergebnissen beruhen, beeinflussen die weitere Entwicklung des wissenschaftlichen Denkens." Einstein calls this chapter "Der philosophische Rahmen" ('philosophical framework').

[114] Inst1740, 13.

It seems perfectly clear to Du Châtelet, therefore, that she would have to defend Leibniz's theories against common ridicule, ridicule which was furthermore mostly based on Voltaire's ironic comments.

> C'est encore un des sentiments de M. de Leibnits, qui a le plus besoin d'être éclairci & *d'être sauvé du ridicule, dont on pourroit le charger*, que cette représentation de l'univers entier, & de tous ces changemens, qu'il prétend être un attribut de notre Ame.[115]

(a) From a mechanical cosmos to a dynamic and energetic structure of the universe

Leibniz's theory of monads together with his concept of force, which Du Châtelet repeatedly defended, facilitate and constitute a dramatic change to the paradigmatic description of universal order. Following Einstein in so far as we are "embedded in an ever-changing and progressive universe",[116] Leibniz's insights are very close to the contemporary view of the universe as being involved in constant change and progress due to its inherent dynamics. Succeeding Newton's discovery of the gravitational laws, which assisted mankind in grasping the infinite of the universe through the means of mathematics, new discoveries in chemistry, physics and microbiology brought about a further paradigm shift in the natural philosophy of the Enlightenment and a decisive change in mankind's understanding of universal order. As opposed to the "world machine" based on mathematical principles, the universe could now be understood as an energetic and dynamic concept; a concept shaped and inherently structured by a range of creative and progressive forces.

Newton's research on the macrocosm had resulted in his own "System of the World". Yet as further research into the true nature of microcosmic structures opened up an entirely new world – "une nouvelle nature", as Maupertuis calls it –, billions of micro-organisms had already started to widen the human horizon to new and unknown dimensions of infinity. Experimental microscopic research is, of course, primarily based on the dynamics of live structures; structures which generally tend to resist mathematical or quantitative approaches, since even the smallest drop of water or grain of sand appears to be impressively active and filled with life: "mille millions de corps mouvans que l'on découvre dans l'eau commune [...] ne sont pas si gros qu'un grain de sable ordinaire."[117] In addition to these changes, new and more experimental approaches to physical research furthermore brought to light dynamic, energetic and regenerative processes and reactions in nature which seemed to defy traditional and more mechanical approaches to terms such as matter and/or nature.[118]

---

[115] Inst1740, 143 (my emphasis – U.W.).

[116] Prigogine and Stengers. 1983. *Dialog mit der Natur*. 16.

[117] "Animalcules". In: *Encyclopédie ou Dictionnaire raisonnée des sciences, des arts et des métiers, par une société de gens de lettres*, vol. I. 475. Daubenton here refers to Leeuwenhoek's microscopic research.

[118] For a more detailed discussion see Winter. 2001. Naturphilosophie und Naturwissenschaften. 173–208.

In order to identify the cause of these processes, Du Châtelet here refers to an energy inherent in the elements of nature, namely to Leibniz's principle of the monads. "Les Etres simples", she writes, "sont donc doués d'une force, [...] *par l'énergie de laquelle ils tendent à agir.*"[119]

In order to grasp all forces of nature in a paradigmatic order, Newton's definition of physical bodies, as laid out in his *Principia*, proves to be inadequate when considered in the light of these more recent scientific discoveries. As opposed to Newton's tenet that the smallest parts of all bodies are extended and hard, impenetrable, and endowed with inertia, Leibniz defines matter as being energetic and nature as being ruled by its inherent forces and actions. For Leibniz, the threefold categories of "énergie", "force" and "activité" constitute the characteristics of all being,[120] and Du Châtelet was convinced that Leibniz's idea of the "force vive" had in fact laid open one of the fundamental secrets of creation.[121]

This, of course, also applies to Leibniz's concept of the monads. In 1740 Du Châtelet describes their inherent force as follows: "Chaque Estre simple est en vertu de sa nature & par sa force interne, dans un mouvement qui produit en lui des changemens perpétuels."[122]

The energy potential of the monads as the elements of being causes constant motion and change in the universe: "[l]a force des Etres simples se déploye continuellement, parce qu'elle produit des changemens sensibles à chaque instant."[123] Impulses of energy, motion and action she equates with force as such: "[l]es Etres simples ont un principe d'action, & c'est ce qu'on appelle *Force*."[124] This inherent force and activity causes motion, change and development in nature. "On observe dans les composés un changement perpétuel; rien ne demeure dans l'état où il est; *tout tend au changement dans la nature*"[125] Each element, no matter how small, possesses different individual characteristics, "comme vous venez de voir que les Etres simples sont tous dissemblables,"[126] Du Châtelet states in an important addition to the 1742 edition. Rather than being defined by a uniform mathematical structure, nature is here constituted by individual and complex systems of infinite diversity.

Du Châtelet outlines some of the most influential of Leibniz's metaphysical theories in the *Institutions*, and she understands them as being of ground-breaking

---

[119] Inst1740, 138 (my emphasis – U.W.). For the concept of energy, compare Heisenberg. 1959. *Physik und Philosophie*. 131: "Alle Elementarteilchen sind aus demselben Stoff gemacht, den wir nun Energie oder universelle Materie nennen können: sie sind nur verschiedene Formen, in denen Materie erscheinen kann."

[120] See Winter. 2004. Diderot und Leibniz. Die Leibniz-Rezeption in der Naturphilosophie der französischen Aufklärung. 57–69.

[121] LetChBI 217, to Maupertuis, 10 February 1738.

[122] Inst1740, 138.

[123] Inst1740, 138.

[124] Inst1740, 137; also in Inst1742Am, 143. (emphasis in the original).

[125] Inst1740, 137, also in Inst1742Am, 143. (my emphasis – U.W.).

[126] Inst1742, 145. This idea is particularly important for it is explicitly added to the 1742 edition and refers to Leibniz's principle of the identity of the Indiscernibles.

importance to the scientific explanation of the universe.[127] The 'Principle of Sufficient Reason' she defines as "fondement de la vérité"[128] and claims that it is "le seul fil qui puisse nous conduire dans ces labyrinthes d'erreurs".[129] Du Châtelet accepts Leibniz's 'Principle of Continuity' as a fundamental contribution to the natural sciences and follows his mathematical evidence. She also adheres to Leibniz's structural nexus of all being in the universe and in the human mind and follows his principle on the identity of the indiscernibles – the infinite complexity and uniqueness of all being – via an elaborate reference to the Leibniz-Clarke correspondence.

Here, too, Emilie du Châtelet supports one of the anticipatory aspects of Leibniz's thoughts. Leibniz's infinite complexity of all being shares considerably more common ground with current research than Newton's ideas, as Prigogine has pointed out in his study of new paths in natural science.

> Wherever we turn, we discern progress, diversification and instability. It is remarkable that this is true for all fundamental levels of research, be it in the realms of atoms, biology, [or] in astrophysics with its view of an ever expanding universe. [...] Natural science has rid itself of the concept of an objective reality in both the macro and the microcosms; an objective reality, which had thought it necessary to deny the new and the diverse in favour of an ever-lasting and non-negotiable law.[130]

(b) Newton's postulate of "Hypotheses non fingo"

Du Châtelet is very critical of Newton's ideas as far as the methodological value of hypothesis for natural science is concerned. Although Newton's well-known postulate of "hypotheses non fingo" in the *Principia* did, in fact, influence a wide range of mathematicians and physicists at the time, Du Châtelet in her work decides to adopt Leibniz's methodological concepts rather than Newton's widely accepted postulate. Newton had claimed that he

> [...] n'imagine point des hypothèses. Car tout ce qui ne se déduit point des phénomènes est une hypothèse: & les hypothèses, soit métaphysiques, soit physiques, soit mécaniques, soit celles des qualités occultes, ne doivent pas être recues dans la philosophie expérimentale. Dans cette philosophie, on tire les propositions des phénomènes, & et on les rend ensuite générales par induction.[131]

---

[127] For an in-depth discussion of Du Châtelet's influence on the reception of Leibniz's thought during the time of the Enlightenment please refer to the author's DFG project *Das vernetzte Universum. Die Leibniz-Rezeption in der Naturphilosophie der französischen Aufklärung* to be published soon by Franz Steiner Verlag Stuttgart as part of the *Studia Leibnitiana*.

[128] Inst1742Am, 99.

[129] Inst1740, 24–25.

[130] Prigogine and Stengers. 1983. Dialog mit der Natur. 284. The German edition reads: "Wohin wir auch blicken, finden wir Entwicklung, Diversifikation, und Instabilitäten. Dies gilt interessanterweise für alle grundlegenden Ebenen – im Bereich der Elementarteilchen, der Biologie, der Astrophysik, die uns ein expandierendes Universum zeigt. [...] Die Naturwissenschaften haben sich somit auf der makroskopischen wie auch auf der mikroskopischen Ebene von einer Konzeption der objektiven Realität befreit, die glaubte, das Neue und das Mannigfaltige im Namen eines unwandelbaren ewigen Gesetzes leugnen zu müssen."

[131] PrincChat1990, vol. II. 179. Also see "Table Alphabétique", vol I, 432: "HYPOTHESE cette philosophie les rejette de quelque espèce qu'elles soient."

However, following Newton in his efforts to reject *all* hypotheses, could only mean seriously endangering scientific progress, Du Châtelet writes. "Ce seroit donc faire un grand tort aux sciences, & retarder infiniment leurs progrès que d'en bannir avec quelques philosophes modernes, les hipotheses."[132]

From the very first pages of her *Institutions*, Du Châtelet decidedly attacks any opponents to scientific hypotheses by simply stating that "[u]n des torts de quelques Philosophes de ce téms, c'est de vouloir bannir les Hypothèses de la Physique; elles y sont aussi nécessaires que les èchafauts dans une maison que l'on bâtit." Even astronomy, she claims, has to be based on hypotheses. "Toute l'Astronomie, par exemple, n'est fondée que sur des Hypotheses."[133] It is here in chapter IV that Newton is identified as the original initiator of this methodological error.

> M. Newton, & surtout ses disciples, [...] se sont elevés contre les hipotheses, & ont tâché de les rendre suspectes & ridicules, en les appellant, *le poison de la raison, & la peste de la Philosophie*.[134]

In contrast to Newton's postulate that "hypotheses non fingo", Du Châtelet repeatedly underlines their importance for all kinds of scientific research. "Les hipotheses doivent donc trouver place dans les sciences, puisqu'elles sont propres à nous faire découvrir la verité, & à nous donner de nouvelles vûes."[135] In doing so, Du Châtelet again accepts one of Leibniz's basic concepts, which had already been mentioned in a letter to Conti and was published later in Desmaizeaux's *Recueil*:

> Cependant, si les Data ne suffisent point, il est permis [...] d'imaginer des hypothèses. Si elles sont heureuses, on s'y tient provisionellement en attendant que de nouvelles expériences nous apportent nova Data.[136]

However, instead of merely pointing out the function of hypotheses, as Leibniz had done before her, Du Châtelet goes even further in understanding them as the indispensable foundation of scientific research.

> Les bonnes hipotheses seront donc toujours l'ouvrage des plus grands hommes. Copernic, Képler, Hughens, Descartes, Leibnits, M. Newton lui même, ont tous imaginé des hipotheses utiles pour expliquer des Phénomenes compliquées & difficiles; & les exemples de ces grands hommes & leurs succés doivent nous faire voir combien ceux qui veulent bannir les hipotheses de la Philosophie, entendent mal les interêts des sciences.[137]

Hypotheses, Du Châtelet states, allow for new insights into scientific phenomena but always have to be validated and adjusted according to subsequent

---

[132] Inst1740, 75.

[133] Inst1742Am, 9.

[134] Inst1740, 75 (emphasis in the original).

[135] Inst1740, 76.

[136] Leibniz. 1768. Brief an Conti. 446. Dutens points out that this letter was taken from Desmaizeaux's *Recueil*. Leibniz also mentions this concept in his correspondence with Clarke, which Du Châtelet had already been familiar with at the time.

[137] Inst1740, 89.

scientific observation, since "[...] l'hipothese n'est vraie et ne mérite d'être adoptée que lorsqu'elle rend raison de toutes les circonstances."[138] They are therefore indispensable to scientific progress.

> Il est donc évident que c'est aux hipotheses successivement faites et corrigées que nous sommes redevables des belles & sublimes connoissances dont l'Astronomie & les sciences qui en dépendent sont à présent remplies.[139]

These remarks in chapter IV of her *Institutions* include groundbreaking analyses of the importance theoretical pre-constructs have for scientific research and empiric experiments in general, and Du Châtelet's ideas greatly influenced further development of methodological concepts. D'Alembert, for instance, quotes several passages from Du Châtelet's *Institutions* in his article on "Hypothèse" in the *Encyclopédie* clearly attributing them to Du Châtelet.

Apart from these transfers of fundamentally important paradigms in the fields of natural science, Du Châtelet's comments on and analyses of Leibniz's and Newton's ideas also provided the basis for several of her own theories on natural philosophy. The importance of theoretical assumptions in physics and natural science in general, however, has come to be an undisputed part of scientific methodology and theory to this very day.[140]

As always when she does not concur with Newton, Du Châtelet consciously avoids these methodological questions in her commentary to Newton's *Principia* in order to allow for a positive and self-contained exposition of his physical and astronomical theories.

Categories of Time and Space in Du Châtelet's Institutions de Physique

The fundamental difference inherent in Leibniz's and Newton's view of cosmic structure and their respective theories of time and space caused great controversy[141] throughout eighteenth-century Europe. The concept of space in particular provided both "a cornerstone of the system of any philosopher in modern times" and a valid meeting point for the two disciplines of philosophy and natural science, as Jammer points out. "The concept of space is, on the one hand one of the fundamental concepts of physics and, on the other – since all of material reality is spatial – one of the basic concepts of philosophy."[142]

---

[138] Ibid., 89.

[139] Inst1740, 79.

[140] See Poser. 2001. *Wissenschaftstheorie. Eine philosophische Einführung*. Also see Heisenberg. 1959. *Physik und Philosophie*. 179.

[141] The notorious controversy between Newton and Leibniz regarding the infinitesimal calculus was, of course, also part of this controversy.

[142] Jammer. 1969. *Concepts of Space: The history of theories of space in physics*. 1.

It was Einstein who finally defined these two divergent theories of the concept of space by stating that "[t]hose two concepts of space may be contrasted as follows: (a) space as positional quality of the world of material objects; (b) space as container of all material objects." In case (b), the notion of absolute space exists independently of its material objects, thus establishing the concept of absolute space as 'superior' to the material world.

> In case (a), space without a material object is inconceivable. In case (b), a material object can only be conceived as existing in space; space then appears as a reality which in a certain sense is superior to the material world.[143]

Newton's principle of absolute space coincides with case number two in Einstein's definition of space. "L'espace absolu, sans relation aux choses externes, demeure toujours similaire et immobile,"[144] Newton claims, since to him space and time as unchangeable entities are primordial to any subsequent contents induced to them.

By contrast, Leibniz posits temporal and spatial relativity as the structuring principles of all being.[145]

> J'ay démontré que l'espace n'est autre chose qu'un ordre de l'existence des choses, qui se remarque dans leur simultaneité. Ainsi la Fiction d'un Univers materiel fini, qui se promene tout entier dans un espace vuide infini, ne sauroit être admise. [...] Ce sont des imaginations des Philosophes à notions incompletes, qui se font de l'espace une réalité absolue.[146]

Leibniz understands the notion of space as a theoretical assumption grounded in physical relations. He also favours a relational theory of time, which he defines as an "ordinal structure of being". In his correspondence with Clarke, he consequently discards Newton's idea of absolute space, "l'Espace reel absolu" as an "idole de quelques Anglois".

> Pour moy, j'ay marqué plus d'une fois, *que je tenois l'Espace pour quelque chose de purement relatif, comme le Temps;* pour un ordre des Coexistences, comme le temps est un ordre de successions.[147]

Regarding the theorems of space and time in their function as a foundation for further philosophical and scientific ideas,[148] Du Châtelet clearly adopts Leibniz's theories on the relativity of space and time as ordinal structures of being, as abstract notions, space being the order of coexistence, time is the order of succession.

---

[143] See Albert Einstein's preface in: Jammer. 1969. *Concepts of Space: The history of theories of space in physics*. XIII.

[144] PrincChat1990 I, 8.

[145] See Poser. 1989. Leibnizens Theorie der Relativität von Raum und Zeit. 125. For a detailed discussion of the reception of Leibniz's concept of time and space during the time of the Enlightenment, see Winter. 2005. Zeitreise in die Unendlichkeit des Universums – Leibniz und Diderot zur Struktur des Kosmos. http://www.inst.at/trans/16Nr/121/winter16.htm

[146] Streitschriften zwischen Leibniz und Clarke, Fünftes Schreiben. 1875–1890. 369.

[147] Streitschriften zwischen Leibniz und Clarke, Drittes Schreiben. 363 (my emphasis – U.W.).

[148] See Jammer. 1969. *Concepts of Space: The history of theories of space in physics*. I. See also Weizsäcker, and Rudolph. 1989. *Zeit und Logik bei Leibniz. Studien zu Problemen der Naturphilosophie, Mathematik, Logik und Metaphysik*.

> Les notions du Tems & de l'Espace ont beaucoup d'analogie entre elles; dans l'Espace, on considère simplement l'ordre des coexistans, entant qu'ils coexistent; & dans la durée, l'ordre des choses successives, entant qu'elles se succèdent.[149]

She explicitly refers to the famous controversy between Leibniz and Clarke/Newton by stating that

> Mr. Clarke s'est donné beaucoup de peine pour soutenir les sentiments de Mr. Newton, & les siens propres sur l'Espace absolu, contre Mr. de Leibnits, qui pretendoit que l'Espace n'étoit que l'ordre des choses coexistantes.

Du Châtelet goes on to state that, in this particular dispute of European relevance and interest, she adopts Leibniz's position, since "[o]n ne peut se dispenser d'avouer que Mr. de Leibnits avoit raison de bannir l'Espace absolu de l'Univers."[150] Newton's ideas on absolute space and time she defines as "idées confuses", since time according to Newton's theory is endowed with divine attributes: like God, it is necessary, eternal, invariable, and of self-contained existence.

> On considère ordinairement le Tems de même que l'Espace [...] par des idées confuses: ainsi, on se le figure comme un Etre qui coule uniformément, indépendamment des choses qui existent dans le Tems qui a été dans un flux continuel de toute éternité.[...] le Tems seroit un Etre nécessaire, immuable, eternel, subsistant par lui-même, par conséquent les atrributs de Dieu lui conviendroient.[151]

Repeatedly, Du Châtelet favours Leibniz's idea of relativity of space. "Ainsi le raisonnement de Mr. de Leibnits contre l'Espace absolu est sans replique, & l'on est forcé d'abandonner cet Espace." Newton's concept of absolute space, she argues on the other hand, implies absurd conclusions, since in his theorem space is endowed with God's "immensite" and could only be understood as God's omnipresent "sensorium".[152]

> L'autorité de M. Newton a fait embrasser l'opinion de vuide absolu à plusieurs Mathematiciens. Ce grand homme croyoit, au rapport de M. Locke, qu'on pouvoit appliquer la création de la matière par l'Espace, en se figurant que Dieu auroit rendu plusieurs parties de l'Espace impénétrables: on voit dans le *Scholium generale* qui est à la fin des Principes de M. Newton, qu'il croyoit que l'Espace étoit l'immensité de Dieu, et il l'appelle dans son Optique le *Sensorium* de Dieu; c'est-à-dire, ce, par le moyen de quoi Dieu est présent à toutes choses.[153]

---

[149] Inst1742Am, 118. Her subtitle in § 102 again emphasises the relativity of time: "Le Tems n'est donc réellement autre chose que l'ordre des Etres successifs; & on s'en forme une idée, entant qu'on ne considère que l'ordre de leur succession. Ainsi, il n'y a point de Tems sans les Etres successifs, rangés dans une suite continue."

[150] Inst1742Am, 97.

[151] Inst1742Am, 118–119. Her subheading reads "Opinion singulière de M. Newton sur l'espace" and she critically underlines that "M. de Leibnits n'eut pas de peine à renverser cette objection du Docteur Anglois."

[152] Ibid. 97. "Il y a encore une grande absurdité à dévorer dans l'opinion de l'Espace absolu."

[153] Inst1740, 92, also in 1742Am. (emphasis in the original).

These particular paragraphs of the *Institutions* constitute Du Châtelet's commentary on both the "Scholium generale" and Newton's philosophical reflections from the beginning of the *Principia*,[154] which were consciously omitted in her Newton commentary. This can be seen as further evidence that the *Institutions de physique*, being first and foremost an analysis of natural philosophy inspired by Leibniz, and her commentary on Newton, a self-contained physical-astronomical treatise, complement each other.

Newton's basic philosophical principles and his paradigm of nature (consciously left aside in her commentary to the *Principia*) are discussed extensively in the *Institutions* to be rejected in favour of Leibniz's understanding of potential energy and of relativity of time and space.

Again, Du Châtelet opts for the anticipatory theories of Leibniz. since it was no other than Einstein himself who stresses physics' departure from the concept of absolute time.

"As a framework for the empirical sciences", Einstein writes, "the one-dimensional time continuum together with the three-dimensional space continuum was no longer valid."[155]

In opposition to the predominant 'ironic' discourse surrounding Leibniz's thoughts among the scholars of the "République des Savants" during the time of the French Enlightenment, Du Châtelet's analytic commentary in the *Institutions de physique* greatly contributed to a programmatic shift in the reception of Leibniz's philosophical theorems and stimulated a critical examination of the authentic thoughts of Leibniz in France.[156] Instead of merely following Voltaire's predominant ironic comments on Leibniz's philosophy, Du Châtelet initiated a philosophic discourse which was aimed at a critical re-evaluation of Leibniz's thoughts towards the development of a more thorough and complex paradigm of nature and cosmos. The universe was no longer understood as a 'machine', working on everlasting and fixed principles and mechanisms; rather, it was established that it possessed an inherent energetic and dynamic structure which was linked to its continuous evolutionary processes, thereby illustrating its complexity, its uniqueness and its perpetual self-development.

"According to our understanding", Prigogine points out, "contemporary science is distancing itself from Newton's myth, having rediscovered diversity and time."[157]

---

[154] Regarding her critique of Newton's concept of absolute space, Du Châtelet explicitly states her sources, namely John Locke's translation and the *Commercium epostolicum*. The margin quotes "Traduction de M. Locke pag. 521 note 2", and "Commercium epistolicum" in relation to Clarke's concepts. See also Inst1740, 124.

[155] Einstein and Infeld. 1950. *Die Evolution der Physik*. 319. The German text reads: "Als Rahmen für das Naturgeschehen wurde fortan nicht mehr das eindimensionale Zeitkontinuum in Verbindung mit dem dreidimensionalen Raumkontinuum angesehen."

[156] A range of other factors regarding the reception of Leibniz's thoughts are the Academy's call for papers on Leibniz's philosophy and several new publications of his work and letters towards the middle of the eighteenth century.

[157] Prigogine and Stengers. 1983. *Dialog mit der Natur*. 62: "Nach unserer Auffassung löst sich die Wissenschaft heute vom Newtonschen Mythos, weil sie die Vielfalt und die Zeit wiederentdeckt hat."

References

1966–1967. Feu. In *Encyclopédie ou Dictionnaire raisonnée des sciences, des arts et des métiers, par une société de gens de lettres*, eds. D. Diderot and J. L. R. d'Alembert, vol. XI. Stuttgart.

Badinter, E., and D. Muzerelle, eds. 2006. *Madame du Châtelet. La femme des Lumières*. Paris: Bibliothèque Nationale de France.

Barber, W. H. 1967. Mme du Châtelet and Leibnizianism: the genesis of the *Institutions de physique*. In *The Age of Enlightenment: Studies presented to Theodore Besterman*, ed. William H. Barber, J.H Brumfitt, R.A Leigh, R. Shakleton, and S.S.B Taylor, 200–222. Edinburgh/London: Oliver and Boyd.

Belaval, Y. 1993. Quand Voltaire rencontre Leibniz. In *Etudes leibniziennes: de Leibniz à Hegel*, ed. Yvon Belaval. Paris: Gallimard.

Biarnais, M.-F. 1982. *Les Principia de Newton. Genèse et structure des chapitres fondamentaux avec traduction nouvelle*. Cahiers d'histoire et de philosophie des sciences 2. Paris: Société française d'histoire des sciences et des techniques.

Carboncini, S. 1988. Vorwort. In *Gesammelte Werke*, ed. Christian Wolff, vol. 28, Abt. 3: Materialien und Dokumente. Hildesheim/Zürich/New York: Olms.

D'Alembert, J. B. 1904. *Discours préliminaire de l'Encyclopédie*, ed. F. Picavet. Paris: Colin.

Darnton, R. 1979. *The business of enlightenment*. Cambridge: Belknap Press.

De la Harpe, J. 1941. *Le Journal des Savants et l'Angleterre, 1702 –1789*. Berkeley: University of California Press.

Des Maizeaux. 1720. *Recueil de Diverses Pièces sur la Philosophie, la Religion naturelle, l'Histoire, les Mathématiques, etc. par Mrs. Leibniz, Clarke, Newton et autres Autheurs célèbres*. Amsterdam.

Diderot, D. 1955–1973. *Correspondance*, ed. Georges Roth. Paris: Minuit.

Diderot, D., and J. L. R. D'Alembert, eds.1966–1967. *Encyclopédie ou Dictionnaire raisonnée des sciences, des arts et des métiers, par une société de gens de lettres (1751–1765)*. Stuttgart: Frommann.

Einstein, A. 1993. Zur Methodik der theoretischen Physik. In *Mein Weltbild*. Frankfurt a.M./Berlin.

Einstein, A., and L. Infeld. 1950. *Die Evolution der Physik*. Berlin/Darmstadt: Deutsche Buchgemeinschaft. Epitre sur l'amour de l'étude, à Madame la Marquise du Chastelet, par un élève de Voltaire. 1815. In *Mélanges*, ed. Fayolle, Paris.

Falkenburg, B. 1989. Kants Einwände gegen Symmetrieargumente bei Leibniz. In *Zeit und Logik bei Leibniz. Studien zu Problemen der Naturphilosophie, Mathematik, Logik und Metaphysik*, ed. Carl Friedrich Weizsäcker, Enno Rudolph, 148–180. Stuttgart.

Freudenthal, G. 1986. *Atom and Individual in the Age of Newton: On the Genesis of the Mechanistic World View*. Boston Studies in the Philosophy of Science 88. Dordrecht/Boston/Tokyo: Reidel Publishing Company.

Gandt, F. de, ed. 2001. *Cirey dans la vie intellectuelle: la réception de Newton en France*. Oxford: Voltaire Foundation.

Guicciardini, N. 1999. *Reading the Principia: The debate on Newton's mathematical methods for natural philosophy from 1687 to 1736*. Cambridge: Cambridge University Press.

Hagengruber, R. 1999. Eine Metaphysik in Briefen. Emilie du Châtelet an Maupertuis. In *Pierre Louis Moreau de Maupertuis (1698–1759). Eine Bilanz nach 300 Jahren*, ed. Hartmut Hecht, 187–206. Berlin: Spitz-Verlag.

Halley, E. 1988. Geleitwort zu*: Isaaci Newtoni Opera quae Extant Omnia*. In *Mathematische Grundlagen der Naturphilosophie*, ed. Ed Dellian, 7–8. Hamburg: Meiner.

Hegel. G. W. F. ³1959. *Einleitung in die Geschichte der Philosophie*, ed. Johannes Hoffmeister. Hamburg: Meiner.

Heisenberg, W. 1959. *Physik und Philosophie*. Berlin: Ullstein.

Jammer, M. 1969. *Concepts of Space: The history of theories of space in physics*. Cambridge: Harvard University Press.

Kant, I. 1996. Allgemeine Naturgeschichte und Theorie des Himmels oder Versuch von der Verfassung und dem mechanischen Ursprunge des ganzen Weltgebäudes nach Newtonischen Grundsätzen abgehandelt. In Immanuel Kant, *Vorkritische Schriften bis 1768*, ed. Wilhelm Weischedel, vol. 1. Frankfurt a. M.: Suhrkamp.

Kant, I. 1996. Metaphysische Anfangsgründe der Naturwissenschaft. In *Werkausgabe*, ed. Wilhelm Weischedel, vol. 9. Frankfurt a.M.: Suhrkamp.

Kant, Immanuel. 1996. Übergang von den metaphysischen Anfangsgründen der Naturwissenschaft zur Physik. Aus dem Opus posthumum, ed. Ingeborg Heidemann. Hildesheim/Zürich/New York: Olms.

Kant, Immanuel. 1996. Gedanken von der wahren Schätzung der lebendigen Kräfte und Beurtheilung der Beweise, derer sich Herr von Leibnitz und andere Mechaniker in dieser Streitsache bedienet haben nebst einigen vorhergehenden Betrachtungen, welche die Kraft der Körper überhaupt betreffen". In *Vorkritische Schriften*, ed. Wilhelm Weischedel, vol. 1. Frankfurt a. M.: Suhrkamp.

Klens, U. 1994. *Mathematikerinnen im 18. Jahrhundert: Maria Gaetana Agnesi, Gabrielle-Emilie Du Châtelet, Sophie Germain: Fallstudien zur Wechselwirkung von Wissenschaft und Philosophie im Zeitalter der Aufklärung*. Forum Frauengeschichte 12. Pfaffenweiler: Centaurus.

Knobloch, E., and S. Probst, eds. 2003. Vorwort. In *Sämtliche Schriften und Briefe*, ed. Leibniz, G.W. Mathematische Schriften. Reihe 7. vol. 3. Berlin: Akademie Verlag.

Kölving, U., and O. Courcelle., eds. 2008. *Émilie du Châtelet: Éclairages & documents nouveaux*. [… du Colloque du Tricentenaire de la Naissance d'Emilie du Châtelet qui s'est tenu du 1er au 3 juin 2006 à la Bibliothèque Nationale de France et à l'ancienne Mairie de Sceaux]. Ferney-Voltaire: Centre International d'Étude du XVIIIe Siècle.

Koyré, A. 1957. *From the Closed World to the Infinite Universe*. Baltimore: Johns Hopkins Press.

Kuhn, T. ^{12}1993. *Die Struktur wissenschaftlicher Revolutionen*. Frankfurt a.M.: Suhrkamp.

La Mettrie, J. O. de. 1970. *Oeuvres philosophiques 1774*. Hildesheim/New York: Olms.

Leibniz, G. W. 1768. Brief an Conti. In *Leibnitii Opera omnia*, ed. Ludovicus Dutens, vol. 3, Hildesheim, Olms.

Leibniz, G. W. 1965. *Die philosophischen Schriften*, ed. Carl Immanuel Gerhardt. 7 vol. Hildesheim. Olms.

Mainzer, K. 1988. *Symmetrien der Natur. Ein Handbuch zur Natur- und Wissenschaftsphilosophie*. Berlin/New York: de Gruyter.

Mainzer, K. 1996. Der Krieg der Philosophen. Zum Verhältnis von Physik, Philosophie und Religion bei Leibniz bis zur Aufklärung. In *Die andere Hälfte der Wahrheit*, ed. Jürgen Andretsch, 72–95. München: Beck.

Mainzer, K. 2003. *KI. Künstliche Intelligenz. Grundlagen intelligenter Systeme*. Darmstadt: Wiss. Buchges.

Pemberton, H. 1728. *A View of Sir Isaac Newton's Philosophy*. London: Palmer.

Pope, A. 1954. Epitaph intended for Sir Isaac Newton. In *Minor Poems*, ed. Norman Ault, 317. London/New Haven: Methuen.

Poser, H. 1989. Leibnizens Theorie der Relativität von Raum und Zeit. In *Philosophie-Physik-Wissenschaftsgeschichte*, ed. W. Muschik and W.R. Shea, 123–138. Berlin.

Poser, H. 2001. *Wissenschaftstheorie. Eine philosophische Einführung*. Stuttgart: Reclam.

Prigogine, I., and I. Stengers. ^{4}1983. *Dialog mit der Natur. Neue Wege naturwissenschaftlichen Denkens*. München/Zürich: Piper.

Rousseau, J.-J. 1971. *Discours sur les Sciences et les Arts*. Paris: Garnier-Flammarion.

Streitschriften zwischen Leibniz und Clarke, 1875–1890. In *Die Philosophischen Schriften von Gottfried Wilhelm Leibniz*, ed. Carl Immanuel Gerhardt, vol. 7, Berlin.

Terrall, M. 2004. Vis viva revisited. *History of Science* 42, no. 2:189–209.

Wade, I. O. 1947, ^{2}1967. *Studies on Voltaire. With some unpublished papers of Mme du Châtelet*. Princeton: Princeton University Press.

Walters, R. L. 2001. La querelle des forces vives et le rôle de Mme du Châtelet. In *Cirey dans la vie intellectuelle: la réception de Newton en France*, ed. François de Gandt, 198–211. Oxford: Voltaire Foundation.

Wehinger, B. 2003. Experimente auf der Suche nach dem Glück. Gabrielle Emilie du Châtelet und Voltaire. In *Das literarische Paar – Le couple litteraire. Intertextualität der Geschlechterdiskurse – Intertextualité et discourses des sexes*, ed. Gislinde Seybert, 89–110. Bielefeld: Aisthesis Verlag.

Wehinger, B., and Brown, H. 2008. Übersetzungskultur im 18. Jahrhundert. Übersetzerinnen in Deutschland, Frankreich und der Schweiz. Hannover-Laatzen: Wehrhahn-Verlag.

Weizsäcker, C. F. von, and E. Rudolph. 1989. *Zeit und Logik bei Leibniz. Studien zu Problemen der Naturphilosophie, Mathematik, Logik und Metaphysik*. Stuttgart: Klett-Cotta.

Winter, U. 1972. *Der Materialismus bei Diderot*, Genf/Paris: Droz et Minard.

Winter, U. 2001. Naturphilosophie und Naturwissenschaften. In *Die Wende von der Aufklärung zur Romantik*, ed. Horst A. Glaser and György M. Vajda, 173–208. Amsterdam/Philadelphia: John Benjamins Publishing.

Winter, U. 2004. Diderot und Leibniz. Die Leibniz-Rezeption in der Naturphilosophie der französischen Aufklärung. *Studia Leibnitiana* XXXVI/1: 57–69.

Winter, U. 2005. Zeitreise in die Unendlichkeit des Universums – Leibniz und Diderot zur Struktur des Kosmos. http://www.inst.at/trans/16Nr/121/winter16.htm

Winter, U. 2006. Vom Salon zur Akademie. Emilie du Châtelet und der Transfer naturwissenschaftlicher und philosophischer Paradigmen innerhalb der europäischen Gelehrtenrepublik des 18. Jahrhunderts. In *Höfe - Salons - Akademien. Kulturtransfer und Gender im Europa der Frühen Neuzeit*, ed. Gesa Stedman and Margarete Zimmermann, 285–306. Hildesheim: Olms.

Winter, U. 2007. Übersetzungsdiskurse der französischen Aufklärung: Die Newton-Übersetzung Emilie du Châtelets. In *Übersetzerinnen im 18. Jahrhundert*, ed. Brunhilde Wehinger and Hilary Brown. Hannover-Laatzen: Wehrhahn Verlag.

Zacher, H. J. 1973. *Die Hauptschriften zur Dydaktik von G. W. Leibniz. Ein Beitrag zur Geschichte des binären Zahlensystems*. Frankfurt a. M.: Klostermann.

Zinsser, J. P. 2001. Translating Newton's 'Principia': The Marquise du Chatelet's Revisions and Additions for a French Audience. *Notes and Records of the Royal Society of London* 55, no. 2:227–45.

Zinsser, J. P. 2006. *La dame d'esprit A Biography of the Marquise du Châtelet*. New York: Viking.

Zinsser, J. P., and J. C. Hayes, eds. 2006. *Emilie du Châtelet: rewriting Enlightenment philosophy and science*. Oxford: Voltaire Foundation.

Emilie du Châtelet, a Bibliography

Ana Rodrigues

Primary Sources

Du Châtelet, E. Discours sur le bonheur. Mazarine: no. 4.344.
Du Châtelet, E. Examen de la Genèse. Manuscripts non autographés. Bibliothèque de Troyes: no. 2376.
Du Châtelet, E. Examen des livres du Nouveau Testament. Manuscripts non autographés. Bibliothèque de Troyes: no. 2377.
Du Châtelet, E. L'Essai sur l'optique. Handschriftenband. UB Basel: L I a 755, fo. 230–265.
Du Châtelet, E. Réflexions sur le bonheur. Bibliothèque Nationale: ffr. 15.331.
Du Châtelet, E. Institutions de Physique. Bibliothèque Nationale: ffr. 12.265.
Du Châtelet, E. Principes de la Philosophie naturelle, par M. Newton, traduits en français par Mme la Marquise du Chastellet, avec un Commentaire sur les propositions qui ont rapport au système du monde. Bibliothèque Nationale: ffr. 12.266–12.268.
Du Châtelet, E. Lettres autographes de la marquise du Châtelet. Bibliothèque Nationale: ffr. 12.269.
Du Châtelet, E. Sur "Descartes" par Mme du Châtelet [Autogr.]. National Library of Russia, St. Petersburg. Voltaire Collection. Vol. IX: 122.
Du Châtelet, E. Notes sur la "physique" par la même [Autogr.]. National Library of Russia, St. Petersburg. Voltaire Collection. Vol. IX: 123.
Du Châtelet, E. Essai inédit de Mme du Châtelet, chap. v: Sur la liberté [Minutes avec corrections autographes.]. National Library of Russia, St. Petersburg. Voltaire Collection. Vol. IX: 126.
Du Châtelet, E. Chap. vi: Des mots en général considérés selon leurs signification grammaticale, chap. vii.: Des verbes auxiliaires; chap. viii.: Des mots qui désignent les opérations de notre entendement sur les objets. [Min. avec corr. autogr.]. National Library of Russia, St. Petersburg. Voltaire Collection. Vol. IX: 133.
Du Châtelet, E. Pensées de Madame du Châtelet [Autogr.]. National Library of Russia, St. Petersburg. Voltaire Collection. Vol. IX: 150.
Du Châtelet, E. Lettre de *** à Mad. du Châtelet, 19 octobre 1747 [Orig.]. National Library of Russia, St. Petersburg. Voltaire Collection. Vol. IX: 152.

A. Rodrigues (✉)
Universität Paderborn, Fach Philosophie, Warburger Str. 100, 33098 Paderborn, Germany
e-mail: ana.rodrigues@upb.de

Du Châtelet, E. Traduction de la Fable des abeilles de Mandeville [Copie avec corr. autogr.]. National Library of Russia, St. Petersburg. Voltaire Collection. Vol. IX: 153.

Du Châtelet, E. Préface de cette traduction [Min. autogr.]. National Library of Russia, St. Petersburg. Voltaire Collection. Vol. IX: 217.

Du Châtelet, E. Préface de Mad. la marquise du Chastellet à la tête de sa traduction de la Fable des abeilles, et dissertation sur la liberté [Copie avec corr. autogr.]. National Library of Russia, St. Petersburg. Voltaire Collection. Vol. IX: 223.

Du Châtelet, E. Traduction de la Fable des abeilles par Mme du Châtelet. [Copie.]. National Library of Russia, St. Petersburg. Voltaire Collection. Vol. IX: 240.

Du Châtelet, E. Institutions de physique par Mme du Châtelet, chap. iv : de la formation des couleurs [Copie.]. National Library of Russia, St. Petersburg. Voltaire Collection. Vol. IX: 286.

Du Châtelet, E. 1738. Lettre sur les *Eléments de la Philosophie de Newton*. Paris. *Journal des sçavans*:534–41.

Du Châtelet, E. 1738. Lettre sur les *Eléments de la Philosophie de Newton*. Amsterdam. *Journal des sçavans*:458–75.

Du Châtelet, E. 1739. Dissertation sur la nature et la propagation du feu. In *Recueil des pièces qui ont remporté le prix de l'Académie royale des Sciences en 1738*, ed. Académie royale des Sciences, 85–168. Paris: Imprimerie royale.

Du Châtelet, E. 1740. *Institutions de physique*. Paris: Prault

Du Châtelet, E. 1741. *Institutions de physique*. Amsterdam: Pierre Mortier

Du Châtelet, E. 1741. *Institutions de physique*. London: Paul Vaillant

Du Châtelet, E. 1741. *Réponse de madame la marquise du Chastellet à la Lettre que M. de Mairan, secrétaire perpétuel de l'Académie royale des Sciences, lui a écrite le 18 février 1741 sur la question des forces vives*. Bruxelles: Foppens.

Du Châtelet, E. 1741. *Zwo Schriften, welche von der Frau Marquise von Chatelet, gebohrener Baronessinn von Breteuil, und dem Herrn von Mairan, beständigem Sekretär bei der französischen Akademie der Wissenschaften, das Maaß der lebendigen Kräfte betreffend, gewechselt worden: aus dem Französischen übersetzt von Louise Adelgunde Victoria Gottsched, geb. Kulmus*. Leipzig: Bernh. Breitkopf.

Du Châtelet, E. 1742. *Institutions physiques de madame la marquise du Chastellet adressés à M. son fils: Nouvelle édition, corrigée et augmentée considérablement par l'auteur*. Amsterdam: Aux dépens de la Compagnie

Du Châtelet, E. 1743. *Der Frau Marquisinn von Chastellet Naturlehre an ihren Sohn. Erster Theil nach der zweyten Französischen Ausgabe übersetzet von Wolfgang Balthasar Adolf von Steinwehr Prof. Publ. Ord. auf der Universitet zu Frankfurt an der Oder, derselben Bibliothecario, und der Königl. Preußischen Societet der Wissenschaften Mitgliede*. Halle/Leipzig: Rengerische Buchhandlung.

Du Châtelet, E. 1743. *Istituzioni di Fisica di Madama la Marchesa du Chastelet indiritte a suo figliuolo. Traduzione dal linguaggio francese nel toscano, accresciuta con la Dissertazione sopra le forze motrizi di M. de Mairan*. Venedig: Presso Giambatista Pascali.

Du Châtelet, E. 1744. *Dissertation sur la nature et la propagation du feu*. Paris: Prault Fils.

Du Châtelet, E. 1747. Mémoire touchant les forces vives adresseè [sic] en forme de lettre à M. Jurin par madame Ureteüil [sic] Du Chastellet. In *Memorie sopra la fisica e istoria naturale di diversi valentuomine*, ed. Carlantonio Giuliani, 75–84.

Du Châtelet, E. 1752. Dissertation sur la nature et la propagation du feu. In *Recueil des pièces qui ont remporté le prix de l'Académie royale des Sciences depuis leur fondation jusqu'à présent, avec les pièces qui y sont concouru, depuis 1738 jusqu'en 1740: Tome quatrième, contenant les pièces depuis 1738 jusqu'en 1740*, ed. Académie royale des Sciences, 87–170. Paris: Gabriel Martin, J. B. Coignard, Hippolyte-Louis Guérin, Charles-Antoine Jombert.

Du Châtelet, E., and I. Newton. 1756. *Principes mathématiques de la philosophie naturelle: par feue Madame la marquise Du Chastellet*. 2 vols. Paris: Desaint & Saillant.

Du Châtelet, E. 1759. Exposition abrégée du Système du monde, et explication des principaux phénomènes astronomiques tirée des Principes de M. Newton. In *Principes mathématiques de la philosophie naturelle de Newton: Par feue madame la marquise Du Chastellet*. 2 vols. Paris: Desaint et Saillant.

Du Châtelet, E., and I. Newton, 1759. *Principes mathématiques de la philosophie naturelle de Newton: Par feue madame la marquise Du Chastellet*. 2 vols. Paris: Desaint et Saillant.
Du Châtelet, E. 1771. *Briefwechsel der Marquise du Châtelet mit Mairan*, ed. Adelheid Gottsched. Leipzig: Bernh. Breitkopf.
Du Châtelet, E. 1777. Idées sur le bonheur. In *Journal étranger de littérature, des spectacles et de politique: Ouvrage périodique*, ed. G. Bigg and P. Elmsly, 61–73. London.
Du Châtelet, E. 1779. Discours sur le bonheur par feue Mme Du Châtelet. In *Huitième recueil philosophique et littéraire de la Société typographique de Bouillon*, ed. Société typographique de Bouillon, 1–36. Bouillon: Aux dépens de la Société typographique.
Du Châtelet, E. 1781. Marquise du Châtelet à Moussinot: 11. sept. 1738, 17. fev. 1741. In *Lettres de M. de Voltaire à M. l'abbé Moussinot son trésorier, écrites depuis 1736 jusqu'en 1742, pendant sa retraite à Cirey, chez madame la marquise du Châtelet, et dans lesquelles on voit quelques détails de sa fortune, ses bienfaits; quelles furent alors ses études, ses querelles avec Desfontaines etc. Publiées par l'abbé D****, ed. l'abbé Duvernet, 165, 184–185. La Haye/Paris: Moutard.
Du Châtelet, E., and Voltaire. 1782. *Lettres de M. de Voltaire et de sa célèbre amie: Suivies d'un petit poème, d'une lettre de J. J. Rousseau & d'un parallèle entre Voltaire et J. J. Rousseau*. Genève/Paris: Cailleau.
Du Châtelet, E. 1792. *Doutes sur les religions révélées adressées à Voltaire, par Emilie du Châtelet: ouvrage posthume*. Paris.
Du Châtelet, E. 1796. Réflexions sur le bonheur. In *Opuscules philosophiques et littéraires, la plupart posthumes ou inédits*, ed. Jean-Baptiste-Antoine Suard and Simon-Jérôme de Bourlet Vauxcelles, 1–40. Paris: De l'Imprimerie de Chevet.
Du Châtelet, E. 1806. *Lettres inédites de Madame la marquise du Châtelet à M. le Comte d'Argental, auxquelles on a joint une Dissertation sur l'existence de Dieu, les Réflexions sur le bonheur, par le même auteur, et deux notices historiques sur madame Du Chastelet et M. d'Argental*. Paris: Xhrouet, Déterville, Lenormand, Petit.
Du Châtelet, E. 1806. Dissertation sur l'existence de Dieu. In *Lettres inédites de Madame la marquise du Châtelet à M. le Comte d'Argental, auxquelles on a joint une Dissertation sur l'existence de Dieu, les Réflexions sur le bonheur, par le même auteur, et deux notices historiques sur madame Du Chastelet et M. d'Argental*. Paris: Xhrouet, Déterville, Lenormand, Petit.
Du Châtelet, E. 1806. Réflexions sur le bonheur. In *Lettres inédites de Madame la marquise du Châtelet à M. le Comte d'Argental, auxquelles on a joint une Dissertation sur l'existence de Dieu, les Réflexions sur le bonheur, par le même auteur, et deux notices historiques sur madame Du Chastelet et M. d'Argental*, 335–78. Paris: Xhrouet, Déterville, Lenormand, Petit.
Du Châtelet, E. 1818. *Lettres inédites de Mme la Mise du Châtelet, et Supplément à la correspondance de Voltaire avec le roi de Prusse, et avec différentes personnes célèbres: On y a joint quelques lettres de cet écrivain, qui n'ont point été recueillies dans les Œuvres complètes, avec des notes historiques et littéraires*. Paris: Lefèvre.
Du Châtelet, E. 1820. Lettre XXVII. Madame la marquise du Châtelet à M. de Thieriot. In *Pièces inédites de Voltaire imprimées d'après les manuscrits originaux, pour faire suite aux différentes éditions publiées jusqu'à ce jour*, 271–72. Paris: Lefèvre.
Du Châtelet, E. 1826. Observation de madame la marquise du Châtelet sur cette lettre (en date du 31 décembre 1738) de Thiriot. In *Mémoires sur Voltaire, et sur ses ouvrages, par Longchamp et Wagnière, ses secrétaires. Suivis de divers écrits inédits de la Marquise du Châtelet, du Président Hénault, de Piron, Darnaud Baculard, Thiriot, etc., tous relatifs à Voltaire*, ed. Sébastien W. J. L. Longchamps, 431–34. 2 vols. Paris: Aimé André.
Du Châtelet, E. 1826. Réponse à une lettre diffamatoire de l'abbé Desfontaines, par Mme la marquise du Châtelet. In *Mémoires sur Voltaire, et sur ses ouvrages, par Longchamp et Wagnière, ses secrétaires. Suivis de divers écrits inédits de la Marquise du Châtelet, du Président Hénault, de Piron, Darnaud Baculard, Thiriot, etc., tous relatifs à Voltaire*, ed. Sébastien W. J. L. Longchamps, 423–34. 2 vols. Paris: Aimé André.
Longchamps, S. W. J. L., ed. 1826. *Mémoires sur Voltaire, et sur ses ouvrages, par Longchamp et Wagnière, ses secrétaires. Suivis de divers écrits inédits de la Marquise du Châtelet, du Président Hénault, de Piron, Darnaud Baculard, Thiriot, etc., tous relatifs à Voltaire*. 2 vols. Paris: Aimé André.

Du Châtelet, E. 1828. Lettre au comte d'Argental du 13 janvier 1737. In *Isographie des hommes célèbres ou collection de fac-similé de lettres autographes et de signatures*. 3 vols. Paris: Alexandre Mesnier.

Du Châtelet, E. 1878. *Lettres de la Marquise du Châtelet: Réunies pour la première fois revues sur les autographes et les éditions originales augmentées de 37 lettres entièrement inédites, de nombreuses notes d'un index et précédées d'une notice biographique par Eugène Assé*. Paris: G. Charpentier.

Du Châtelet, E. 1896. In *The Collection of autograph letters and historical documents formed by Alfred Morrison*, ed. Alfred Morrisson, 167–243. London: Strangeways & Sons.

Du Châtelet, E. 1906. *Quelques lettres inédites de la marquise du Châtelet et de la duchesse du Choiseul*. Paris: Henry Leclerc.

Du Châtelet, E. 1939. Dissertation sur la nature et la propagation du feu. In *Pièces qui int remporté le prix de l'Académie royale des sciences, en M.DCCXXXVIII*, ed. Académie royale des Sciences, 85–168. Paris: Imprimerie royale.

Du Châtelet, E. 1941. Examen de la Genèse. In *Voltaire and Madame du Châtelet: An essay on the intellectual activity at Cirey*, ed. Ira O. Wade, 48–107. Princeton: Princeton University Press. Repr.: 1967. New York: Octagon Books.

Du Châtelet, E. 1947. De la liberté. In *Studies on Voltaire. With some unpublished papers of Mme. du Châtelet*, ed. Ira O. Wade, 92–108. Princeton: Princeton University Press. Repr.: 1967. New York: Russell & Russell.

Du Châtelet, E. 1947. Grammaire Raisonnée: Chap. 6, Des mots en général considérés selon leur signification grammaticale. Chap. 7, Des mots qui représentent les objets de nos perceptions. Chap. 8, Des mots qui désignent les opérations de notre entendement sur les objets. In *Studies on Voltaire. With some unpublished papers of Mme. du Châtelet*, ed. Ira O. Wade, 209–41. Princeton: Princeton University Press. Repr.: 1967. New York: Russell & Russell.

Du Châtelet, E. 1947. L'Essai sur l'optique: Chapitre IV. De la formation des couleurs par Me du Chastellet. In *Studies on Voltaire. With some unpublished papers of Mme. du Châtelet*, ed. Ira O. Wade, 188–208. Princeton: Princeton University Press. Repr.: 1967. New York: Russell & Russell.

Du Châtelet, E. 1947. Mme. du Châtelet's Translation of the *Fable of the Bees*. In *Studies on Voltaire. With some unpublished papers of Mme. du Châtelet*, ed. Ira O. Wade, 131–87. Princeton: Princeton University Press. Repr.: 1967. New York: Russell & Russell.

Du Châtelet, E. 1954. Réponse à une libelle. In *Voltaire's Correspondence*. App 30. 102 vols. Genève: Institut et Musée Voltaire.

Du Châtelet, E. 1958. *Discours sur le bonheur*. Genève: Institut et Musée Voltaire.

Du Châtelet, E. 1958. *Les Lettres de la marquise Du Châtelet: publiées par Theodore Besterman*. 2 vols. Genève: Institut et Musée Voltaire.

Du Châtelet, E. 1961. *Discours sur le bonheur: introduction et notes de Robert Mauzi*. Paris: Les Belles-Lettres.

Du Châtelet, E., and I. Newton. 1966. *Principes mathématiques de la philosophie naturelle de Newton: par feue madame la marquise Du Chastellet*. Paris: Albert Blanchard.

Du Châtelet, E. 1969. Réponse à un libelle. In *Oeuvres complètes de Voltaire/Complete Works of Voltaire*, ed. Theodor Bestermann, 508–12. Genève/Oxford: Institut et Musée Voltaire.

Du Châtelet, E. 1970. Mme Du Châtelet's inventory, 1749. In *Oeuvres complètes de Voltaire/ Complete Works of Voltaire*, ed. Theodor Bestermann, 410–76. Genève/Oxford: Institut et Musée Voltaire.

Du Châtelet, E. 1978. *Lettres de M. de Voltaire et de sa célèbre amie (la Marquise du Chatelêt) (Microforme): suivies d'un Petit poëme, d'une lettre de J. J. Rousseau et d'un paralelle entre Voltaire et J. J. Rosseau*. Paris: Microéditions Hachette.

Du Châtelet, E. 1988. Institutions physiques: Nouvelle édition. In *Gesammelte Werke*, ed. Jean Ecole 28, Abt. 3: Materialien und Dokumente. Hildesheim/Zürich/New York: Olms.

Du Châtelet, E., and I. Newton. 1990. *Principes mathématiques de la philosophie naturelle*. Sceaux: Jacques Gabay.

Du Châtelet, E. 1991. *Réflexions sur le bonheur: manuscrit inconnu*. Paris: F. et R. Chamonal.

Du Châtelet, E. 1992, ²1993. *Discorso sulla felicità: A cura di Maria Cristina Leuzzi. Con una nota di Giuseppe Scaraffia*. Palermo: Sellerio editore.

Du Châtelet, E. 1994. *Disertación sobre la naturaleza y la propagación del fuego*. Traducción de Carmen Mataix. Madrid: Universidad Complutense.
Du Châtelet, E. 1994. Dissertation sur la nature et la propagation du feu. In *"De la nature et de la propagation du feu": cinq mémoires couronnés par l'académie royale des sciences, Paris 1738*, ed. ASPM Association pou la sauvegarde et la promotion du patrimoine métallurgique haut-marnais, 90–151. Wassy.
Du Châtelet, E. 1996. *Discorso sobre la felicidad y correspondencia: Edición de Isabel Morant Deusa. Traduction par Alicia Martorelli*. Madrid: Cátedra.
Du Châtelet, E. 1997. *Discours sur le bonheur. Préface d'Elisabeth Badinter.* Paris: Edition Payot et Rivages.
Du Châtelet, E. 1997, ²2000. *Lettres d'amour au marquis de Saint-Lambert*. Paris: Editions Paris-Méditerrannée.
Du Châtelet, E. 1998. *Correspondance autographe de Gabrielle-Émilie Le Tonnelier de Breteuil, Marquise du Chastelet avec le Marquis de St Lambert et le Mae. Duc de Richelieu*. New York: Pierpont Morgan Library.
Du Châtelet, E. 1998, ²1999. Rede vom Glück. In *Klassische philosophische Texte von Frauen*, ed. Ruth Hagengruber. München: Deutscher Taschenbuchverlag.
Du Châtelet, E. 1999. *Rede vom Glück. Discours sur le bonheur. Übers. und hrsg. von Iris Roebling, mit einer Anzahl Briefe der Mme du Châtelet an den Marquis de Saint-Lambert*. Berlin: Friedenauer Presse
Du Châtelet, E. 1999. Zwo Schriften, welche von der Frau Marquise von Chatelet und dem Herrn von Mairan, das Maaß der lebendigen Kräfte betreffend, gewechselt worden sind. In *Louise Gottsched – "mit der Feder in der Hand", Briefe aus den Jahren 1730–1762*, ed. Inka Kording, 104–08. Darmstadt: Wissenschaftliche Buchgesellschaft.
Du Châtelet, E. 2000. *Acerca de la felicidad*. Buenos Aires: Grupo Imaginador de ediciones.
Du Châtelet, E. 2001. Extrait d'un livre intitulé Discours sur les miracles de Iesus traduit de l'anglois. In *Six discours sur les miracles de notre Sauveur. Deux traductions manuscrits du XVIIIe siècle dont unde de Mme Du Châtelet*, ed. William Trapnell, 329–72. Paris: Champion.
Du Châtelet, E. 2001. *Lettres inédites de madame la marquise du Chastelet à M. le comte d'Argental: auxquelles on a joint une Dissertation sur l'existance de Dieu, les réflexions... sur madame du Chastelet et M. d'Argental*. Boston: Adamant Media Corporation.
Du Châtelet, E. 2001. *Lettres inédites de madame la marquise du Chastelet à M. le comte d'Argental: Auxquelles on a joint une Dissertation sur l'existance de Dieu, Les ... ... sur madame du Chastelet et M. d'Argental*. Boston: Adamant Media Corporation.
Du Châtelet, E. 2002. *Discurso sobre a felicidade: Prefácio de Elisabeth Badinter. Tradução de Marina Appenzeller.* São Paulo: Martins Fontes.
Du Châtelet, E., and I. Newton. 2005. *Principia. Principes mathématiques de la philosophie naturelle. Traduit de l'anglais par la marquise Du Châtelet. Préface de Voltaire.* Paris: Dunod.
Du Châtelet, E. 2006. *Examens de la Bible*. Paris: Champion.
Du Châtelet, E. 2008. Discours sur le bonheur. In *L'art de vivre d'une femme au XVIIIe siècle*, ed. Robert Mauzi. Paris: Desjonquères.
Du Châtelet, E. 2009. *Dissertation Sur La Nature Et La Propagation Du Feu (1744):* Kessinger Publishing.
Du Châtelet, E. 2009. *Institutions Physiques VI (1742)*. Whitefish/MT: Kessinger Publishing.
Du Châtelet, E. 2009. *Emilie Du Châtelet: Selected Philosophical and Scientific Writings*. Selected and edited by Judith P. Zinsser. Translated by Isabelle Bour and Judith P. Zinsser. Chicago: University of Chicago Press.
Du Châtelet, E. 2010. *Lettres de la Mse Du Chatelet*. Charleston/SC: Nabu Press.
Du Châtelet, E. 2010. *Lettres Inedites de Mme. La Mise. Du Chatelet: Et Supplement a la Correspondance de Voltaire (1818)*. Whitefish/MT: Kessinger Publishing.
Du Châtelet, E. 2010. *Zwo Schriften, Welche Von Der Frau Marquis Von Chatelet (1741)*. Whitefish/MT: Kessinger Pub Co.
Du Châtelet, E. Forthcoming. *Correspondance. Sous la direction d'Ulla Kölving et André Magnan*. Genève: Ferney-Voltaire, Centre international d'étude du XVIIIe siècle.
Du Châtelet, E. Forthcoming. *Naturlehre an ihren Sohn*. Hildesheim/Zürich/New York: Olms.

Secondary Sources

1736–1742. *Observations sur les écrits modernes*. Paris: Chaubert.
1739. *Göttingische Zeitungen von gelehrten Sachen*:883–84.
1739. *Mémoires pour l'histoire des sciences et des beaux-arts (de Trévoux)*. Paris: Chaubert.
1740. *Journal des savants*. Compte rendu des *Institutions de physique*.
1741. Les Thuilleries. In *Le Perroquet, ou mélanges de diverses pièces intéressantes pour l'esprit et pour le cœur*, 707–49 2. Francfort sur le Meyn: François Varrentrapp.
1741. *Journal des savants*:65–107.
1741. *Journal des savants* 3:135–53.
1741. *Göttingische Zeitungen von gelehrten Sachen*:233 ff.
1741. Mémoires pour l'Histoire des sciences et des beaux arts, no. August:1381–402.
1741. *Journal des savants, Amsterdam* 3:291–331.
1741. *Le Perroquet, ou mélanges de diverses pièces intéressantes pour l'esprit et pour le cœur* 2. Francfort sur le Meyn: François Varrentrapp.
1741. *Mémoires pour l'histoire des sciences et des beaux-arts (de Trévoux)*. Paris: Chaubert.
1741. *Mémoires pour l'Histoire des Sciences & des Beaux Arts (de Trévoux)*.
1741. *Mémoires por l'Histoire des Sciences et des beaux Arts*. Paris: Chaubert.
1742. *Le Perroquet, ou mélanges de diverses pièces intéressantes, pour l'esprit et pour le cœur* 2. Francfort sur le Meyn: François Varrentrapp.
1742. *Göttingische Zeitungen von gelehrten Sachen*:67 ff.
1745. Le Portrait de Madame la Marquise du Châtelet, tenant le livre de L'Institution physique, qu'elle a composé. *Salon de 1745*.
1745. *Histoire de Mlle d'Attily, par Mme de***, dédiée à Mme la marquise Du Chatellet*. La Haye: J. Neaulme.
1746. *Journal Universel*, no. 10:421.
1746. *Mémoires pour l'histoire des sciences et des beaux-arts (de Trévoux)*.
1759. *La Feuille nécessaire, contenant divers deétails sur les sciences, les lettres et les arts*:301.
1759. *L'Année littéraire*, no. 6:13–27.
1759. *L'Année littéraire*, no. 1:327.
1764. *Le Journal universel ou mémoires pour servir à l'histoire civile, politique, ecclésiastique et littéraire du XVIIIe siècle*, no. 10:411–21.
1765. Repos (Phys.). In *Encyclopédie, ou dictionnaire raisonné des sciences, des arts et des métiers*, ed. D. Diderot and J. B. D'Alembert, 138b-139b. Neufchastel: Samuel Faulche.
1765. Pésanteur (Phys.). In *Encyclopédie, ou dictionnaire raisonné des sciences, des arts et des métiers*, ed. D. Diderot and J. B. D'Alembert, 443b-446a. Neufchastel: Samuel Faulche.
1765. Suffisante raison (Métaphysique). In *Encyclopédie, ou dictionnaire raisonné des sciences, des arts et des métiers*, ed. D. Diderot and J. B. D'Alembert, 634b-635a. Neufchastel: Samuel Faulche.
1765. Impossible (Métaphysique.). In *Encyclopédie, ou dictionnaire raisonné des sciences, des arts et des métiers*, ed. D. Diderot and J. B. D'Alembert, 600a-600b. Neufchastel: Samuel Faulche.
1765. Hypothèse (Métaphysique.). In *Encyclopédie, ou dictionnaire raisonné des sciences, des arts et des métiers*, ed. D. Diderot and J. B. D'Alembert, 417a-418a. Neufchastel: Samuel Faulche.
1765. Pendule (Métaphysique.). In *Encyclopédie, ou dictionnaire raisonné des sciences, des arts et des métiers*, ed. D. Diderot and J. B. D'Alembert, 294a-294b. Neufchastel: Samuel Faulche.
1765. Temps (Métaphysique.). In *Encyclopédie, ou dictionnaire raisonné des sciences, des arts et des métiers*, ed. D. Diderot and J. B. D'Alembert, 93b-96a. Neufchastel: Samuel Faulche.
1767. *Laïs et Phriné, poème en quatre chants*. Orléans, Londres, Paris, Panckoucke, Delalain, Couret de Villeneuve.
1777. *Journal étranger de littérature, des spectacles et de politique*, no. 6.
1778. Complainte sur la mort de Mme la marquise du Châtelet, morte en couches, ou dialogue entre son mari, M. de Voltaire et M. de Saint-Lambert. In *Correspondance littéraire*, ed. Jacob H. Meister.

1792. *Life of Madame Du Châtelet*: C. Forster.
1821–1823. Chastelet, Gabrielle-Émilie Le Tonnelier de Breteuil, marquise du. In *Dictionnaire historique, critique et bibliographique, contenant les vies des hommes illustres (...) suivi d'un dictionnaire abrégé des mythologies et d'un tableau chronologique des évènements les plus remarquables qui ont eu lieu depuis le commencement du monde jusqu'à nos jours. Par une société de gens de lettres*. Sous la direction de Jean-Daniel Goigoux, ed. Lois-Mayeul Chaudon and Antoine-François Delandine, 427–29. 30 vols. Paris: Ménard et Desenne.
1828. *Isographie des hommes célèbres ou collection de fac-similé de lettres autographes et de signatures*. 3 vols. Paris: Alexandre Mesnier.
1854. Le château de Cirey. *Revue étrangère*, no. 51:210–21.
1863. *L'amateur d'autographes* 2, no. 46:347–49.
1865. *L'amateur d'autographes* 4, no. 77:78.
1874. *Parnasse satyrique, XVIIe siècle, pièces trop libres échapées dans les débauches d'esprit à quelques gens de lettres connus et inconnus*. Neuchâtel: Presses de la Société des bibliophiles cosmopolites.
1892. Madame Du Châtelet. *Temple Bar (...) A London magazine for town and country readers* 95:75–84.
1949. Voltaire et les Du Châtelet. *Cahier haut-marnais* 19–20:198.
1952. L'hôtel Lambert. *Réalités* 81:92–99.
Académie royale des Sciences, ed. 1739. *Recueil des pièces qui ont remporté le prix de l'Académie royale des Sciences en 1738*. Paris: Imprimerie royale.
Académie royale des Sciences, ed. 1752. *Recueil des pièces qui ont remporté les prix de l'Académie royale des Sciences, depuis leur fondation jusqu'à présent, avec les pièces qui y ont concouru: Tome quatrième, contenant les pièces depuis 1738 jusqu'en 1740*. Paris: Martin, Coignard, Guérin, Jombert.
Académie royale des Sciences, ed. 1765. *Histoire et mémoires de l'Académie royale des sciences*.
Académie royale des Sciences, ed. 1939. *Pièces qui int remporté le prix de l'Académie royale des sciences, en M.DCCXXXVIII*. Paris: Imprimerie royale.
Adelson, R. 2008. La belle Issé: Mme Du Châtelet musicienne. In *Émilie Du Châtelet: Éclairages & documents nouveaux*, ed. Ulla Kölving and Olivier Courcelle, 127–34. Ferney-Voltaire: Centre International d'Étude du XVIIIe Siècle.
Aimery Pierrebourg, M. T. G. de. 1948. *Madame du Châtelet, une maîtresse de Voltaire*. Paris: Fayard.
Alanen, L., and C. Witt, eds. 2004, ²2005. *Feminist Reflections on the History of Philosophy*. Boston: Kluwer Academic Publishers Group.
Albus, V. 2001. *Weltbild und Metapher: Untersuchungen zur Philosophie im 18. Jahrhundert*. Würzburg: Königshausen & Neumann.
Aldridge, A. O. 1997. The art of autobiography in Voltaire's Mémoires. In *Voltaire et ses combats. Actes du congrès international Oxford-Paris 1994*, ed. Ulla Kölving and Christiane Mervaud, 319–27. 2 vols. Oxford: Voltaire Foundation.
Algarotti, F. 1737. *Il Newtonianismo per le dame, overo dialoghi sopra la luce*. Napoli (Milano).
Algarotti, F. 1794. *Opere del conte Algarotti* 16. Venedig: C. Palese.
Alic, M. 1986. *Hypatia's Heritage. A history of women in science from Antiquity to the late nineteenth century*. Boston: Beacon.
Allen, L. D. 1998. *Physics, Frivolity, and 'Madame Pompon-Newton': The Historical Reception of the Marquise du Châtelet from 1750 to 1996*. Ph.D. Dissertation, University of Cincinnati.
Allen, B. 2007. The Multi-Tasking Marquise. *The Hudson Review* 60, no. 2.
American Association of Physics Teachers, ed. 1984. *Making Contributions: an historical overview of women's role in physics*. College Park/ Md: American Association of Physics Teachers.
American Historical Association, ed. 2005. *Womens History in Global Perspective*. Chicago: University of Illinois Press.

Ancelot, J.-A.-P.-F. 1832. *Madame du Châtelet, ou point de lendemain, comédie en un acte, mêlée de couplets, par MM. Angelot et Gustave, représentée pour la première fois, sur le Théâtre de vaudeville, le 5 mais 1832*. Paris: J.N. Barba.

Anderson, B. S. 1995. *Eine eigene Geschichte - Frauen in Europa*. Frankfurt/M.: Fischer-Taschenbuch-Verl.

Anderson, B. S. 2000. *A history of their own: Women in Europe from prehistory to the present*. New York: Oxford Univ. Press.

Anderson, B. S., and J. P. Zinsser. 1995. *Aufbruch - Vom Absolutismus zur Gegenwart*. Frankfurt/M.: Fischer-Taschenbuch-Verl.

Andrieux, L. 1930. *Une grande dame sous le règne du Bien-Aimé. Ceci est un roman vécu. Il n'y faut pas chercher les grâces de la fiction*. Paris: Georges Andrieux.

Argens, J. B. B. d', ed. 1736–1742. *Nouvelle bibliothèque ou histoire littéraire des principaux écrits qui se publient*. 12 vols. La Haye: Pierre Poupie.

Argenson, M.-P. 1825. *Mémoires du marquis d'Argenson, ministre sous Louis XV; avec une notice sur la vie et les ouvrages de l'auteur; publié par René d'Argenson*. Paris: Baudouin Frères.

Argenson, R.-L. 1857–1858. *Mémoires et journal inédit du marquis d'Argenson, ministre des Affaires étrangères sous Louis XV*. 5 vols. Paris: P. Jannet.

Argenson, R.-L. 1859–1867. *Journal et mémoires du marquis d'Argenson: publiés pour la première fois d'après les manuscrits autographes de la Bibliothèque du Louvre, pour la Société de l'histoire de France, par E.J.B. Rathery*. 9 vols. Paris: Yve Jules Renouard.

Argenson, R.-L. 1923. *Autour d'un ministre de Louis XV: lettres intimes inédites*. Paris: Albert Messein.

Armandi, G. 1968. Voltaire innamorato. *Ausoniaer*, no. 5:37–46.

Ascoli, G. 1924. Voltaire et la marquise du Châtelet. *Revue des cours et conférences* 25:302–15.

ASPM Association pou la sauvegarde et la promotion du patrimoine métallurgique haut-marnais, ed. 1994. *"De la nature et de la propagation du feu": cinq mémoires couronnés par l'académie royale des sciences, Paris 1738*. Wassy.

Assé, E. 1878. Notice biographique. In *Lettres de la Mse du Châtelet: Réunies pour la première fois revue sur les autographes et les éditions originales augmentées de 37 lettres entièrement inédites, de nombreuses notes, d'un index et précédées d'une notice biographique par Eugène Assé*, ed. Eugène Assé. Paris: G. Charpentier.

Aubaud, C. 1993. *Lire les femmes de lettres*. Paris: Dunod.

Bachelard, G. 1938. *La Formation de l'esprit scientifique. Contribution à une psychanalyse de la connaissance objective*. Paris: J. Vrin.

Bachelard, G. 1938. *La Psychanalyse du feu*. Paris: Gallimard.

Badilescu, S. 1996. "Lady Newton" – an eighteenth century marquise. *Physics education* 31, no. 4:242–45.

Badinter, E. 1980. *L'amour en plus*. Paris: Club français du livre.

Badinter, E. 1981, ²1983, ³1984, ⁴1993. *Emilie, Emilie, l'ambition féminine au XVIIIieme siècle*. Paris: Flammarion.

Badinter, E. 1984. *Emilie, Emilie: weibl. Lebensentwurf im 18. Jhdt*. München: Pieper.

Badinter, E. 1999. *Les Passions intellectuelles. Désirs de gloire. 1735–1751*. Paris: Fayard

Badinter, E. 2006. Les relations sociales et amicales de la marquise. In *Madame du Châtelet. La femme des Lumières*, ed. Elisabeth Badinter and Danielle Muzerelle, 27–29. Paris: Bibliothèque Nationale de France.

Badinter, E. 2006. ²2007. *Madame du Châtelet, Madame d'Epinay: ou l'Ambition féminine au XIIIe siècle*. Paris: Flammarion.

Badinter, E. 2006. Une femme dans tous ses états. In *Madame du Châtelet. La femme des Lumières*, ed. Elisabeth Badinter and Danielle Muzerelle, 9–11. Paris: Bibliothèque Nationale de France.

Badinter, E. 2006. Une intellectuelle hors pair. In *Madame du Châtelet. La femme des Lumières*, ed. Elisabeth Badinter and Danielle Muzerelle, 85–87. Paris: Bibliothèque Nationale de France.

Badinter, E. 2007. *Les Passions intellectuelles I: Désirs de gloire (1735-1751)*. Paris: Fayard.

Badinter, E. 2008. Portrait de Mme Du Châtelet. In *Émilie Du Châtelet: Éclairages & documents nouveaux*, ed. Ulla Kölving and Olivier Courcelle, 13–23. Ferney-Voltaire: Centre International d'Étude du XVIIIe Siècle.
Badinter, E., and J. Duhême. 2006. *Les passions d'Emilie: la marquise du Châtelet, une femme d'exception.* Paris: Gallimard Jeunesse.
Badinter, E., and D. Muzerelle, eds. 2006. *Madame du Châtelet. La femme des Lumières.* Paris: Bibliothèque Nationale de France.
Baeyer, H. C. von. 2007. La Dame d'Esprit. A Biography of the Marquise Du Châtelet by Judith P. Zinsser. *American journal of physics* 75, no. 6:575
Baldensperger, F. 1934. Voltaire et la Lorraine. *Pays lorrain* 26:209–28.
Baldensperger, F. 1934. Voltaire et la Lorraine. *Le Pays lorrain et le pays messin* 26:1–20.
Barber, W. H. 1955. *Leibniz in France, from Arnauld to Voltaire, a study in french reactions to leibnizianism, 1670–1760.* Oxford: Clarendon Press.
Barber, W. H. 1967. Mme du Châtelet and Leibnizianism: the genesis of the *Institutions de physique*. In *The Age of Enlightenment: Studies presented to Theodore Besterman*, ed. William H. Barber, J.H Brumfitt, R.A Leigh, R. Shakleton, and S.S.B Taylor, 200–22. Edinburgh/London: Oliver and Boyd.
Barber, W. H. 1975. Voltaire at Cirey: art and thought. In *Studies in eighteenth-century French literature presented to Robert Niklas*, ed. J. H. Fox, M. H. Waddicor, and D. A. Watts, 1–13. Exeter: University of Exeter.
Barber, W. H.. 1978. Penny plain, twopence coloured: Longchamp's memoirs of Voltaire. In *Studies in the French eighteenth century presented to John Lough by colleagues and pupils*, ed. D. J. Mossop, G. E. Rodmell, and D. B. Wilson, 9–21. Durham: University of Durham.
Barber, W. H.. 1979. Voltaire and Samuel Clarke. *SVEC* 179:47–61.
Barber, W. H.. 2003. Exposition du livre des *Institutions de physique*. In *1739–1741*, ed. Robert L. Walters. Oxford: Voltaire Foundation.
Barber, W. H.. 2006. Mme Du Châtelet and Leibnizianism: The genesis of the *Institutions de physique*. In *Emilie Du Châtelet: rewriting Enlightenment philosophy and science*, ed. Judith P. Zinsser and Julie C. Hayes, 5–23. Oxford: Voltaire Foundation.
Barber, W. H., J. Brumfitt, R. Leigh, R. Shakleton, and S. Taylor, eds. 1967. *The Age of Enlightenment: Studies presented to Theodore Besterman.* Edinburgh/London: Oliver and Boyd.
Barbier de Montault, X. 1887. Épitaphe d'une princesse de la maison Du Châtelet, à Naples. *Journal de la Société d'archéologie et du Musée historique lorrain* 36:193–95.
Barry, J. A. 1985. *À la française. Le couple à travers l'histoire: Traduit de l'américain par Christine Blanchet et Liliane Lassen.* Paris: Seuil.
Barry, J. A. 1987. *French lovers: from Héloïse and Abelard to Beauvoir and Sartre.* New York: Arbour House.
Bart, J. 2000. *Women succeeding in Science: Theories and Practices across Discipline.* West Lafayette/IN: Purdue University Press.
Barth, E. M. 1992. *Women Philosophers: A Bibliography of Books through 1990:* Philosophy Documentation Center.
Beaulieu, J.-P., and D. Desrosiers-Bonin, eds. 1998. *Dans les miroirs de l'écriture: La réflexivité chez les femmes écrivains d'Ancien Régime.* Montréal: Département d'études françaises, Université de Montréal.
Becker, B. 2000. *Dynamics: Madame du Châtelet and Voltaire: student reader.* Dubuque/ IA: Kendall/ Hunt Pub. Co.
Bedel, C. 1964. *Enseignement et diffusion des sciences en France au XVIIIe siècle.* Paris: Hermann.
Beeson, D. 1997. Il n'y a pas d'amour heureux: Voltaire, Émilie and the debate on forces vives. In *Voltaire et ses combats.* Actes du congrès international Oxford-Paris 1994, ed. Ulla Kölving and Christiane Mervaud, 901–13. 2 vols. Oxford: Voltaire Foundation.
Beeson, D. 2006. *Maupertuis: An intellectual biography.* Studies on Voltaire and the eighteenth century 299. Oxford: Voltaire Foundation.

Belaval, Y. 1993. Quand Voltaire rencontre Leibniz. In *Etudes leibniziennes: de Leibniz à Hegel*, ed. Yvon Belaval. Paris: Gallimard.
Belaval, Y., ed. 1993. *Etudes leibniziennes: de Leibniz à Hegel*. Paris: Gallimard.
Belin, J. P. 1913. Une invitée de Voltaire à Cirey. *Le journal des débats, édition hébdomadaire* 20:700–01.
Bellessort, A. 1925. *Essai sur Voltaire, cours professé à la Société des conférences*. Paris: Perrin et Cie.
Bellugou, H. 1962. *Voltaire et Frédéric au temps de la marquise du Châtelet. Un trio singulier.* Paris: Marcel Rivière.
Bernis, F.-J. P. 1878. *Mémoires et lettres de François-Joachim Pierre cardinal de Bernis (1715–1758) publiés avec l'autorisation de sa famille d'après les manuscrits inédits*. 2 vols. Paris: E. Plon.
Bertaut, J. 1952. L'infortune sentimentale de Voltaire. *Figaro littéraire*:1–8.
Bertaut, J. 1978. Pregnancy of Gabrielle Emilie de Breteuil, marquise of Chatelet. *Historia* 381: 94–101.
Bérubé, G. 2000. Mme de Graffigny à Cirey: écrire pour exister "par procuration". In *Femmes en toutes lettres: les épistolaires du XVIIIe siècle*, ed. Marie-France Silver and Marie-Laure Giroud Swiderski, 23–32. Oxford: Voltaire Foundation.
Bervens-Stevelinck, C., and J. Häseler, eds. 2005. *Les grands intermédiaires culturels de la République des Lettres: études de réseaux de correspondances du XVIe au XVIIIe siècles*. Paris: Honoré Champion.
Bessire, F. 2008. Mme Du Châtelet épistolière. In *Émilie Du Châtelet: Éclairages & documents nouveaux,* ed. Ulla Kölving and Olivier Courcelle, 25–35. Ferney-Voltaire: Centre International d'Étude du XVIIIe Siècle.
Besterman, T. 1962. Émilie du Châtelet: portrait of a unknown woman. In *Voltaire Essays and another*, ed. Theodore Besterman. London: Oxford University Press. Repr.: 1980. Santa Barbara/Cal.: Greenwood Pub. Group Inc.
Besterman, T., ed. 1962. *Voltaire Essays and another*. London: Oxford University Press. Repr.: 1980. Santa Barbara/Cal.: Greenwood Pub. Group Inc.
Biagioli, M., ed. 2003. *Scientific Authorship: Credit and Intellectual Property in Science*. New York: Routledge.
Biarnais, M.-F. 1981. *Les Principia de Newton et "leurs traductions" françaises au milieu du XVIIIe Siècle*. Dissertation.
Biarnais, M.-F. 1982. *Les Principia de Newton. Genèse et structure des chapitres fondamentaux avec traduction nouvelle*. Paris: Société française d'histoire des sciences et des techniques.
Bingham, A. J. 1961. The Recueil philosophique et littéraire (1769–1779). *SVEC* 18:113–28.
Bioche, C. 1926. Madame du Châtelet et la querelle des forces vives. *La Nature, revue des sciences et de leurs applications à l'art et à l'industrie* 54, no. 2720:332–33.
Bioche, C. 1927. Les aventures scientifiques de Mme du Châtelet. *Revue des deux mondes*, no. 6:689–97.
Black, L. 1993. Gabrielle-Émilie Le Tonnelier de Bréteuil, marquise du Châtelet (1706–1749). In *Women in chemistry and physics: a biobibliogpaphical sourcebook*, ed. Louise S. Grinstein, K. Rose, and Miriam H. Rafailovich, 101–05. Westport/Conn.: Greenwood Press.
Blancocorujo, O. 1993. Noble Dissemination: The contribution of Emilie du Châtelet and Mary Wortley Montagu to the diffusion of scientific ideas during the Enlightenment. *Arbor – Ciencia Pensamiento y Cultura* 144, no. 565:65–78.
Blay, M. 1987. Le traitement newtonien du mouvement des projectiles dans les milieux résistants. *Revue d'Histoire des sciences* 40, no. 3–4:325–55.
Blay, M., and R. Halleux. 1998. *La science classique, XVIème-XVIIIème siècle, dictionnaire critique*. Paris: Flammarion.

Blay, M., and M. Toulmonde. 2008. Vers une nouvelle édition des *Principes mathématiques*. In *Émilie Du Châtelet: Éclairages & documents nouveaux*, ed. Ulla Kölving and Olivier Courcelle, 333–39. Ferney-Voltaire: Centre International d'Étude du XVIIIe Siècle.

Bléchet, F. 2008. La marquise Du Châtelet et les institutions: l'Académie royale des sciences et la Bibliothèque du roi. In *Émilie Du Châtelet: Éclairages & documents nouveaux*, ed. Ulla Kölving and Olivier Courcelle, 99–109. Ferney-Voltaire: Centre International d'Étude du XVIIIe Siècle.

Bloch, O., ed. 1982. *Le matérialisme du XVIIIe siècle et la littérature clandestine*. Paris: J. Vrin.

Bocquillon, M. 2004. Échanger ou (se) donner le change: la correspondance d'Émilie du Châtelet et de Jean-François de Saint-Lambert. *Lumen. Travaux choisis de la Société canadienne d'étude du dix-huitième siècle. Selected Proceedings from the Canadian society for eighteenth-century studies* 23:151–63.

Bodanis, D. 2000. ²2005. *E=mc2: A biography of the world's most famous equation*. New York: Walker.

Bodanis, D. 2006. *Passionate minds: The great love affair of the Enlightenment featuring the scientist Emilie Du Châtelet the poet Voltaire sword fights book burnings assorted kings seditious verse and the birth of the modern world*. New York: Crown Publishers.

Boerhaave, H. 1754. *Eléments de chimie*. Paris: Guillyn.

Boijolin, J. de, and G. Mossé. 1905. Mme du Châtelet. In *Quelques meneuses d'hommes au XVIIe siècle*, ed. J. de Boisjolin and G. Mossé, 12–20. Paris: Éditions de la Nouvelle Revue.

Boisjolin, J. de, and G. Mossé, eds. 1905. *Quelques meneuses d'hommes au XVIIe siècle*. Paris: Éditions de la Nouvelle Revue.

Boncompagni, p. B. de. 1894. Lettere di Alessio Claudio Clairaut. *Atti dell'Accademia pontifica di Nuovi Lincei* 45:233–91.

Bonnefon, P. 1902. Une inimitié littéraire au XVIIIe siècle d'après des documents inédits. Voltaire et Jean-Baptiste Rousseau. *Revue d'histoire littéraire de la France* 9:547–95.

Bonnel, R. 2000. La correspondance scientifique de la marquise du Châtelet: la 'lettre laboratoire'. In *Femmes en toutes lettres: les épistolaires du XVIIIe siècle*, ed. Marie-France Silver and Marie-Laure Giroud Swiderski, 79–85. Oxford: Voltaire Foundation.

Bos, H. J. M. 1997. *Lectures in the history of mathematics*. Providence/RI: American Math. Soc.

Brosshart-Pfluger, C., D. Grisard, and C. Späthi, eds. 2005. *Geschlecht und Wissen = Genre et savoir = Gender and knowledge. Beiträge der 10. Schweizerischen Historikerinnentagung 2002*. Zürich: Chronos.

Böttcher, F. 2006. Emilie du Châtelet – 'die Nachwelt wird sie mit Erstaunen betrachten'. *Mathematische Semesterberichte* 53, no. 2:245–57.

Böttcher, F. 2008. La réception des *Institutions de physique* en Allemagne. In *Émilie Du Châtelet: Éclairages & documents nouveaux*, ed. Ulla Kölving and Olivier Courcelle, 243–54. Ferney-Voltaire: Centre International d'Étude du XVIIIe Siècle.

Bottiglia, W. F., ed. 1968. *Voltaire: a collection of critical essays*. Englewood Cliffs/N.J.: Prentice-Hall.

Bourdi, C. J. 2002. *What was mechanical about mechanics? The concept of force between metaphysics and mechanics from Newton to Lagrange*. Dordrecht: Kluwer.

Boyé, P. 1791. *Cour de Lunéville en 1748 ou Voltaire chez le roi Stanislas*. Nancy: G. Crépin-Leblond.

Brackenridge, J. B. 1985. The defective diagram as an analytical device in Newton's *Principia*. In *Religion, science and worldview*, ed. Margaret J. Osler and Paul L. Farber, 61–93. Cambridge: CUP.

Braidotti, R. 1990. *Women's studies and philosophy: an international bibliography on feminist sources with an accent on sexual differences and theories of the subject*. Utrecht.

Braine, S. E. 1897. Gabrielle Émilie, marquise du Châtelet, Voltaire's marquise. *Englishwoman* 6:280–83.

Bray, B., and J. Varloot. 1976. *La Correspondance littéraire de Grimm et de Meister (1754–1813): Colloque de Sarrebruck (22–24 février 1974)*. Paris: Klincksieck.

Bredow, R. von. 2005. Revier des Weibes. *Der Spiegel* 25:154–55.
Breger, H. 1991. Der mechanistische Denkstil in der Mathematik des 17. Jahrhunderts. In *Gottfried Wilhelm Leibniz im philosophischen Diskurs über Geometrie und Erfahrung*, ed. Hartmut Hecht, 15–46. Berlin: Akademie Verlag.
Breger, H., ed. 1994. *Leibniz und Europa: VI. Internationaler Leibniz-Kongreß*. Hannover: G.-W.-Leibniz-Gesellschaft.
Breger, H. 1999. Über den von Samuel König veröffentlichten Brief zum Prinzip der kleinsten Wirkung. In *Pierre Louis Moreau de Maupertuis (1698–1759). Eine Bilanz nach 300 Jahren*, ed. Hartmut Hecht, 363–81. Berlin: Spitz-Verlag.
Breteuil, L.-N. b. de. 1992. *Mémoires*. Paris: F. Bourin.
Breteuil, H.-F. de. 2006. Le cercle de famille d'Émilie de Breteuil. In *Madame du Châtelet. La femme des Lumières*, ed. Elisabeth Badinter and Danielle Muzerelle, 15–17. Paris: Bibliothèque Nationale de France.
Brewer, H. 1986. *Madame du Châtelet and the search for a metaphysics: Cartesian, Leibnizian and ultimately Newtonian*. Dissertation, Harvard University.
Brockmeier, P., R. Desné, and E. Voss, eds. 1979. *Voltaire und Deutschland. Quellen und Untersuchungen zur Rezeption der französischen Aufklärung: Internat. Kolloquium der Universität Mannheim zum 200. Todestag Voltaires*. Stuttgart: Metzler.
Brosshart-Pfluger, C., D. Grisard, and C. Späthi, eds. 2005. *Geschlecht und Wissen = Genre et savoir = Gender and knowledge. Beiträge der 10. Schweizerischen Historikerinnentagung 2002*. Zürich: Chronos.
Brown, A., and U. Kölving. 2003. Qui est l'auteur du Traité de métaphysique? *Cahiers Voltaire* 2:85–94.
Brown, A., and U. Kölving. 2008. À la recherche des livres d'Émilie Du Châtelet. In *Émilie Du Châtelet: Éclairages & documents nouveaux*, ed. Ulla Kölving and Olivier Courcelle, 111–20. Ferney-Voltaire: Centre International d'Étude du XVIIIe Siècle.
Brucker, J. J. 1745. *Bilder-Sal heutiges Tages lebender und durch Gelahrtheit berühmter Schrifftsteller, in welchem derselbigen nach wahren Original-malereyen entworfene Bildnisse in schwartzer Kunst in natürlicher Aehnlichkeit vorgestellt, und ihre Lebensumstände, Verdienste um Wissenschafften, und Schriften aus glaubewürdigen Nachrichten erzählt werden*. Augsburg: Joh. Jacob Haid.
Brunet, P. 1929. *Maupertuis: étude biographique*. Paris: A. Blanchard.
Brunet, P. 1931. *L'Introduction des théories de Newton en France au XVIIIe siècle*. Paris: Blanchard.
Brunet, P. 1951. *La vie et l'œuvre de Clairaut (1713–1765)*. Paris: PUF.
Buchdahl, G. 1969. *Metaphysics and the philosophy of science*. Oxford: Blackwell.
Burns, W. E. 2003. *Science in the Enlightenment: an encyclopedia*. Santa Barbara/Cal.: ABC-CLIO.
Buschmann, C. 1991. Connubium rationis et experientiae: das Problem von Erfahrung und Theorie in seiner Bedeutung für den Denkeinsatz der Philosophie Christian Wolffs. In *Gottfried Wilhelm Leibniz im philosophischen Diskurs über Geometrie und Erfahrung*, ed. Hartmut Hecht, 186–207. Berlin: Akademie Verlag.
Gillispie, C. C., ed. 1971. *Dictionary of scientific biography*. New York: Charles Scribner's Sons.
Cadilhac, P.-E., ed. 1955. *Demeures inspirées et sites romanesques*, Vol. II. Paris: Snep - Illustration.
Cadilhac, P.-E. 1955. Voltaire et Mme du Châtelet à Cirey et en Lorraine. In *Demeures inspirées et sites romanesques*, ed. Paul-Emile Cadilhac, 53–64. Vol. II. Paris: Snep - Illustration.
Cajori, F. 1926. Madame du Châtelet on Fluxions. *Mathematical Gazette* 13:252.
Calinger, R. 1979. Kant and Newtonian Science: The Pre-Critical Period. *Isis* 70, no. 3:349–62.
Calmet, D. A. 1741. *Histoire généalogique de la maison du Châtelet, branche puînée de la maison de Lorraine*. Nancy: Cusson.
Candaux, J.-D. 2008. La vie à Cirey et la mort à Lunéville de Mme Du Châtelet: deux reportages du *Journal helvétique* de Neuchâtel. In *Émilie Du Châtelet: Éclairages & documents nouveaux*, ed. Ulla Kölving and Olivier Courcelle, 85–91. Ferney-Voltaire: Centre International d'Étude du XVIIIe Siècle.

Canterla, C. 1997. Newtonianism et anticartésianism dans l'*Essai sur la nature du feu, et sur sa propagation*. In *Voltaire et ses combats*. Actes du congrès international Oxford-Paris 1994, ed. Ulla Kölving and Christiane Mervaud, 47–53. 2 vols. Oxford: Voltaire Foundation.

Capefigue, J.-B.-H.-R. 1868. *La Marquise du Châtelet et les amies des philosophes du XVIIIe siècle*. Paris: Amyot.

Carboncini, S. 1984. Lumière e Aufklärung – A proposito della presenza della filosofia di Christian Wolff nell'Encyclopédie. *Annali della Scuola Normale Superiore di Pisa* 14, no. 4:1297–301.

Carboncini-Gavanelli, S. 1993. Christian Wolff in Frankreich. Zum Verhältnis von französischer und deutscher Aufklärung. In *Aufklärung als Mission: Akzeptanzprobleme und Kommunikationsdefizite = La mission des Lumières: accueil réciproque et difficultés de communication*, ed. W. Schneiders. Marburg: Hitzeroth.

Casini, P. 1995. Voltaire, la luce e il fuoco. *Micromégas* 61/62, no. 1/2:123–27.

Cassirer, E. 1969. Newton und Leibniz. In *Ernst Cassirer, Philosophie und exakte Wissenschaft*, ed. Wilhelm Krampf. Frankfurt/M.: Klostermann.

Caussi, F. 1911. La mission diplomatique de Voltaire d'après des documents inédits. *La Grande Revue* 15, no. 3:547–63.

Cavazza, M. 2000. Les femmes à l'académie: le cas de Bologne. In *Académies et sociétés savantes en Europe (1650–1800)*, ed. D. O. Hurel and G. Laudin, 161–75. Paris: H. Champion.

Cayrol, L.-N.-J.-J. de. 1836. *Voltaire étrangement défiguré par l'auteur des Souvenirs de Mme de Créqui*. Compiègne: J. Escuyer.

Chambat, F., and D. Varry. 2008. Faut-il faire une description bibliographique des Principes mathématiques? In *Émilie Du Châtelet: Éclairages & documents nouveaux*, ed. Ulla Kölving and Olivier Courcelle, 317–32. Ferney-Voltaire: Centre International d'Étude du XVIIIe Siècle.

Chandrasekhar, S. 1995. *Newton's Principia for the common reader*. Oxford: Oxford University Press.

Charpentier, J. 1962. *Voltaire en menage*. Monaco: Editions LEP.

Châtelier, L. 2000/2001. Madame du Châtelet et la diffusion du newtonisme en France. *Mémoires de l'académie de Stanislas* 15:135–44.

Chaudon, L.-M., and A.-F. Delandine, eds. 1821–1823. *Dictionnaire historique, critique et bibliographique, contenant les vies des hommes illustres (…) suivi d'un dictionnaire abrégé des mythologies et d'un tableau chronologique des évènements les plus remarquables qui ont eu lieu depuis le commencement du monde jusqu'à nos jours. Par une société de gens de lettres*. Sous la direction de Jean-Daniel Goigoux. 30 vols. Paris: Ménard et Desenne.

Cheve, J. 2006. Emilie de Breteuil, marquise de Châtelet. *Historia* 712:7.

Chvojan, C. 1997. *Le genre et le savoir: Madame du Châtelet et son temps*. Diplomarbeit. Wien.

Clark, W., J. Gollinski, and S. Schaffer, eds. 1999. *The Sciences in Enlightened Europe*. Chicago: University of Chicago Press.

Clarke, E. 1878. Voltaire and Madame du Châtelet at Cirey. *The Nineteenth Century, a monthly review* 3, no. 16:1052–73.

Clément, P. 1967. *Les Cinq années littéraires ou lettres de M. de Clément sur les ouvrages de littérature qui ont paru dans les années 1748–1752*. Genève: Slatkine.

Clinton, K. B. 1975. Femme et Philosophe: Enlightenment Origins of Feminism. *Eighteenth-Century Studies* 8, no. 3:283–99.

Clogenson, J. 1859–1860. Voltaire, jardinier à Cirey et aux Délices. *Précis analytique des travaux de l'Acad. impériale des sciences, belles-lettres et arts de Rouen*:5–24.

Cluzel, E. 1955. La *Dissertation sur la nature du feu* de la marquise du Châtelet. *Bulletin du bibliophile et du bibliothécaire*:1–10.

Cohen, I. B. 1967. Newton's use of 'force', or, Cajori versus Newton: A Note on Translations of the *Principia*. *Isis* 58:226–30.

Cohen, I. B. 1968. The French Translation of Isaac Newton's *Philosophiae Naturalis Principia Mathematica* (1756, 1759, 1966). *Archives internationales d'histoire des sciences* 21:261–90.

Cohen, E. 1997. "What the Women at All Times Would Laugh At": Redefining Equality and Difference, Circa 1660–1760. *Osiris*, no. 12:121–42.

Cohen, I. B. 1999. A guide to Newton's *Principia*. In *The Principia: mathematical principles of natural philosophy*, ed. I. B. Cohen and Anne Whitman. Berkeley: University of California Press.

Cohen, I. B., and G. W. Smith, eds. 2002. *The Cambridge Companion to Newton*. Cambridge: Cambridge University Press.
Cohen, I. B., and A. Whitman, eds. 1999. *The Principia: mathematical principles of natural philosophy*. Berkeley: University of California Press.
Colet, L. 1845. Lettres inédites de madame du Châtelet au maréchal de Richelieu et à Saint-Lambert. *Revue des deux mondes*, no. 11:1011–53.
Colet, L. 1847. *Deux femmes célèbres, par madame Louise Colet*. 2 vols. Paris: Pétion.
Colet, L. 1854. *Madame du Châtelet, par Mme Louise Colet*.
Colet, L. 1863. Mme Du Châtelet. *Romans populaires illustrés*, no. 28.
Collé, C. 1805. *Journal historique, ou mémoires critiques et littéraires, sur les ouvrages dramatiques et sur les évènements les plus mémorables, depuis 1748 jusqu'en 1751 inclusivement. Par Charles Collé, auteur de la Partie de chasse de Henri IV. Imprimés sur le manuscrit de l'auteur, et précédés d'une notice sur sa vie et ses écrits*. 3 vols. Paris: Imprimerie bibliographique.
Colnet Du Ravel, C.-J., ed. 1825. *L'Hermite du faubourg Saint-Germain, ou observations sur les mœurs et les usages français au commencement du XIXe siècle par M. Colnet, auteur de l'Art de dîner en ville, faisant suite à la Collection des mœurs françaises de M. de Jouy*. 2 vols. Paris: Pillet aîné.
Colnet Du Ravel, C.-J. 1825. Vie privée de Voltaire et de Mme du Châtelet. In *L'Hermite du faubourg Saint-Germain, ou observations sur les mœurs et les usages français au commencement du XIXe siècle par M. Colnet, auteur de l'Art de dîner en ville, faisant suite à la Collection des mœurs françaises de M. de Jouy*, ed. Charles-Joseph Colnet Du Ravel, 234–44. 2 vols. Paris: Pillet aîné.
Condorcet, J. A. N. C., and E. Badinter. 1994. *Vie de Voltaire*. Paris: Quai Voltaire.
Conti, A., ed. 1739. *Prose, e poesie del signor abate Antonio Conti, patrizio veneto*. 2 vols. Venezia: Giambatista Pasquali.
Conti, A., ed. 1739. Annotazione sul secondo. In *Prose, e poesie del signor abate Antonio Conti, patrizio veneto*, ed. Antonio Conti, 115–16. Vol. I. Venezia: Giambatista Pasquali.
Cook, A. 1997. Ladies in the scientific revolution: Lady Katherine Ranelagh, Queen Christina of Sweden, Elisabeth Hevelius, Catherine Barton, Lady Masham, Queen Caroline, Emilie du Chatelet, Nicole-Reine Etable de la Biere Lepaute. *Notes and records of the Royal Society*, no. 51/1:1–12.
Cook, A. 1997. Ladies in the Scientific Revolution. *Notes and Records of the Royal Society of London* 51, no. 1:1–12.
Coolidge, J. L. 1951. Six Female Mathematicians. *Scripta Mathematica* 17:20–31.
Cooney, M. P., ed. 1996. *Celebrating women in mathematics and science*. Reston/Va.
Costabel, P. 1984. *La signification d'un débat sur trente ans (1728–1758): la question des forces vives*. Paris: CNRS.
Costabel, P., and E. Winter, eds. 1986. *Correspondance de Leonhard Euler avec P.-L. M. de Maupertuis et Frédéric II*. Basel: Birkhäuser Verlag.
Countess of Blessington. 1854. A Glance at the Life and Times of Gabrielle Emilie, Marchioness Du Châtelet. *The ladies' companion* 2, no. 6:29–32.
Courcelle, O. 1997. Clairaut la Comète. *Quadrature. Magazine de mathématiques pures et appliquées*, no. 27:15–20.
Courcelle, O. 2008. La publication tardive des *Principes mathématiques*. In *Émilie Du Châtelet: Éclairages & documents nouveaux*, ed. Ulla Kölving and Olivier Courcelle, 301–08. Ferney-Voltaire: Centre International d'Étude du XVIIIe Siècle.
Cox, J. F. 1950. Hommage à la marquise du Châtelet. *Ciel Terre, bulletin de la société belge d'astronomie, de météreologie et de physique du globe* 66:1–11.
Créqui, R.-C.-V. F. 1834–1835. *Souvenirs de la marquise de Créqui 1710–1800*. 7 vols. Paris: Fournier jeune.
Crocker, L. G. 1963. *Nature and Culture. Ethical Thought in the French Enlightenment*. Baltimore: The Johns Hopkins Press.

D'Alembert, J. B. 1765. Mouvement (Méchan.). In *Encyclopédie, ou dictionnaire raisonné des sciences, des arts et des métiers*, ed. Denis Diderot and Jean Baptiste Le Rond D'Alembert, 830b-840b. Neufchastel: Samuel Faulche.

D'Alembert, J. B. 1765. Newtonianisme. In *Encyclopédie, ou dictionnaire raisonné des sciences, des arts et des métiers*, ed. Denis Diderot and Jean Baptiste Le Rond D'Alembert, 122b-125b. Neufchastel: Samuel Faulche.

D'Alembert, J. B. 1904. *Discours préliminaire de l'Encyclopédie*. Paris: Armand Colin.

D'Alembert, J.B. 1997. *Abhandlung über die Dynamik*. Frankfurt/M.: Thun.

Dampier-Whetham, W. C. 1929. *A history of science and its relations with philosophy and religion*. Cambridge: Cambridge University Press.

Dannreuther, H. 1886. Le protestantisme dans la maison du Châtelet. *Journal de la Société d'archéologie et du Musée historique lorrain* 35:135-41.

Davidson, A. 1956. *A study of the influence of Madame du Châtelet on Voltaire*. Dissertation, Ohio University.

Deakin, M. 1992. Women in mathematics: fact versus fabulation. *The Australian Mathematical Society Gazette* 19:105-14.

Debever, R. 1987. La marquise du Châtelet traduit et commente les *Principia* de Newton. Académie royale des sciences, des lettres et des beaux-arts de Belgique. *Bulletin de la Classe des Sciences* 73, no. 12:509-27.

Desfontaines, P.-F. G. 1736-1742. In *Observations sur les écrits modernes*, 42-44. Paris: Chaubert.

Deidier, l., ed. 1741. *La Méchanique générale contenant la statique, l'airométrie, l'hydrostatique et l'hydraulique, pour servir d'introduction aux sciences physicomathématiques*. Paris: Charles-Antoine Jombert.

Deidier, l., ed. 1741. Nouvelle réfutacion de l'hypothèse des forces vives. In *La Méchanique générale contenant la statique, l'airométrie, l'hydrostatique et l'hydraulique, pour servir d'introduction aux sciences physicomathématiques*, ed. l'abbé Deidier. Paris: Charles-Antoine Jombert.

Deidier, l., ed. 1741. Remarque en forme de dissertation touchant les forces vives. In *La Méchanique générale contenant la statique, l'airométrie, l'hydrostatique et l'hydraulique, pour servir d'introduction aux sciences physicomathématiques*, ed. l'abbé Deidier. Paris: Charles-Antoine Jombert.

Deligiorgi, K. 2001. The Public Tribunal of Political Practical Reason: Kant and the Culture of Enlightenment. In *Kant und die Berliner Aufklärung: Akten des IX. Internationalen Kant-Kongresses*, ed. Volker Gerhard, 148-55. Berlin: de Gruyter.

Deslisle, J., ed. 2002. *Portraits de traductrices*. Ottawa: PU d'Ottawa.

Delon, M. 1981. La marquise et le philosophe. *Revue des sciences humaines* 182:65-78.

Deschamps, J. 1743-1747. *Cours abrégé de la philosophie Wolfienne, en forme de lettres*. 3 vols. Amsterdam/Leipzig: Arkstée und Merkus.

Deschamps, J. 1991. Cours abrégé de la philosophie Wolfienne, en forme de lettres. In *Wolff, Christian, Gesammelte Werke*, ed. Jean École. Hildesheim/Zürich/New York: Olms.

Deslisle, J., and J. Woodsworth, eds. 1995. *Les traducteurs dans l'histoire*. Ottawa: PU d'Ottawa.

Desnoiresterres, G., ed. 1855. Du Châtelet (Gabrielle-Émilie Le Tonnelier de Breteuil, marquise). In *Nouvelle biographie générale depuis les temps les plus reeculés jusqu'à nos jours, avec les renseignements bibliographiques et l'indication des sources à consulter: publié par MM. Firmin Didot frères, sous la direction de M. le Dr Hoefer*, ed. D. Hoefer, 940-44. Paris: Firmin Didot frères.

Desnoiresterres, G., ed. 1867-1876. Voltaire au château de Cirey. In *Voltaire et la société au XVIIIème siècle*, ed. Gustave Desnoiresterres. 8 vols. Paris: Didier et Cie.

Desnoiresterres, G., ed. 1867-1876. *Voltaire et la société au XVIIIème siècle*. 8 vols. Paris: Didier et Cie.

Desnoiresterres, G. 1869. *Voltaire à la Cour*. Paris.

Desnoiresterres, G., ed. 1891. La marquise du Châtelet un cochon. *Intermédiaire des chercheurs et des curieux* 24:178-80.

Diderot, D. 1955 ff. *Correspondance*. 11 vols. Paris: Éditions de Minuit.
Diderot, D., and J. B. D'Alembert, eds. 1751–1780. *Encyclopédie, ou dictionnaire raisonné des sciences, des arts et des métiers*. 35 vols. Genève, Neufchastel, Paris: Briasson, David, Le Breton, Durand.
Diderot, D., and J. B. D'Alembert, eds. 1765. *Encyclopédie, ou dictionnaire raisonné des sciences, des arts et des métiers*. Neufchastel: Samuel Faulche.
Didier, B. 2008. La correspondance de Mme Du Châtelet, un journal intime? In *Émilie Du Châtelet: Éclairages & documents nouveaux*, ed. Ulla Kölving and Olivier Courcelle, 53–60. Ferney-Voltaire: Centre International d'Étude du XVIIIe Siècle.
Dijksterhuis, E. J. 1956. *Die Mechanisierung des Weltbildes*. Berlin.
Dilthey, W. 1927. Friedrich der Große und die deutsche Aufklärung. In *Gesammelte Schriften*. Leipzig/Berlin: B. G. Teubner.
Dilthey, W. 1927. *Gesammelte Schriften*. Leipzig/Berlin: B. G. Teubner.
Dodson, K. E., ed. 2004. *Enlightenment and Revolution, 1690–1815: An Interdisciplinary Biographical Dictionary*. Santa Barbara/Cal.: Greenwood Pub Group Inc.
Doetsch, B., ed. 2004. *Philosophinnen im 3. Jahrtausend*. Bielefeld: Kleine Verlag.
Dorat, C.-J. 1874. Les trois pères. In *Parnasse satyrique, XVIIe siècle, pièces trop libres échapées dans les débauches d'esprit à quelques gens de lettres connus et inconnus*, 56–58. Neuchâtel: Presses de la Société des bibliophiles cosmopolites.
Douay-Soublin, F. 2008. Nouvel examen de la *Grammaire raisonnée*. In *Émilie Du Châtelet: Éclairages & documents nouveaux*, ed. Ulla Kölving and Olivier Courcelle, 173–96. Ferney-Voltaire: Centre International d'Étude du XVIIIe Siècle.
Droysen, H. 1910. Die Marquise du Châtelet, Voltaire und der Philosoph Christian Wolff. *Zeitschrift für französische Sprache und Literatur* 35:226–48.
Du Deffand, M. V.-C. 1778. Portrait de feue madame la marquise du Châtelet, par Mme la marquise du Deffand. In *Correspondance littéraire*, ed. Jacob H. Meister.
Du Deffand, M. V.-C. 1809. *Correspondance inédite de Mme du Deffand avec d'Alembert, Montesquieu, le président Hénault, la duchesse du Maine; mesdames de Choiseul, de Staal; le marquis d'Argens, le chevalier d'Aydie, etc. suivie des lettres de M. de Voltaire à Mme du Deffand*. 2 vols. Paris: Léopold Collin.
Du Deffand, M. V.-C. 1812. Portrait de la marquise du Châtelet. *Journal des arts, des sciences et de la littérature* 8, no. 128:91–92.
Du Deffand, M. V.-C. 1865. *Correspondance complète de la marquise du Deffand avec ses amis le président Hénault, Montesquieu, d'Alembert, Voltaire, Horace Walpole, classée dans l'ordre chronologique et sans suppressions, augmentée des lettres inédites au chevalier de L'Isle, précédé d'une histoire de sa vie, de son salon, de ses amis, suivie de ses œuvres diverses et éclairée de nombreuses notes par M. de Lescure*. 2 vols. Paris: H. Plon.
Du Deffand, M. V.-C. 1865. Lettre de la madame de Staal. In *Correspondance complète de la marquise du Deffand avec ses amis le président Hénault, Montesquieu, d'Alembert, Voltaire, Horace Walpole, classée dans l'ordre chronologique et sans suppressions, augmentée des lettres inédites au chevalier de L'Isle, précédé d'une histoire de sa vie, de son salon, de ses amis, suivie de ses œuvres diverses et éclairée de nombreuses notes par M. de Lescure*. 2 vols. Paris: H. Plon.
Dufrenoy, M.-L. 1963. La diffusion des théories de Newton: Maupertuis, Voltaire et Mme du Châtelet. *SVEC* 25:531–48.
Dumas Mascarel, B. de. 1990. Cirey dans l'intimité des Lumières. *Point de vue – Images du monde*, no. 42/2200:18–20.
Eberhard, J. A. 1789. *Philosophisches Magazin*. Halle: Gebauer.
Eberhard, J. A. 1789. Über den wesentlichen Unterschied der Erkennnis durch die Sinne und durch den Verstand. In *Philosophisches Magazin*, ed. Johann A. Eberhard, 290–306. Halle: Gebauer.
École, J., ed. 1991. *Wolff, Christian, Gesammelte Werke*. Hildesheim/Zürich/New York: Olms.
Edeen, S., J. Edeen, and V. Slachman. 1990. *Women Mathematicians*. Palo Alto/ Calif.: Dale Seymour Publ.
Edwards, S. 1970. *The Divine Mistress: A Biography of Emilie du Châtelet, the Beloved of Voltaire*. New York: David McKay.

Ehrman, E. 1986. *Mme du Châtelet*. Leamington Spa: Berg.
Emch, G., and A. Emch-Deriaz. 1999. Is Madame du Châtelet's a fair presentation of Newton's *Principia?* In *Conférence faite au Dixième Congrès international des Lumières*, ed. François de Gandt, 25–31. Dublin.
Emch, G., and A. Emch-Deriaz. 2006. Mathématicienne, et comment. In *Madame du Châtelet. La femme des Lumières*, ed. Elisabeth Badinter and Danielle Muzerelle, 90–92. Paris: Bibliothèque Nationale de France.
Emch, G., and A. Emch-Deriaz. 2006. On Newton's French Translator: How Faithful was Mme Du Châtelet? In *Emilie Du Châtelet: rewriting Enlightenment philosophy and science* ed. Judith P. Zinsser and Julie C. Hayes, 226–51. Oxford: Voltaire Foundation.
Henriot, E., ed. 1931. *Épistoliers et mémorialistes*. Paris: Éditions de la nouvelle revue critique.
Euler, L. 1957. De universali principio aequilibrii et motus in vi viva reperto deque nexu inter vim vivam et actionem, utriusque minimo, dissertatio, autore Sam. Koenigio Profess. Franequer. In *Opera omnia sub auspici Societatis Scientiarium Naturalium Helveticae*, 304–24 5, Ser. 2. Basel: Birkhäuser.
Euler, L. 1957. *Opera omnia sub auspici Societatis Scientiarium Naturalium Helveticae* 5, Ser. 2. Basel: Birkhäuser.
Euler, L. 1963. *Briefe an Gelehrte*. Moskau.
Euler, L. 1986. Euler à Maupertuis 19.02.1740. In *Correspondance de Leonhard Euler avec P.-L. M. de Maupertuis et Frédéric II*, ed. Pierre Costabel and Eduard Winter. Basel: Birkhäuser Verlag.
Euler, L. 1986. Euler à Maupertuis 21.10.1740 and 10.12.1745. In *Correspondance de Leonhard Euler avec P.-L. M. de Maupertuis et Frédéric II.*, ed. Pierre Costabel and Eduard Winter. Basel: Birkhäuser Verlag.
Everett, K. L. 1935. *La Marquise du Châtelet: her life and influence on Voltaire*. Dissertation, George Washington University.
Eves, H. 1990. *An Introduction to the History of Mathematics*. Philadelphia/Pa: Saunders College Publishing, Saunders Series.
Pichler Linz, F., ed. 2004. *Von den Planetentheorien zur Himmelsmechanik: die Newtonsche Revolution*. Linz: Trauner.
Faguet, É. 1907. *Amours d'hommes de lettres*. Paris: Société française d'imprimerie et de librairie.
Falvey, J. 1979. Women and sexuality in the thought of La Mettrie. In *Women and society in eighteenth-cetury France: Essays in honour of John Stephenson Spink*, ed. E. Le Jacobs, William H. Barber, J. H. Bloch, F. W. Leakey, and E. Breton, 55–68. London: The Athlone Press.
Fang, J. 1976/77. Mathematicians, man or woman: exercises in a 'Verstehen-Approach'. *Philosophia Mathematica* 13/14:15–71.
Fara, P. 2002. Images of Emilie du Châtelet. *Endeavour* 26, no. 2:39–40.
Fara, P. 2004. Emilie du Châtelet: the genius without a beard. *Physics World* 17, no. 6:14–15.
Fara, P. 2004. *Pandora's breeches: women, science, power in the Enlightenment*. London: Pimlico.
Farquhar, D., and M.-R. Lynn. 1989. *Women sum it up: biographical sketches of women mathematicians*. Christchurch: Hazard Press.
Faur, L.-F. 1791. *Vie privée du maréchal de Richelieu, contenant ses amours et intrigues et tout ce qui a rapport aux divers rôles qu'a joués cet hommes célèbre pendant plus de 80 ans*. 3 vols. Paris: Buisson.
Fayolle, R. 1972. *Sainte-Beuve et le XVIIIe siècle, ou comment les révolutions arrivent*. Paris: Armand Collin.
Feltman, J. 1978. *Voltaire's World*. Muncie: Ball State University.
Ferval, C. A., and M.-T.-G. Pierrebourg. 1948. *Madame du Châtelet, une maîtresse de Voltaire*. Paris: Arthème Fayard.
Firode, A. 2001. Locke et les philosophes français. In *Cirey dans la vie intellectuelle: la réception de Newton en France*, ed. François de Gandt, 57–72. Oxford: Voltaire Foundation.
Fleuriot de Langle. 1967. Une amie de Voltaire: la comtesse de Graffigny. *Revue générale belge* 123, no. 4:67–76.
Fontenelle, B., 1818. *Œuvres complètes*. 3 vols. Paris.
Fontenelle, B. 1818. Du Bonheur. In *Œuvres complètes*. 3 vols. Paris.

Fontenelle, B. 1818. Traité de la liberté de l'âme. In *Œuvres complètes*. 3 vols. Paris.
Formey, J. H. S. 1767. Continu. In *Dictionnaire instructif, où l'on trouve les principaux termes des sciences et des arts dont l'explication peut être utile ou agréable aux personnes qui n'ont pas fait des études approfondies*, ed. Jean H. S. Formey. Halle: Gebauer.
Formey, J. H. S. 1767. Contradiction. In *Dictionnaire instructif, où l'on trouve les principaux termes des sciences et des arts dont l'explication peut être utile ou agréable aux personnes qui n'ont pas fait des études approfondies*, ed. Jean H. S. Formey. Halle: Gebauer.
Formey, J. H. S. 1767. *Dictionnaire instructif, où l'on trouve les principaux termes des sciences et des arts dont l'explication peut être utile ou agréable aux personnes qui n'ont pas fait des études approfondies*. Halle: Gebauer.
Formey, J. H. S. 1767. Espace. In *Dictionnaire instructif, où l'on trouve les principaux termes des sciences et des arts dont l'explication peut être utile ou agréable aux personnes qui n'ont pas fait des études approfondies*, ed. Jean H. S. Formey. Halle: Gebauer.
Formey, J. H. S. 1767. Loi de continuité. In *Dictionnaire instructif, où l'on trouve les principaux termes des sciences et des arts dont l'explication peut être utile ou agréable aux personnes qui n'ont pas fait des études approfondies*, ed. Jean H. S. Formey. Halle: Gebauer.
Formey, J. H. S. 1767. Mouvement. In *Dictionnaire instructif, où l'on trouve les principaux termes des sciences et des arts dont l'explication peut être utile ou agréable aux personnes qui n'ont pas fait des études approfondies*, ed. Jean H. S. Formey. Halle: Gebauer.
Formey, J. H. S. 1767. Pendule. In *Dictionnaire instructif, où l'on trouve les principaux termes des sciences et des arts dont l'explication peut être utile ou agréable aux personnes qui n'ont pas fait des études approfondies*, ed. Jean H. S. Formey. Halle: Gebauer.
Formey, J. H. S. 1767. Pesanteur. In *Dictionnaire instructif, où l'on trouve les principaux termes des sciences et des arts dont l'explication peut être utile ou agréable aux personnes qui n'ont pas fait des études approfondies*, ed. Jean H. S. Formey. Halle: Gebauer.
Formey, J. H. S. 1767. Repos. In *Dictionnaire instructif, où l'on trouve les principaux termes des sciences et des arts dont l'explication peut être utile ou agréable aux personnes qui n'ont pas fait des études approfondies*, ed. Jean H. S. Formey. Halle: Gebauer.
Formey, J. H. S. 1767. Temps. In *Dictionnaire instructif, où l'on trouve les principaux termes des sciences et des arts dont l'explication peut être utile ou agréable aux personnes qui n'ont pas fait des études approfondies*, ed. Jean H. S. Formey. Halle: Gebauer.
Formey, J. H. S., ed. 1789. *Souvenirs d'un citoyen*. 2 vols. Berlin: Lagarde.
Formey, J. H. S. 1983. La belle Wolfienne. In *Gesammelte Werke*, ed. Christian von Wolff. Hildesheim: Olms.
Fouquier, H. 1884. *Au siècle dernier*. Brüssel: H. Kistemaeckers.
Fox, J. H., M. H. Waddicor, and D. A. Watts, eds. 1975. *Studies in eighteenth-century French literature presented to Robert Niklas*. Exeter: University of Exeter.
Francis, L. 1948. *La vie privée de Voltaire*. Paris: Hachette.
Freadman, A. 1989. Émilie: a cautionary tale. *Australian feminist studies* 10:97–102.
Fremont, H. 1966. Gabrielle-Emilie le Tonnelier de Breteuil (du Châtelet). In *Dictionnaire de biographie française*, ed. Michel Prevost, 1191–97. Paris: Letouzey et Ané.
Frédéric II. 1788. Lettres à la marquise du Châtelet. In *Œuvres posthumes de Frédéric II. roi de Prusse*. Berlin: Voss&Söhne/Decker&Söhne.
Frédéric II. 1789. Correspondance de Frédéric II, Roi de Prusse. In *Œuvres posthumes de Frédéric Roi de Prusse*, ed. J.P Heubach. Lausanne.
Frédéric II. 1846–1857. *Oeuvres de Frédéric le Grand*. Berlin: R. Decker.
Lottes, G., and I. D'Aprile, eds. 2006. *Hofkultur und aufgeklärte Öffentlichkeit: Potsdam im 18. Jahrhundert im europäischen Kontext*. Berlin: Akademie Verlag.
Gaffiot, J. C., ed. 2003. *Lunéville: fastes du Versailles lorrain*. Paris: D. Carpentier.
Galitzin, A. 1860. Une lettre inédite de Crousaz à la marquise du Châtelet. *Bulletin du bibliophile et du bibliothécaire*, no. 14:1621–25.
Gandt, F. de. 1995. La réception de Newton: philosophes et géomètres. *Revue du Nord* 77, no. 312:845–57.

Gandt, F. de, ed. 1999. *Conférence faite au Dixième Congrès international des Lumières*. Dublin.
Gandt, F. de, ed. 2001. *Cirey dans la vie intellectuelle: la réception de Newton en France*. Oxford: Voltaire Foundation.
Gandt, F. de. 2001. Qu'est-ce qu'être newtonien en 1740? In *Cirey dans la vie intellectuelle: la réception de Newton en France*, ed. François de Gandt, 126–47. Oxford: Voltaire Foundation.
Gardiner, L. J. 1982. Searching for the Metaphysics of Science: the Structure and composition of Madame du Châtelet's *Institutions de physique*, 1737–40. *SVEC* 201:85–113.
Gardiner, L. J. 1984. Women in Science. In *French Women an the Age of Enligthenment*, ed. Samia I. Spencer. Bloomington/ IN: Indiana University Press.
Gardiner, L. J. 2008. Mme Du Châtelet traductrice. In *Émilie Du Châtelet: Éclairages & documents nouveaux*, ed. Ulla Kölving and Olivier Courcelle, 167–72. Ferney-Voltaire: Centre International d'Étude du XVIIIe Siècle.
Gastineau, B. 1878. *Voltaire en exil, sa vie et son oeuvre en France et à l'étranger (Angleterre, Hollande, Belgique, Prusse, Suisse) avec des lettres inédites de Voltaire et de Mme du Châtelet (Centennaire de Voltaire)*. Paris: Germer Baillière et Cie.
Gastineau, B. 2006. *Voltaire in Exile: His Life and Works in France and Abroad with unpublished Letters of Voltaire and Mme Du Châtelet*. Whitefish/MT: Kessinger Pub Co.
Gautier, J. 1752. *Réfutation du Celse moderne, ou objections contre le christianisme avec des reponses*. Lunéville: Goebel und Messuy.
Gauvin, J.-F. 2006. Le cabinet de physique du château de Cirey et la philosophie naturelle de Mme Du Châtelet et Voltaire. In *Emilie Du Châtelet: rewriting Enlightenment philosophy and science*, ed. Judith P. Zinsser and Julie C. Hayes, 165–202. Oxford: Voltaire Foundation.
Gawlina, M. 2001. Der Ansatz von Kants politischer Philosophie, betrachtet in seiner Gegenüberstellung zu Fichte und Hegel. In *Kant und die Berliner Aufklärung: Akten des IX. Internationalen Kant-Kongresses*, ed. Volker Gerhard, 262–70. Berlin: de Gruyter.
Gehler, J. S. T. 1787. *Physikalisches Wörterbuch oder Versuch einer Erklärung der vornehmsten Begriffe und Kunstwörter der Naturlehre mit kurzen Nachrichten von der Geschichte der Erfindungen und Beschreibungen der Werkzeuge begleitet in alphabetischer Ordnung*. Leipzig.
Genlis, S.-F. D. C. c. d. 1825. *Mémoires inédits de Mme la comtesse de Genlis, sur le dix-huitième siècle et la Révolution française, depuis 1756 jusqu'à nos jours*. 10 vols. Paris: Ladvocat.
Gerhardt, V., ed. 2001. *Kant und die Berliner Aufklärung: Akten des IX. Internationalen Kant-Kongresses*. Berlin: de Gruyter.
Gerlach, H.-M., ed. 1989. *Descartes und das Problem der wissenschaftlichen Methode*. Halle/Saale: Abt. Wiss.-Publ. D. Martin-Luther-Univ.
Gerson, N. B. 1989. *Die göttliche Geliebte Voltaires: das Leben der Emilie du Châtelet*. Stuttgart: Engelhorn-Verlag.
Gigot, J.-G. 1961. Autographes inédits de Voltaire et de la marquise du Châtelet à l'occasion d'une affaire de vol commis au château de Cirey (1736). *Cahiers haut-marnais* 66, no. 3:123–31.
Gillispie, C. C. 1980. *Science and polity in France: the end of the old regime*. Princeton: Princeton Univ. Press.
Gipper, A. 2002. *Wunderbare Wissenschaft. Literarische Strategien naturwissenschaftlicher Vulgarisierung in Frankreich von Cyrano de Bergerac bis zur Encyclopédie*. München: Fink.
Gireau-Geneaux, A. 2001. Mme du Châtelet entre Leibniz et Newton. In *Cirey dans la vie intellectuelle: la réception de Newton en France*, ed. François de Gandt, 173–86. Oxford: Voltaire Foundation.
Gjertsen, D. 1986. *The Newton handbook*. London/New York: Routledge and Kegan Paul.
Glaser, H. A., and G. M. Vayda, eds. 2001. *Die Wende von der Aufklärung zur Romantik*. Amsterdam/Philadelphia: John Benjamins Publishing.
Goncourt, E., and Jules de. 1862. *La femme au XVIIIe siècle*. Paris: F. Didot frères, fils et Cie.
Gooch, G. P. 1961. Mme du Châtelet an her Lovers. *Contemporary Review* 200:648–53.
Gooch, G. P. 1962. Mme du Châtelet an her Lovers. *Contemporary Review* 201:44–48.
Gooch, G. P. 1962. Mme du Châtelet an her Lovers. *Contemporary Review* 201:203–07.
Goodman, D. 1994. *The Republic of Letters: A Cultural History of the French Enlightenment*. Ithaka/N.Y: Cornell University Press.

Goodman, E. 2000. *The Portraits of Madame de Pompadour: celebrating the femme savante*. Berkeley/Los Angeles: University of California Press.

Gordon, D. 1994. *Citizens without Sovereignty: Equality and Sociability in French Thought, 1670–1780*. Princeton: Princeton University Press.

Gottsched, L. A. 1771. *Briefe der Frau Luise Adelgunde Viktorie Gottsched geb. Kulmus. Zweyter Theil*. Dresden: Harpeter.

Graffigny, F. de. 1820. *Vie privée de Voltaire et de Mme du Châtelet, pendant un séjour de six mois à Cirey; par l'auteur des Lettres péruviennes: suivie de cinquantes lettres inédites, en vers et en prose, de Voltaire*. Paris: Treuttel et Wurz, Pelicier, Delaunay, Mongie.

Graffigny, F. de. 1879. *Lettres de Mme de Graffigny suivies de celle de Mmes de Staal, d'Épinay, du Boccage, Suar, du chevalier de Boufflers, du marquis de Vilette, etc., des Relations de Marmontel, de Gibbon, de Chabanon, du prince de Ligne, de Grétry, de Genlis sur leur séjour près de Voltaire*. Revues sur les éditions originales augmentées de nombreuses notes, d'un index et précédées d'une notice biographique par Eugène Assé. Paris: C. Charpentier.

Graffigny, F. de. 1985. *Correspondance de Madame de Graffigny*. Oxford: The Voltaire Foundation.

Grand, S., ed. 1985. *Ces bonnes femmes du XVIIIe siècle: flâneries à travers les salons littéraires*. Paris: Pierre Horay.

Grand, S., ed. 1985. Une rencontre décisive: Madame du Châtelet et Voltaire. In *Ces bonnes femmes du XVIIIe siècle: flâneries à travers les salons littéraires*, ed. Serge Grand, 131–46. Paris: Pierre Horay.

Granet, F., ed. 1737. *Reflexions sur les ouvrages de literature*. Paris: Briasson.

Greckol, S. 2005. Emilie explains Newton to Voltaire – Gabrielle Emilie le Tonnelier de Breteuil du Chatelet (1706–1749). *Dalhousie Review* 85, no. 1:119.

Greenberg, J. 1986. Mathematical physics in eighteenth-century France. *Isis* 77:59–78.

Grimm, Diderot, Raynal, and Meister. 1877–1882. *Correspondance littéraire, philosophique et critique par Grimm, Diderot, Raynal, Meister, etc. revue sur les textes originaux comprenant outre ce qui a été publié à divers époques les fragments supprimés en 1813 par la censure, les parties inédites conservées à la bibliothèque ducale de Gotha et à l'Arsenal à Paris*. Paris: Garnier frères.

Grinstein, L. S., and P. J. Campbell, eds. 1987. *Women of Mathematics. A Bibliographic Sourcebook*. New York: Greenwood Press.

Grinstein, L. S., K. Rose, and M. H. Rafailovich, eds. 1993. *Women in chemistry and physics: a biobibliogpaphical sourcebook*. Westport/Conn.: Greenwood Press.

Guerlac, H. 1981. *Newton on the Continent*. Ithaka: Cornell University Press.

Gunderson, L. 2010. *Emilie: La marquise du Châtelet defends her life tonight*. Samuel French acting. New York: Samuel French.

Guyot, P. 2008. La pédagogie des *Institutions de physique*. In *Émilie Du Châtelet: Éclairages & documents nouveaux*, ed. Ulla Kölving and Olivier Courcelle, 267–81. Ferney-Voltaire: Centre International d'Étude du XVIIIe Siècle.

Haase-Dubosc, D., and E. Viennot, eds. 1991. *Femmes et pouvoirs sous l'Ancien Régime*. Paris: Rivages.

Hagengruber, R., ed. 1998, ²1999. *Klassische philosophische Texte von Frauen*. München: Deutscher Taschenbuchverlag.

Hagengruber, R., ed. 1998. Philosophinnen. In *Feministische Hochschuldidaktik. Materialien zu den Koblenzer Frauenstudien*, ed. Elisabeth de Sotelo, 185–246. Münster: Lit-Verlag.

Hagengruber, R., ed. 1998. Über die Vervollständigung des Wissens. Philosophinnen in der Wissenschaft. In *Sie und Er. Frauenmacht und Männerherrschaft im Kulturvergleich*, ed. G. Völger, 105–09. Köln: Rautenstrauch-Joest-Museum.

Hagengruber, R., ed. 1999. Eine Metaphysik in Briefen. Emilie du Châtelet an Maupertuis. In *Pierre Louis Moreau de Maupertuis (1698–1759). Eine Bilanz nach 300 Jahren*, ed. Hartmut Hecht, 187–206. Berlin: Spitz-Verlag.

Hagengruber, R., ed. 2002. Emilie du Châtelet. Gegen Rousseau und für die Physik. Wissenschaft im Zeitalter der Aufklärung. *Konsens* 3, no. 18:27–30.

Hagengruber, R., ed. 2003. Tradition und Wandel. Frauen in der Wissenschaft. *Forschung und Lehre* 5:249–51.

Hagengruber, R., ed. 2004. Vom Besonderen zum Allgemeinen – zu einer Neuorientierung im Verhältnis von Philosophie, Wissenschaft und Feminismus. In *Philosophinnen im 3. Jahrtausend*, ed. B. Doetsch, 17–29. Bielefeld: Kleine Verlag.

Hagengruber, R., ed. 2005. Frauen in der Geschichte der Wissenschaft. In *Frauenbilder*, ed. Ulrike Schultz, 11–13.

Hagengruber, R., ed. 2010. Das Glück der Vernunft - Emilie du Châtelets Reflexionen über die Moral. In *Von Diana zu Minerva. Philosophierende Aristokratinnen des 17. und 18. Jahrhunderts.*, ed. Ruth Hagengruber and Ana Rodrigues, 109–28. Berlin: Akademie Verlag.

Hagengruber, R., ed. 2010. Von Diana zu Minerva. Philosophierende Aristokratinnen des 17. und 18. Jahrhunderts. In *Von Diana zu Minerva. Philosophierende Aristokratinnen des 17. und 18. Jahrhunderts.*, ed. Ruth Hagengruber and Ana Rodrigues, 11–32. Berlin: Akademie Verlag.

Hagengruber, R., and A. Rodrigues, eds. 2010. *Von Diana zu Minerva. Philosophierende Aristokratinnen des 17. und 18. Jahrhunderts.* Berlin: Akademie Verlag.

Hagengruber, R., and H. Hecht, eds. Forthcoming. *Emilie du Châtelet und die deutsche Aufklärung.* Hildesheim/New York: Olms.

Hager, A. J. 2006. *Euclid, the father of geometry: and, selected women in mathematics.*

Hall, A. R. 1988. Newtonianism after 300 Years. *Notes and Records of the Royal Society of London* 42, no. 1:5–9

Hamel, F. 1910. *An Eighteenth-Century Marquise. A study of Emilie du Châtelet and her Times: With frontispiece and illustrations.* London: Stanley Paul & Co.

Hamel, F. 1911. *An eighteenth century marquise: a study of Emilie du Châtelet and her times, etc. [With portraits].* New York.

Hamel, F. 2009. *An Eighteenth Century Marquise: A Study of Émilie Du Châtelet and her Times.* Charleston/SC: BiblioBazaar.

Hamou, P. 2001. Algarotti vulgarisateur. In *Cirey dans la vie intellectuelle: la réception de Newton en France*, ed. François de Gandt, 73–89. Oxford: Voltaire Foundation.

Hankins, T. L. 1965. Eighteenth-century attempts to solve the vis viva controversy. *Isis* 56, no. 3:281–97.

Haroche-Bouzinac, G. 2001. Voltaire à Cirey, poète et philosophe, d'après sa correspondance, 1735–1738. In *Cirey dans la vie intellectuelle: la réception de Newton en France*, ed. François de Gandt, 16–25. Oxford: Voltaire Foundation.

Hartemann, J. 1966–1967. La malheureuse grossesse de madame du Châtelet. Discours de réception. *Mémoires de l'académie de Stanislas* 63, no. 47:83–101.

Hartemann, J. 1984. *Eros et grandes dames. La folle Émilie ou la malheureuse grossesse de madame Du Châtelet.* Nancy: A. Hartemann

Harth, E. 1992. *Cartesian Women: Versions and Subversion of Rational Discourse in the Old Regime.* Ithaca/London: Cornell Univ. Press.

Havard, J.-A, ed. 1863. *Voltaire et Mme du Châtelet. Révélations d'un serviteur attaché à leurs personnes. Manuscrit et pièces inédites avec commentaires et notes historiques par d'Albanès Havard.* Paris: E. Dentu

Havard, J.-A. 1873. *Voltaire et Mme. du Châtelet.* Paris.

Hayes, J. C., ed.. 1999, ²2006. *Rereading the French Enlightenment: System and Subversion.* Cambridge: Cambridge University Press.

Hayes, J. C., ed.. 2006. Physics and figuration in du Châtelet's *Institutions de physique*. In *Rereadind the french enlightenment: system and subversion*, ed. Julie C. Hayes. Cambridge: CUP.

Hayes, J. C., ed. 2006. *Rereading the french enlightenment: system and subversion.* Cambridge: Cambridge University Press.

Hecht, H. 1989. Der Erfahrungsbegriff im Philosophie- und Wissenschaftsverständnis bei Descartes und Leibniz. In *Descartes und das Problem der wissenschaftlichen Methode*, ed. Hans-Martin Gerlach, 52–57. Halle/Saale: Abt. Wiss.-Publ. D. Martin-Luther-Univ.

Hecht, H., ed. 1991. *Gottfried Wilhelm Leibniz im philosophischen Diskurs über Geometrie und Erfahrung.* Berlin: Akademie Verlag.

Hecht, H. 1994. Pierre Louis Moreau de Maupertuis oder die Schwierigkeit der Unterscheidung von Newtonianern und Leibnizianern. In *Leibniz und Europa: VI. Internationaler Leibniz-Kongreß*, ed. Herbert Breger, 331–38. Hannover: G.-W.-Leibniz-Gesellschaft.

Hecht, H., ed. 1999. *Pierre Louis Moreau de Maupertuis (1698–1759). Eine Bilanz nach 300 Jahren.* Berlin: Spitz-Verlag.

Hecht, H. 2004. Das Triumvirat Euler, Maupertuis, Merian in den Leibniz-Debatten der Berliner Akademie. In *Leibnizbilder im 18. und 19. Jahrhundert,*, ed. Alexandra Lewendowski, 147–68. Stuttgart: Steiner.

Hecht, H. 2006. Maupertuis und die Leibniz-Tradition an der Berliner Akademie. In *Hofkultur und aufgeklärte Öffentlichkeit: Potsdam im 18. Jahrhundert im europäischen Kontext,* ed. Günther Lottes and Iwan D'Aprile. Berlin: Akademie Verlag.

Hellegouarc'h, J. 1997. Finalité des Mémoires pour servir à la vie de M. de Voltaire: règlement de compte ou mise au point? In *Voltaire et ses combats.* Actes du congrès international Oxford-Paris 1994, ed. Ulla Kölving and Christiane Mervaud, 329–37. 2 vols. Oxford: Voltaire Foundation.

Helvétius, C.-A. 1773. *De l'homme.* London.

Helvétius, C.-A. 1814. Épître sur l'amour de l'étude, à madame la marquise du Chastelet, par un élève de Voltaire, avec des notes du maître. *Magasin encyclopédique ou journal des sciences, des lettres et des arts* 6:273–85.

Helvétius, C.-A. 1815. *Épître sur l'amour de l'étude, à madame la marquise du Chastelet, par un élève de Voltaire, avec des notes du maître.* Paris: J.-B. Sajou.

Hénault, C.-J.-F. 1855. *Mémoires du président Hénault, de l'Académie française, écrits par lui-même, recueillis et mis en ordre par son arrière-neveu, M. le baron de Vigan.* Paris: E. Dentu.

Henriot, E. 1931. Émilie. In *Épistoliers et mémorialistes,* ed. Emile Henriot, 139–46. Paris: Éditions de la nouvelle revue critique.

Henriot, E. 1937. *De Marie de France à Katherine Mansfield, portraits de femmes.* Paris: Plon.

Hérard, L. 1962. *Voltaie à Sémur.* Dijon: l'Auteur.

Hermes, L., A. Hirschen, and I. Meißner, eds. 2003. *Ausgewählte Beiträge der 3. Fachtagung Frauen-/Gender-Forschung in Rheinland-Pfalz.* Tübingen: Stauffenburg-Verlag.

Heubach, J., ed. 1789. *Œuvres posthumes de Frédéric Roi de Prusse.* Lausanne.

Heydemann, M.-C. 1974. Histoire de quelques mathématiciennes. In *Mathématiques, Mathématiciens et Société,* ed. Pierre Samuel. Orsay: Université de Paris-Sud.

Hochet, 1806. Notice historique sur Madame du Châtelet. In *Lettres inédites de Madame la marquise du Châtelet à M. le Comte d'Argental, auxquelles on a joint une Dissertation sur l'existence de Dieu, les Réflexions sur le bonheur, par le même auteur, et deux notices historiques sur madame Du Chastelet et M. d'Argental.* Paris: Xhrouet, Déterville, Lenormand, Petit.

Hoefer, D., ed. 1855. *Nouvelle biographie générale depuis les temps les plus reeculés jusqu'à nos jours, avec les renseignements bibliographiques et l'indication des sources à consulter: publié par MM. Firmin Didot frères, sous la direction de M. le Dr Hoefer.* Paris: Firmin Didot frères.

Horne, H. 1957. Voltaire amoureux. *Miroir de l'histoire* 95, no. 8:583–85.

Horst Baader, ed. 1980. *Voltaire.* Darmstadt: Wissenschaftliche Buchgesellschaft.

Howells, C., ed. 2004. French Women Philosophers: A Contemporary Reader: Subjectivity, Identity, Alterity. London: Routledge.

Hurel, D., and G. Laudin, eds. 2000. *Académies et sociétés savantes en Europe (1650–1800).* Paris: H. Champion.

Hutton, S. 2004. Emilie du Châtelet's *Institutions de physique* as a Document in the History of French Newtonianism. *Studies in history and philosophy of science* 35 A, no. 3:515–31.

Hutton, S. 2004. Women, Sience, and Newtonianism: Emilie du Châtelet versus Francesco Algarotti. In *Newton and Newtonianism: new studies,* ed. Sarah Hutton and James E. Force, 183–203. Boston: Kluwer.

Hutton, S., and J. E. Force, eds. 2004. *Newton and Newtonianism: new studies.* Boston: Kluwer.

Iltis, C. 1973. The Leibnizian-Newtonian Debates: Natural Philosophy and Social Psychology. *The British Journal for the History of Science* 6, no. 4:343–77

Iltis, C. 1977. Leibniz and the Vis viva Controversy. *Isis* 62:21–35.

Iltis, C. 1977. Madame du Châtelet's Metaphysics and Mechanics. *Studies in history and philosophy of science* 8, no. 1:29–48.

Inguenaud, M.-T. 2006. La Grosse et le Monstre: histoire d'une haine. In *Emilie Du Châtelet: rewriting Enlightenment philosophy and science*, ed. Judith P. Zinsser and Julie C. Hayes, 65–90. Oxford: Voltaire Foundation.
International Congress on the History of Science, ed. 1971. *Science et Philosophie: XVIIe et XVIIe siècles*. Actes XIIe Congrès Internat. d'Histoire des Sciences, Paris 1968 III B. Paris: A. Blanchard.
Isely, L. 1894. *Les femmes mathématiciennes: Hypathie, Emilie du Châtelet, Marie Agnesi, Sophie Germain, Sophie Kowalewska*. Neuchâtel: Impr. Nouvelle.
Iverson, J. R. 2006. A female member of the Republic of Letters: Du Châtelet's Portrait in Bilder-Sal (…) berühmter Schrifftsteller. In *Emilie Du Châtelet: rewriting Enlightenment philosophy and science*, ed. Judith P. Zinsser and Julie C. Hayes, 35–51. Oxford: Voltaire Foundation.
Iverson, J. 2008. Émilie Du Châtelet, Luise Gottsched et la Société des Alétophiles: une traduction allemande de l'échange au sujet des forces vives. In *Émilie Du Châtelet: Éclairages & documents nouveaux*, ed. Ulla Kölving and Olivier Courcelle, 283–99. Ferney-Voltaire: Centre International d'Étude du XVIIIe Siècle.
Iverson, J., and M.-P. Pieretti. 1998. Une gloire réfléchie: du Châtelet et les stratégies de la traductrice. In *Dans les miroirs de l'écriture: La réflexivité chez les femmes écrivains d'Ancien Régime*, ed. Jean-Philippe &. Desrosiers-Bonin, Diane Beaulieu, 135–44. Montréal: Département d'études françaises, Université de Montréal.
Iverson, J., and M.-P. Pieretti. 2004. Toutes personnes (…) seront admises à concourir: la participation des femmes au concours académiques. *Dix-huitième siècle* 36:313–32.
Jaggar, A. M., and I. M. Young, eds. 1998. *A Companion to Feminist Philosophy*. Oxford: Blackwell.
Jameson, A. B. 1857. *Memoirs of the loves of the poets*. Boston: Ticknor & Fields.
Janssens, J. 1963. Voltaire à Bruxelles. *Revues des deux mondes*:263–73.
Jauch, U. P. 1988. *Immanuel Kant zur Geschlechterdifferenz: Aufklärerische Vorurteilskritik und bürgerliche Geschlechtsvormundschaft*. Wien: Passagen Verlag.
Jauch, U. P. 1990, ²1991. *Damenphilosophie & Männermoral*. Wien: Passagen Verlag.
Jauch, U. P. 1992. Emilie, 'Marquisin von Chastellet'. Aufklärung als gelebte Geschichte. *NZZ*, April 25.
Jauch, U. P. 1998. *Jenseits der Maschine: Philosophie, Ironie und Ästhetik bei Julien Offray de La Mettrie (1709–1751)*. München: Hanser.
Jauch, U. P. 1999. Adieu, Madame Newton-Pompon.Vor 250 Jahren starb Emilie du Châtelet. *NZZ*, September 11.
Jauch, U. P. 2006. Emilie, Exzellenz und Exzentrik. Endstation Lumières – in Paris wird die Aufklärung wiederbelebt. *NZZ*, May 6.
Jaumann, H. 2004. *Handbuch Gelehrtenkultur der frühen Neuzeit*. Berlin/New York: de Gruyter.
Jindráková, T., and J. Folta. 2004. Die Rolle von Mme du Châtelet in der Anwendung des Infinitesimalkalküls in Newtons *Principia*. In *Von den Planetentheorien zur Himmelsmechanik: die Newtonsche Revolution*, ed. Franz Pichler Linz, 155–62. Linz: Trauner.
Johnson, W. 1994. Some women in the history of mathematics, physics, astronomy and engineering. *Journal of materials processing technology* 40:33–71.
Johnson, J. 1994. Voltaire after 300 Years. *Notes and Records of the Royal Society of London* 48, no. 2:215–20.
Joly, B. 1995. Voltaire chimiste: l'influence des théories de Boerhaave sur sa doctrine du feu. *Revue du Nord* 77, no. 312:817–43.
Joly, B. 2001. Les théories du feu de Voltaire et de Mme du Châtelet. In *Cirey dans la vie intellectuelle: la réception de Newton en France*, ed. François de Gandt, 212–37. Oxford: Voltaire Foundation.
Jordan, L. 1915. Ein zeitgenössisches Manuskript der *Réflexions sur le bonheur* der Marquise du Châtelet. *Archiv* 133:115–19.
Jordanova, L. 1989. *Sexual Visions: Images of Gender in Science and Medicine between the Eighteenth and Twentieth Centuries*. Madison/Wis.: The University of Wisconsin Press.

Jovy, E. 1906. *Quelques Lettres inédites de da Mise du Châtelet et de la Duchesse de Choiseul.* Paris: H. Leclerc.
Jovy, E. 1922. *Une illustration scientifique vitryate: le P. François Jacquier et ses correspondants, 116 lettres inédites conservées, pour la plupart, à la bibliothèque de la ville de Vitry-le-François et publiées par Ernest Jovy.* Vitry-le-François: Société des sciences et arts de Vitry-le-François.
Juškevič, A., and E. Winter. 1965. *Euler, Leonhard - Goldbach, Christian: Briefwechsel (1729–64).* Berlin: Akademie Verlag.
Kahan, M. 1967. *The unreported first French edition of Newtons Principia, 1756. A bibliographical essay.* Vaxholms Antikvariat.
Kant, I. 1746. *Gedanken von der wahren Schätzung der lebendigen Kräfte und Beurtheilung der Beweise, derer sich Herr von Leibnits und andere Mechaniker in dieser Streitsache bedient haben: nebst einigen vorhergehenden Betrachtungen, welche die Kraft der Körper überhaupt betreffen.* Königsberg: Martin Eberhard Dorn.
Kawashima, K. 1990. La participation de madame du Châtelet à la querelle sur les forces vives. *Historia scientiarum: international journal of the History of Science Society of Japan* 40:9–28.
Kawashima, K. 1993. Les idées scientifiques de Madame du Châtelet dans ses *Institutions de physique*: un rêve de femme de la haute société dans la culture scientifique au Siècle des Lumières. *Historia scientiarum: international journal of the History of Science Society of Japan* 3, no. 1:63–82.
Kawashima, K. 1993. Les idées scientifiques de Madame du Châtelet dans ses *Institutions de physique*: un rêve de femme de la haute société dans la culture scientifique au Siècle des Lumières. *Historia scientiarum: international journal of the History of Science Society of Japan* 3, no. 2:137–55.
Kawashima, K. 1995. Madame du Châtelet dans le journalisme. *LLULL. Revista de la Sociedad española de Historia de las ciencias y las técnicas* 18:471–91.
Kawashima, K. 1998. Madame du Châtelet et Madame Lavoisier, deux femmes de science. *La Revue. Conservatoire National des Arts et Métiers, Musée National des Techniques* 22:22–29.
Kawashima, K. 2004. Birth of ambition: Madame du Châtelet's *Institutions de physique*. *Historia scientiarum: international journal of the History of Science Society of Japan* 14, no. 1:49–66.
Kawashima, K. 2005. *Emilie du Châtelet and Marie-Anne Lavoisier: The Issue of Gender and Science in 18th Century France.* Tokyo: University of Tokyo Press.
Kawashima, K. 2005. The issue of gender and science: a case study of Madame du Châtelet's *Dissertation sur le feu*. *Historia scientiarum: international journal of the History of Science Society of Japan* 15, no. 1:23–43.
Kawashima, K. 2006. Le débat entre Mme du Châtelet et Dortous de Mairan. In *Madame du Châtelet. La femme des Lumières*, ed. Elisabeth Badinter and Danielle Muzerelle, 99. Paris: Bibliothèque Nationale de France.
Kawashima, K. 2007. Two popular accounts of Émilie du Châtelet and the gender problem. *Historia scientiarum: international journal of the History of Science Society of Japan* 17, no. 2:121–33
Kelley, L. 1996. Why were so few mathematicians female? *Mathematics Teacher* 89, no. 7:592–96.
Kersey, E. M., and C. O. Schrag. 1989. *Women philosophers: A bio-critical source book.* New York: Greenwood Press.
Kiernan, C. 1968. *Science and the Enlightenment in Eighteenth Century France.* Genèva: Institut et Musée Voltaire.
Kim, Y. L. 1999. *Les lettres de Cirey. Etude de la correspondance de Voltaire (1734–1749).* Dissertation.
Kittay, E. F., and L. M.-N. Alcoff, eds. 2006. *The Blackwell Guide to Feminist Philosophy.* Oxford: Blackwell.
Kleinert, A. 1974. *Die allgemeinverständlichen Physikbücher der französischen Aufklärung.* Aarau: Verlag Sauerländer.
Klens, U. 1994. Châtelet-Lomont, Gabrielle-Emilie du. In *Philosophinnen-Lexikon*, ed. Ursula Meyer, 141–45. Aachen: Ein-Fach-Verlag.

Klens, U. 1994. *Mathematikerinnen im 18. Jahrhundert: Maria Gaetana Agnesi, Gabrielle-Emilie Du Châtelet, Sophie Germain: Fallstudien zur Wechselwirkung von Wissenschaft und Philosophie im Zeitalter der Aufklärung*. Pfaffenweiler: Centaurus.

Kölving, U. 2002. Deux lettres inédites d'Émilie du Châtelet. *Cahiers Voltaire* 1.

Kölving, U. 2008. Bibliographie chronologique d'Émilie Du Châtelet. In *Émilie Du Châtelet: Éclairages & documents nouveaux*, ed. Ulla Kölving and Olivier Courcelle, 341–85. Ferney-Voltaire: Centre International d'Étude du XVIIIe Siècle.

Kölving, U. 2008. Émilie Du Châtelet devant l'histoire. In *Émilie Du Châtelet: Éclairages & documents nouveaux*, ed. Ulla Kölving and Olivier Courcelle, 1–12. Ferney-Voltaire: Centre International d'Étude du XVIIIe Siècle.

Kölving, U., and A. Brown. 2008. Émilie Du Châtelet, lectrice d'une Apologie d'Homère. In *Émilie Du Châtelet: Éclairages & documents nouveaux*, ed. Ulla Kölving and Olivier Courcelle, 135–65. Ferney-Voltaire: Centre International d'Étude du XVIIIe Siècle.

Kölving, U., and O. Courcelle, eds. 2008. *Émilie Du Châtelet: Éclairages & documents nouveaux*. [... du Colloque du Tricentenaire de la Naissance d'Emilie Du Châtelet qui s'est tenu du 1er au 3 juin 2006 à la Bibliothèque Nationale de France et à l'ancienne Mairie de Sceaux]. Ferney-Voltaire: Centre International d'Étude du XVIIIe Siècle.

Kölving, U., and C. Mervaud, eds. 1997. *Voltaire et ses combats*. Actes du congrès international Oxford-Paris 1994. 2 vols. Oxford: Voltaire Foundation.

Kording, I., ed. 1999. *Louise Gottsched – "mit der Feder in der Hand", Briefe aus den Jahren 1730–1762*. Darmstadt: Wissenschaftliche Buchgesellschaft.

Kourany, J. 1998. *Philosophy in a Feminist Voice*. Princeton: Princeton University Press.

Koyré, A. 1965. *Newtonian Studies*. Chicago: University of Chicago Press.

Koyré, A. 1968. *Etudes newtoniennes*. Paris: Gallimard.

Krampf, W., ed. 1969. *Ernst Cassirer, Philosophie und exakte Wissenschaft*. Frankfurt/M.: Klostermann.

Kraus, G. 2006. *Bedeutende Französinnen*. Mühlheim/M.: Schröder.

Kraus, G. 2007. Bedeutende Französinnen. Oldenburg: Antiquariat Schröder.

La Beaumelle, L. A. de. 1755. *Die Antwort des aus der Bastille befryten Anglivie (sic) de la Beaumelle an den Herrn von Voltaire auf dessen Supplement Der Zeiten Ludewigs des XIV. aus dem Französischen übersetzet mit einem Anhange zwoer sinnreichen Grab-Schrifften, deren eine auf den Herrn von Voltaire und die andere auf die Marquise du Chastelet gemacht worden*. Colmar (Jena): Fischer.

La Mettrie, J. O. de. 1747. *Histoire naturelle de l'ame, traduite de l'Anglois de M. Charp, par feu M.H.*** de l'Académie des Sciences, et. Nouvelle Edition ... & augmentée par la Lettre Critique de M. de la Mettrie à Madame la Marquise Du Châtelet*. Oxford.

La Mettrie, J. O. de. 1747. Lettre critique de M. de la Mettrie à Madame la marquise du Châtelet. In *Histoire naturelle de l'âme*. Oxford.

La Mettrie, J. O. de. 1975. *Discours sur le bonheur*. Critical edition by John Falvey. Banbury/Oxfordshire: Voltaire Foundation.

La Porte, J., and J.-F. de Lacroix, eds. 1769. *Histoire littéraire des femmes françaises, ou lettres historiques et critiques sur la vie et les ouvrages des femmes qui se sont distinguées dans la littérature français; par une société de gens de lettres*. 5 vols. Paris: Lacombe.

La Porte, D. 1970. *Theories of Fire and Heat in the first half of the 18th century*. Doctoral Dissertation.

La Rougère. 1746. *Le génie ombre*. Chimerie.

Le Coat, N. 2006. 'Le génie de la sécheresse': Mme Du Châtelet in the eyes of her Second Empire critics. In *Emilie Du Châtelet: rewriting Enlightenment philosophy and science*, ed. Judith P. Zinsser and Julie C. Hayes, 292–307. Oxford: Voltaire Foundation.

Le Coat, G., and A. Eggimann-Besancon. 1986. Le portrait de madame du Châtelet par Marie-Anne Loir – emblématique et émancipation féminine au XVIIIe siecle. *Colóquio artes* 68, no. 2:30–39.

Le Jacobs, E., W. H. Barber, J. H. Bloch, F. W. Leakey, and E. Breton, eds. 1979. *Women and society in eighteenth-cetury France: Essays in honour of John Stephenson Spink*. London: The Athlone Press.

Le Lalande, J. J. F. de. 1803. *Bibliographie astronomique; avec l'histoire de l'astronomie depuis 1781 jusqu'à 1802*. Paris: Imprimerie de la République.
Le Lay, C. 2004. Astronomie des dames. *Dix-huitième siècle* 36:303–12.
Le Ru, V. 2001. La conception sceptique de la matière au temps de Cirey. In *Cirey dans la vie intellectuelle: la réception de Newton en France*, ed. François de Gandt, 148–58. Oxford: Voltaire Foundation.
Le Ru, V. 2008. Quand Voltaire et la marquise parlent métaphysique. In *Émilie Du Châtelet: Éclairages & documents nouveaux*, ed. Ulla Kölving and Olivier Courcelle, 213–18. Ferney-Voltaire: Centre International d'Étude du XVIIIe Siècle.
Le Sueur, A. 1896. *Maupertuis et ses correspondants: lettres inédits du grand Frédéric, du prince Henri de Prusse, de Labeaumelle, du président Hénault, du comte de Tressan, d'Euler, de Kaestner, de Koenig, de Haller, de Condillac, de l'abbé d'Olivet, du maréchal d'Écosse etc.* Montreuil-sur-Mer.
Ledeuil d'Enquin, J. 1892. *La Marquise du Châtelet à Sémur et le passage de Voltaire*. Semur: T. Millon.
Lee, J. P. 2006. Le Receuil de Poésies: manuscrit de Mme Du Châtelet. In *Emilie Du Châtelet: rewriting Enlightenment philosophy and science*, ed. Judith P. Zinsser and Julie C. Hayes, 105–23. Oxford: Voltaire Foundation.
Lefèvre, W., ed. 2001. *Between Leibniz, Newton, and Kant. Philosophy and Sciences in the Eighteenth Century*. Dordrecht/Boston/London: Kluwer.
Lehnert, G. 2006. *Frauen, die man kennen muß*. Berlin: Aufbau TB.
Lenôtre, G. 1956. Une curieuse amie de Voltaire: la marquise du Châtelet. *Historia* 121:572–75.
Lescure, M.-F.-A. de. 1881. *Les Femmes philosophes*. Paris: E. Dentu.
Lescure, M.-F.-A. de. 1881. Mme du Châtelet. In *Les Femmes philosophes*, 210–30. Paris: E. Dentu.
Lescure, M.-F.-A. de. 1878. Les femmes philosophes, le couvent de Voltaire: la marquise Du Deffand, Mlle Lespinasse, la baronne de Staal, la marquise Du Châtelet. *Le Correspondant. Recueil périodique*, no. 77:683–726.
Lescure, M.-F.-A. de. 1878. Les femmes philosophes, le couvent de Voltaire; la Marquise du Deffand, Mlle de Lespinasse, la baronne de Staal, la Marquise Du Châtelet. *Le Correspondant. Recueil périodique*, no. 77:999–1038.
Lewendowski, A., ed. 2004. *Leibnizbilder im 18. und 19. Jahrhundert*. Stuttgart: Steiner.
Libby, M. S. 1935. *The attitude of Voltaire to magic and sciences*. New York: Columbia University Press.
Liebrecht, H. 1948. Un séjour de Voltaire à Bruxelles. *Bulletin de l'Académie de langue et de littérature françaises* 6:57–68.
Lizé, É. 1976. Voltaire 'collaborateur' de la CL. In *La Correspondance littéraire de Grimm et de Meister (1754–1813): Colloque de Sarrebruck (22–24 février 1974)*, 49–67. Paris: Klincksieck.
Lizé, É. 1979. *Voltaire, Grimm et la Correspondance littéraire*. Oxford: Voltaire Foundation.
Lizé, É. 1983. Glanures voltairiennes. *Revue d'histoire littéraire de la France* 83:237–41.
Lleaud, G. 1998. Emilie de Breteuil du Châtelet. In *Notable Women in Mathematics: A Biographical Dictionary*, ed. Charlene Morrow and Teri Perl, 38–43. London: Greenwood Press.
Lloyd, G. 1993. *"Male" and "Female" in Western Philosophy*. Minneapolis: University of Minnesota Press.
Locqueneux, R. 1995. Les *Institutions de physique* de Madame du Châtelet ou d'un traité de paix entre Descartes, Leibniz et Newton. *Revue du Nord* 77, no. 312:859–92.
Locqueneux, R. 2001. La physique expérimentale vers 1740: expérience, systèmes et hypothèses. In *Cirey dans la vie intellectuelle: la réception de Newton en France*, ed. François de Gandt, 90–111. Oxford: Voltaire Foundation.
Loménie, L. de. 1870. *La Comtesse de Rochefort et ses amis. Études sur les mœurs en France au XVIIIe siècle, avec des documents inédits*. Paris: Michel Lévy frères.
Longchamp, S. G., and J.-L. Wagnière. 1826. *Mémoires sur Voltaire, et sur ses ouvrages, par Longchamp et Wagnière, ses secrétaires; Suivis de divers écrits inédits de la marquise Du*

Chatelet, du président Hénault, de Piron, Darnaud Baculard, Thiriot, etc., tous relatifs à Voltaire. Publié par L.-P. Decroix et A. Beuchot. 2 vols. Paris: Aimé André.

Longchamp, S. G., and J.-L. Wagnière. 1838. *Mémoires anecdotiques, très-curieux et inconnus jusqu'à ce jour sur Voltaire, réflexions sur ses ouvrages, suivis de divers écrits inédits de la marquise Du Châtelet, du président Hénault, de Piron, etc.* Paris: Bethune et Plon.

Lowenthal. 1928. Époux et amants au XVIIIe siècle. *Revue mondiale* 185:234–40.

Luynes, C.-P. d'A. 1860–1865. *Mémoires du duc de Luynes sur la cour Louis XV, 1735–1758.* 17 vols. Paris: Firmin Didot frères.

Mach, E. 1991. *Die Mechanik in ihrer Entwicklung.* Darmstadt: Wissenschaftliche Buchgesellschaft.

Machado, A. M. 2009. *Ciencia, vida y metafísica en Madame du Châtelet:* Fundación Canaria Orotava de Historia de la Ciencia.

Machado, A. M. 2009. Madame du Châtelet, leibniziana *malgré* Voltaire. *Thémata. Revista de Filosofía* 42:51–75.

Maglo, K. 2008. Mme Du Châtelet, *l'Encyclopédie,* et la philosophie des sciences. In *Émilie Du Châtelet: Éclairages & documents nouveaux,* ed. Ulla Kölving and Olivier Courcelle, 255–66. Ferney-Voltaire: Centre International d'Étude du XVIIIe Siècle.

Mairan, J. J. D. de. 1741. *Lettre de M. de Mairan, secrétaire perpétuel del'Académie royale des sciences, etc. à madame*** sur la question des forces vives, en réponse aux objections qu'elle lui a faites sur ce sujet dans ses Institutions de physique.* Paris: Charles-Antoine Jombert.

Mandic, S. 1995. *Emilie du Châtelet.* Atlanta/GA: Agnes Scott College.

Mangeot, G. 1914. Une biographie de madame de Graffigny. *Le Pays lorrain et le pays messin. Revue mensuelle illustré* 11:66–77.

Mangeot, G. 1922. Les *Réflexions sur le bonheur* de la marquise du Châtelet. In *Mélanges offerts par ses amis et ses élèves à M. Gustave Lanson,* 275–83. Paris: Hachette.

Mangeot, G. 1922. *Mélanges offerts par ses amis et ses élèves à M. Gustave Lanson.* Paris: Hachette.

Marmontel, J.-F. 1804. *Œuvres posthumes de Marmontel, historiographe de France, secrétaire perpétuel de l'Académie française. Imprimés sur le manuscrit autographe de l'auteur. Mémoires.* 4 vols. Paris: Xhrouet.

Maroy, C. 1905. Les séjours de Voltaire à Bruxelles. *Annales de la société d'archéologie de Bruxelles* 19:288–302.

Martin, M.-M. 1950. *Le "Génie" des femmes.* Paris: Editions du Conquistador.

Martinprey, E. de. 1888. Note sur la famille du Châtelet. *Journal de la Société d'archéologie et du Musée historique lorrain,* no. 37:65–67.

Mascarel, B. D. de. 1990. Cirey dans l'intimité des Lumières. *Point de vue – Images du monde* 42:18–20.

Mason, A. 2006. 'L'air du climat et le goût du terroir': translation as a cultural capital in the writings of Mme Du Châtelet. In *Emilie Du Châtelet: rewriting Enlightenment philosophy and science,* ed. Judith P. Zinsser and Julie C. Hayes, 124–41. Oxford: Voltaire Foundation.

Masson, G. 1880. Épaves du XVIIIe siècle. *Le Cabinet historique,* no. 26:250–51.

Mataix Loma, C. 1993. Madame du Châtelet: un fuego encendido. *Mujer y ciencia* 144, no. 565:79–90.

Maugras, G. 1904. *La Cour de Lunéville au XVIIIe siècle: Les marquises de Boufflers et Du Châtelet, Voltaire, Devau, Saint-Lambert etc.* Paris: Plon.

Maupertuis, P. L. M. de. 1768. Discours sur les différentes figures des astres. In *Oeuvres et lettres,* 4 vols. Lyon: Bruyset.

Maupertuis, P. L. M. de. 1768. Essay de Cosmologie. In *Oeuvres et lettres,* 4 vols. Lyon: Bruyset.

Maupertuis, P. L. M. de. 1768. Lettre sur la comète qui paroissoit en 1742. In *Oeuvres et lettres,* 4 vols. Lyon: Bruyset.

Maupertuis, P. L. M. de. 1768. *Oeuvres et lettres.* 4 vols. Lyon: Bruyset.

Maupertuis, P. L. M. de. 1828. Fragment d'une lettre de Maupertuis. In *Isographie des hommes célèbres ou collection de fac-similé de lettres autographes et de signatures.* 3 vols. Paris: Alexandre Mesnier.

Maupertuis, P. L. M. de. 1974. *Œuvres*. Hildesheim: Olms.
Maupertuis, P. L. M. de. 1974. Essai de philosophie morale. In *Œuvres*. Hildesheim: Olms.
Maurel, A. 1930. *La Marquise du Châtelet, amie de Voltaire*. Paris: Hachette.
Maurel, A. 1930. *The Romance of Mme Du Châtelet & Voltaire: Translated by Walter Mostyn, with some pages from Voltaire's Memoirs by way of preface*. London: Hutchinson & Co. Ltd.
Mauro, F. 2006. *Émilie du Châtelet*. Paris: Plon.
Mauzi, R. 1960, ²1969. *L' idée du bonheur dans la littérature et la pensée française au XVIIIe siècle*. Paris: Armand Colin.
Mauzi, R. 1979. *L'idée du bonheur dans la littérature et la pensée françaises au XVIII siècle*. Genève: Slatkine Repr.
Mauzi, R. 1994. *L'idée du bonheur dans la littérature et la pensée françaises au XVIIIe siècle*. Reprod. photomécanique, éd. au format de poche. Paris: Albin Michel.
Mauzi, R., and H. Coulet. 2008. *L'art de vivre d'une femme au XVIIe siècle : Suivi du "Discours sur le bonheur" de madame du Châtelet*. Paris: Desjonquères.
Mazauric, S. 2008. En passant par la Lorraine. In *Émilie Du Châtelet: Éclairages & documents nouveaux*, ed. Ulla Kölving and Olivier Courcelle, 93–98. Ferney-Voltaire: Centre International d'Étude du XVIIIe Siècle.
Mazzotti, M. 2004. Newton for the ladies: gentility, gender, and radical culture. *British Journal for the History of Science* 37, no. 2:119–46
Mazzotti, M. 2008. Mme Du Châtelet académicienne de Bologne. In *Émilie Du Châtelet: Éclairages & documents nouveaux*, ed. Ulla Kölving and Olivier Courcelle, 121–26. Ferney-Voltaire: Centre International d'Étude du XVIIIe Siècle.
McAlister, L. L., ed. 1996. *Hypatia's Daughters: 1500 Years of Women Philosophers*. Bloomington/IN: Indiana University Press.
McDonald, L. 1994. *The Women Founders of the Social Sciences*. Ottawa: Carleton University Press.
McDonald, L. 1996. *The Early Origins of the Social Sciences*. Montreal: McGill - Queen's University Press.
McDonald, L. 1998. *Women theorists on society and politics*. Waterloo/Ontario: Wilfrid Laurier University Press.
McHugh, N. A. 2007. *Feminist Philosophies A-Z*. Edinburgh: Edinburgh University Press.
McKenna, A., and A. Mothu, eds. 1997. *La philosophie clandestine à l'âge classique: actes du colloque de l'Université Jean Monnet Saint-Etienne du 29 septembre au 2 octobre 1993*. Oxford: Voltaire Foundation.
McNiven Hine, E. 1996. *Jean-Jacques Dortous de Mairan and the Geneva connection: scientific networking in the eighteenth century*. Oxford: Voltaire Foundation.
Menage, G. 1984. *The History of Women Philosophers*. Lanham/ MD: University Press of America.
Mercier, G. 1985. *Bébé, le nain de Stanislas. Ou les amours mouvementées d'Emilie du Châtelet et de Voltaire*. illustr. von Philippe Delestre. Sarreguemines: Editions Pierron.
Mercier, G. 2001. *Madame Voltaire*. Paris: Ed. de Fallois.
Mercier, G. 2003. Quand le roi Stanislas accueillait le roi Voltaire. In *Lunéville: fastes du Versailles lorrain*, ed. J. Charles Gaffiot, 201–05. Paris: D. Carpentier.
Mercier, G. 2004. *Femmes des Lumières à la cour de Stanislas*. Préface de Jeanne Cressanges. Nancy: Éditions de l'Est.
Metzger, F. 2006. Emilie du Châtelet: Femme fatale der Aufklärung. *G-Geschichte* 12:58–61.
Meyer, U. I. 2009. *Aufklärerinnen*. Aachen: Ein-Fach-Verl.
Meyer, U., ed. 1994. *Philosophinnen-Lexikon*. Aachen: Ein-Fach-Verlag.
Michel, F. 1955. La marquise et le philosophe. *Nouvelle critique* 66, no. 7:71–87.
Mitford, N. 1957. A visit to Voltaire. *History today*:18–27.
Mitford, N. 1957. *Voltaire in Love*. London: Hamish Hamilton.
Mòllica, C. 1982. *Un'amica di Voltaire: Emilia du Châtelet 1706–1749*. Rom.
Montucla, J.-É. 1802. *Histoire des mathématiques: Nouvelle édition considérablement augmentée, et prolongée jusque vers l'époque actuelle.Par J. F. Montucla, de l'Institut national de France*. 4 vols. Paris: Henry Agasse.

Montucla, J.-É. 1966. *Histoire des mathématiques*. Paris: Albert Blanchard.
Moriarty, P. V. 2006. The principle of sufficient reason in Du Châtelet's *Institutions*. In *Emilie Du Châtelet: rewriting Enlightenment philosophy and science*, ed. Judith P. Zinsser and Julie C. Hayes, 203–25. Oxford: Voltaire Foundation.
Mornet, D. 1969. *La pensée française au XVIIIe siècle*. Paris: Armand Colin.
Morrow, C., and T. Perl, eds. 1998. *Notable Women in Mathematics: A Biographical Dictionary*. London: Greenwood Press.
Mosebach, M. 2006. Die Kunst, etwas weniger zu lieben. Madame du Châtelet: Rede vom Glück – Discours sur le bonheur. In *Schöne Literatur*, 145–50. München/ Wien: Hanser.
Mosebach, M. 2006. *Schöne Literatur*. München/ Wien: Hanser.
Moser, F. 1998. *Alles am Weibe ist ein Rätsel ...: Der philosophische Blick auf die Frau*. Frankfurt/M.: Eichborn.
Mossop, D. J., G. E. Rodmell, and D. B. Wilson, eds. 1978. *Studies in the French eighteenth century presented to John Lough by colleagues and pupils*. Durham: University of Durham.
Moulin, H. 1885. *Voltaire et le premier président Fyot de la Marche; la marquised du Châtelet; le président de Brosses; les Calas; Marie Corneille; les Fyot de La Marche père et fils: (15 lettres inédites)*. Caen: Le Blanc-Hardel.
Moureau, F., ed. 1993. *De bonne main. La communication manuscrite au XVIIe siècle*. Oxford/ Paris: Voltaire Foundation.
Mozans, H. J. 1913. *Women in science: with an introductory chapter on on woman's long struggle for things of the mind by H.J. Mozans; with a preface by Cynthia Russett and an introduction by Thomas P. Gariepy*. New York/ London: D. Appleton.
Munslow, A., and R. A. Rosenstone, eds. 2004. *Experiments in rethinking history*. London/New York: Routledge.
Murdoch, R. T. 1958. Voltaire, James Thomson and a poem for the marquise du Châtelet. *Studies on Voltaire and the eighteenth century* 6:147–53.
Muzerelle, D. 2006. Académies et femmes savantes. In *Madame du Châtelet. La femme des Lumières*, ed. Elisabeth Badinter and Danielle Muzerelle, 112. Paris: Bibliothèque Nationale de France.
Muzerelle, D. 2006. La bibliothèque de Mme du Châtelet. In *Madame du Châtelet. La femme des Lumières*, ed. Elisabeth Badinter and Danielle Muzerelle, 111. Paris: Bibliothèque Nationale de France.
Muzerelle, D. 2006. Le cabinet de physique de Cirey. In *Madame du Châtelet. La femme des Lumières*, ed. Elisabeth Badinter and Danielle Muzerelle, 103. Paris: Bibliothèque Nationale de France.
Muzerelle, D. 2006. Regards sur Mme du Châtelet. In *Madame du Châtelet. La femme des Lumières*, ed. Elisabeth Badinter and Danielle Muzerelle, 114. Paris: Bibliothèque Nationale de France.
Nedeljkovič, D. 1966. La continuité chez Leibniz, Madame du Châtelet and R. J. Boscovich. *Dijalektika* 3:65–68.
Neumeister, S., and C. Wiedemann, eds. 1987. *Res publica litteraria. Die Institutionen der Gelehrsamkeit in der frühen Neuzeit*. Wiesbaden: Harrassowitz.
Newton, I., ed. 1687. *Philosophiae naturalis principia mathematica*. London.
Newton, I., ed. 1759. *Principes mathématiques de la philosophie naturelle de Newton: traduit du latin par Mme. du Châtelet, préface de Costes, et Éloge historique de Voltaire*. 2 vols. Paris: Desaint et Saillant.
Newton, I., ed. 1972. *Isaac Newton's Philosophiae naturalis principia mathematica*. 2 vols. Cambridge: Cambridge University Press.
Newton, I., ed. 1988. *Mathematische Grundlagen der Naturphilosophie*. Hamburg: Meiner.
Nielsen, N. 1935. *Géomètres français du XVIIIe siècle*. Copenhague: Levin & Munksgaard.
Noordraven, A. 2001. Leibniz' Onto-Logik und die transzendentale Logik Kants. In *Kant und die Berliner Aufklärung: Akten des IX. Internationalen Kant-Kongresses*, ed. Volker Gerhardt, 55–64. Berlin: de Gruyter.
Noury, J. 1894. Voltaire inédit; billets à Cideville; une contrefaçon de ses œuvres à Rouen, la correspondance de Madame du Châtelet avec Cideville et de Cidevile avec Voltaire. *Bulletin historique et philologique du Comité de travaux historiques et scientifiques*:352–56.

Ogilvie, M. B. 1986. *Women in Science: Antiquity through the Nineteenth Century.* Cambridge: MIT Press.
Ogilvie, M. B., and K. L. Meek. 1996. *Women and Science.* New York: Garland.
Orieux, J. 1965. Les enchantements de l'exil: Voltaire à Cirey. *Revue de Paris* 72:68–79.
Orieux, J. 1966. *Voltaire ou la royauté de l'esprit.* Paris: Flammarion.
Osen, L. M. 1974, ²1977. *Women in Mathematics.* Cambridge/MA: MIT Press.
Osler, M. J., and P. L. Farber, eds. 1985. *Religion, science and worldview.* Cambridge: Cambridge University Press.
Oulmont, C. 1934. Au Château de Cirey avec Voltaire. *Le Temps*:4.
Oulmont, C. 1936. *Voltaire en robe de chambre.* Paris: Calmann-Lévy.
Paillet de Warcy, L. 1824. *Histoire de la vie et des ouvrages de Voltaire, suivie des jugements qu'ont portés de cet homme célèbre divers auteurs estimés.* 2 vols. Paris: Mme Dufriche, Ponthieu, Delaunay.
Papineau, D. 1977. The vis viva controversy: do meanings matter? *Studies in history and philosophy of science* 8, no. 1:111–42
Paraf, P. 1928. Un soir avec Voltaire. *La Nouvelle revue* 98:260–70.
Parnes, R. de. 1882. *Anecdotes secrètes du règne de Louis XV: Portefeuille d'un petit-maître publié par Roger de Parnes avec préface par Georges d'Heylli.* Paris: Ed. Rouveyre et G. Blond.
Pasini, M. 1997. Jean-Jacques Dortous de Mairan e Madame du Châtelet: frammenti di una 'querelle'. *Studi settecenteschi* 17:63–82.
Passeron, I. 2001. Muse ou élève? sur les lettres de Clairaut à Mme du Châtelet. In *Cirey dans la vie intellectuelle: la réception de Newton en France*, ed. François de Gandt, 187–97. Oxford: Voltaire Foundation.
Perl, T. 1978. *Math equals. Biographies of Women Mathematicians and Related Activities.* Menlo Park/ California: Addison-Wesley Publishing Company.
Petrovich, V. C. 1999. Women and the Paris Academy of Sciences. *Eighteenth-Century Studies* 32, no. 3:383–90.
Peyrefitte, R. 1992. *Voltaire et Frédéric II.* 2 vols. Paris: Albin Michel.
Pfeiffer, J. 1991. L'engouement des femmes pour les sciences. In *Femmes et pouvoirs sous l'Ancien Régime*, ed. D. Haase-Dubosc and E. Viennot. Paris: Rivages.
Philips, E. 1942. Madame du Châtelet, Voltaire and Plato. *Romanic Review* 33:250–63.
Pieretti, M.-P. 2002. Women Writers and Translation in Eighteenth-Century France. *The French Review* 75, no. 3:474–88.
Pignet, G. 1938. *La vérité sur la vie amoureuse de M. de Voltaire.* Paris: Fasquelle.
Piot, l. 1894. *Cirey-Le-Château: suivi de La marquise du Châtelet (sa liaison avac Voltaire): par M. l'abbé Piot, curé de Chancenay, membre titulaire de la Société des lettres de Saint-Dizier.* Saint-Diziers: O. Godard.
Plavinskaia, N. 2005. Trois lettres d'Émilie du Châtelet retrouvées dans les archives moscovites. *Cahiers Voltaire* 4:37–82.
Poggendorff, J. C. 1863. *Biographisch-literarisches Handwörterbuch zur Geschichte der exacten Wissenschaften enthaltend Nachweisungen über Lebensverhältnisse und Leistungen von Mathematikern, Astronomen, Phyysikern, Chemikern, Mineralogen, Geologen, Geographen usw. aller Völker und Zeiten.* 2 vols. Leipzig: Johann Ambrosius Barth.
Poirier, R. 1991. Une Lettre inédite de Saint-Lambert à Madame du Châtelet. *Revue d'histoire littéraire de la France* 91, no. 4–5:747–55.
Poirier, J.-P. 2002. *Histoire des femmes de science en France: du Moyen Âge à la Révolution.* Paris: Pygmalion.
Pomeau, R. 1956, ²1969, ³1994. *La religion de Voltaire.* Paris: Nizet.
Pomeau, R. 1985. *D'Arouet à Voltaire 1694–1734.* Oxford: Voltaire Foundation.
Pomeau, R. 1995. *Voltaire et son temps. Voltaire et la Marquise du Châtelet.* Paris: Jean Touzot.
Pomeau, R. 2001. Voltaire en passant par la Lorraine… *Revue Voltaire*, no. 1:113–18.
Pomeau, R. 2001. Voltaire et Mme du Châtelet à Cirey: amour et travail. In *Cirey dans la vie intellectuelle: la réception de Newton en France*, ed. François de Gandt, 7–15. Oxford: Voltaire Foundation.

Porset, C. 1997. Position de la philosophie de Voltaire. In *Voltaire et ses combats*. Actes du congrès international Oxford-Paris 1994, ed. Ulla Kölving and Christiane Mervaud, 727–38. 2 vols. Oxford: Voltaire Foundation.
Pound, E., R. Wilbur, and Voltaire. 1992. *A madame du Châtelet*. Amherst/Mass. Amherst College.
Powell, J., ed. 2006. *Great Lives from History. The 18th Century 1701–1800*. Pasadena/CA: Salem Press.
Pratt, T., and D. McCallam, eds. 2004. *The Enterprise of Enlightenment. A Tribute to David Williams from his friends*. Oxford: Peter Lang.
Prevost, M., ed. 1966. *Dictionnaire de biographie française*. Paris: Letouzey et Ané.
Pulte, H. 1993. Die Newton-Rezeption in der rationalen Mechanik des 18. Jahrhunderts. *Beiträge zur Geschichte von Technik und technischer Bildung* 7:33–59.
Summer, M., ed. 1900. *Quelques salons de Paris au XVIIIe siècle*. Paris: L.-Henry May.
Quentin, H. 1906. Les surprises d'une perquisition: lettres inédites de Voltaire. *Revue d'histoire littéraire de la France* 103:332–36.
Quinn, T. 2007. Emilie du Châtelet, John Freind, Robert Hooke, Charles Darwin and John Stanley Gardner. *Notes and Records of the Royal Society of London* 61, no. 2:85.
Ravaisson-Mollien, L. J. F., ed. 1866. *Archives de la Bastille*. Paris: A. Durant et Pedone-Lauriel.
Rebière, A. 1897. *Les Femmes dans la science, notes recueillies par A. Rebière*. 2e édition, très augmentée et ornée de portraits et d'autographes. Paris: Nony & Cie.
Redien-Collot, R. 2006. Émilie Du Châtelet et les femmes: entre l'attitude prométhéenne et la pleine assomption du statut de minoritaire. In *Emilie Du Châtelet: rewriting Enlightenment philosophy and science*, ed. Judith P. Zinsser and Julie C. Hayes, 277–96. Oxford: Voltaire Foundation.
Reiffenberg, F.-A.-F.-T. d. 1845. Encore un manuscrit de la Bibliothèque royale - Séjour de madame du Chastellet et de Voltaire à Bruxelles - Les Esclaves de Marie. *Le Bibliophile belge*, no. 1:14–20.
Rema, E. 1913,²1920. *Voltaires Geliebte*. Dresden/Leipzig: Reißner.
Rey, A.-L. 2008. La figure du leibnizianisme dans les *Institutions de physique*. In *Émilie Du Châtelet: Éclairages & documents nouveaux*, ed. Ulla Kölving and Olivier Courcelle, 231–42. Ferney-Voltaire: Centre International d'Étude du XVIIIe Siècle.
Rodrigues, A. 2008. Emilie du Châtelet - Glück zwischen Leidenschaft und Vernunft. *Spirale der Zeit* 4:36–39.
Rodrigues, A. 2010. Emilie du Châtelet, Julien Offray de La Mettrie und Pierre Louis Moreau de Maupertuis im Zwiegespräch über das Glück. In *Von Diana zu Minerva. Philosophierende Aristokratinnen des 17. und 18. Jahrhunderts.*, ed. Ruth Hagengruber and Ana Rodrigues, 151–60. Berlin: Akademie Verlag.
Rodrigues, A. 2010. Emilie du Châtelet - Vom glücklichen Leben zur Freiheit des Denkens. In *Von Diana zu Minerva. Philosophierende Aristokratinnen des 17. und 18. Jahrhunderts.*, ed. Ruth Hagengruber and Ana Rodrigues, 95–108. Berlin: Akademie Verlag.
Roger, J. 1963. *Les sciences de la vie dans la pensée française du XVIIIe siècle: La génération des animaux de Descartes à l'Encyclopédie*. Paris: Armand Colin.
Rolka, G. M. 1994. *100 women who shaped world history*. San Francisco/CA: Blue wood books.
Rondot, B. 2006. Le goût d'une femme de son temps. In *Madame du Châtelet. La femme des Lumières*, ed. Elisabeth Badinter and Danielle Muzerelle, 55–67. Paris: Bibliothèque Nationale de France.
Rothman, P. 1996. *Women in the history of mathematics: from antiquity to the nineteenth century*. London: Department of Mathematics, University College London.
Rullmann, M. 1993. *Philosophinnen: von der Antike bis zur Aufklärung*. Zürich/Dortmund: Ebersbach im eFeF-Verlag.
Sade, D.-A.-F. de. 1994. Correspondance de madame Du Châtelet et de l'abbé de Sade. In *Correspondance du marquis de Sade et de ses proches enrichies de documents, notes et commentaires*, ed. Alice M. Laborde, 39–64. Genève: Slatkine.
Sade, D.-A.-F. de. 1994. *Correspondance du marquis de Sade et de ses proches enrichies de documents, notes et commentaires*. Genève: Slatkine.

Saget, H. 1993, ²2005. *Voltaire à Cirey.* Chaumont: Le Pythagore.
Saget, H. 1994. Cirey-sur-Blaise, le phlogistique et les idées sur le feu. In *"De la nature et de la propagation du feu" : cinq mémoires couronnés par l'académie royale des sciences, Paris 1738*, ed. ASPM Association pou la sauvegarde et la promotion du patrimoine métallurgique haut-marnais, 7–18. Wassy.
Sainte-Beuve, C.-A. 1850. Lettres de Mme de Graffigny, ou Voltaire à Cirey. *Causeries du Lundi*, no. 2:208–25.
Sainte-Beuve, C.-A. 1850. Madame du Châtelet, suite de Voltaire à Cirey. *Causeries du Lundi*, no. 2:266–85.
Sainte-Beuve, C.-A. 1932. *Les grands écrivains français par Sainte-Beuve. XVIIIe siècle. Voltaire, sa vie et sa correspondance. Etudes des Lundis et des Portraits classées selon un ordre nouveau et annotées par Maurice Allem.* Paris: Garnier frères.
Saint-Marc-Girardin. 1869. Mme du Châtelet. *Revue des cours littéraires de la France et de l'etranger*, no. 6:154–60.
Saint-Marc-Girardin. 1869. Voltaire à Cirey. *Revue des cours littéraires de la France et de l'étranger*, no. 6:106–12.
Saint-Victor. 1810. Lettres de madame la marquise Du Châtelet à monsieur le comte d'Argental. *Le Spectateur français au XIXe siècle*, no. 7:29–34.
Saisselin, R. G. 2006. Portraiture and the ambiguity of being. In *Emilie Du Châtelet: rewriting Enlightenment philosophy and science*, ed. Judith P. Zinsser and Julie C. Hayes, 91–102. Oxford: Voltaire Foundation.
Salama-Carr, M. 1995. Les Traducteurs, diffuseurs des connaissances. In *Les traducteurs dans l'histoire*, ed. Jean Deslisle and Judith Woodsworth. Ottawa: PU Ottawa.
Salvador, A., and M. B. Molero Aparicio. 2003. *Gabrielle Émilie de Breteuil, Marquesa du Châtelet (1706–1749).* Madrid: Ediciones del Orto.
Salzman, P., ed. 2000. *Early Modern Women's Writing: An Anthology 1560–1700.* Oxford: Oxford World's Classics.
Samuel, P., ed. 1974. *Mathématiques, Mathématiciens et Société.* Orsay: Université de Paris-Sud.
Sankey, M., and J. Fornasiero. 2003. Women Philosophers and the History of Philosophy. *Australian Journal of French Studies XL* 3:257–74.
Santana, A. 1993. *Women mathematicians.*
Sareil, J. 1970. Quelques lettres de Voltaire et de ses amis. *Revue d'histoire littéraire de la France* 70:653–58.
Sartori, E. 2006. *Histoire des femmes scientifiques de l'Antiquité au XIXe siècle, les filles d'Hypatie.* Paris: Plon.
Schad, M. 1997. *Frauen, die die Welt bewegten: geniale Frauen, der Vergangenheit entrissen.* Augsburg: Pattloch.
Schiebinger, L. 1988. Feminine Icons: The Face of Early Modern Science. *Critical Inquiry* 14, no. 4:661–91.
Schiebinger, L. 1989. *The Mind has no Sex? Women in the Origins of Modern Science.* Cambridge/MA: Harvard Univ. Press.
Schiebinger, L. 1995. *Nature's Body: Gender in the Making of Modern Science.* Boston: Beacon Press.
Schirmacher, K., ed. 1897. *Aus aller Herren Länder. Gesammelte Studien und Aufsätze.* Leipzig: H. Welter.
Schirmacher, K., ed. 1897. Die Marquise du Châtelet. In *Aus aller Herren Länder. Gesammelte Studien und Aufsätze*, ed. Käthe Schirmacher, 152–79. Leipzig: H. Welter.
Schlaffer, H. 2007. Die Minerva Frankreichs. *FAZ*, 2007.
Schmidt, A.-H. 1999. *Adieu, Madame Newton-Pompon: vor 250 Jahren starb Emilie du Châtelet.* Aigen-Voglhub: A.-H. Schmidt.
Schmidt, A.-H. 1999. *Die Rolle der Frau in der Französischen Revolution und die Ursprünge der Frauenbewegung samt Kurzporträt der Emilie du Châtelet: der Versuch einer analytischen Darstellung samt einer umfangreichen Bibliographie zu diesen Themen.* Aigen-Voglhub: A.-H. Schmidt.

Schneiders, W., ed. 1993. *Aufklärung als Mission: Akzeptanzprobleme und Kommunikationsdefizite = La mission des Lumières: accueil réciproque et difficultés de communication.* Marburg: Hitzeroth.

Schwarzbach, B. E. 1993. Une légende en quête d'un manuscrit : Le *Commentaire sur la Bible* de Madame du Châtelet. In *De bonne main. La communication manuscrite au XVIIe siècle*, ed. F. Moureau, 97–116. Oxford: Voltaire Foundation.

Schwarzbach, B. E. 1995. La critique biblique dans les *Examens de la Bible* et dans certains autres traités clandestins. *La Lettre Clandestine* 4:577–612.

Schwarzbach, B. E. 1997. Le profil littéraire de l'auteur des *Examens de la Bible*. In *La philosophie clandestine à l'âge classique: actes du colloque de l'Université Jean Monnet Saint-Etienne du 29 septembre au 2 octobre 1993*, ed. Anthony McKenna and Alain Mothu, 223–32. Oxford: Voltaire Foundation.

Schwarzbach, B. E. 2001. Les études bibliques à Cirey. In *Cirey dans la vie intellectuelle: la réception de Newton en France*, ed. François de Gandt, 26–54. Oxford: Voltaire Foundation.

Schwarzbach, B. E. 2005. Aux éditeurs de la revue voltaire. *Revue Voltaire* 5:395.

Schwarzbach, B. E. 2006. Mme du Châtelet et la Bible. In *Madame du Châtelet. La femme des Lumières*, ed. Elisabeth Badinter and Danielle Muzerelle, 105–09. Paris: Bibliothèque Nationale de France.

Schwarzbach, B. E. 2006. Mme du Châtelet's *Examens de la Bible* and Voltaire's *La Bible enfin expliquée*. In *Emilie Du Châtelet: rewriting Enlightenment philosophy and science*, ed. Judith P. Zinsser and Julie C. Hayes, 142–64. Oxford: Voltaire Foundation.

Schwarzbach, B. E. 2008. Mme Du Châtelet et la Bible. In *Émilie Du Châtelet: Éclairages & documents nouveaux,* ed. Ulla Kölving and Olivier Courcelle, 197–211. Ferney-Voltaire: Centre International d'Étude du XVIIIe Siècle.

Scott, W. L. 1959. The Significance of "Hard Bodies" in the History of Scientific Thought. *Isis* 50, no. 3:199–210.

Scriba, C. J. 1971. The French edition of Newton's *Principia* (translation of the Marquise du Châtelet): 1759 or 1756? In *Science et Philosophie: XVIIe et XVIIe siècles*, ed. International Congress on the History of Science, 117–19. Actes XIIe Congrès Internat. d'Histoire des Sciences, Paris 1968 III B. Paris: A. Blanchard.

Seth, C. 2003. Deux lettres inédites de Mme du Châtelet. *Revue Voltaire*, no. 3:355–66.

Seybert, G., ed. 2003. *Das literarische Paar – Le couple litteraire. Intertextualität der Geschlechterdiskurse – Intertextualité et discourses des sexes.* Bielefeld: Aisthesis Verlag.

Showalter, E. 1975. Sensibility at Cirey: Mme du Châtelet, Mme du Graffigny, and the Voltaireomanie. *SVEC* 135:181–92.

Showalter, E. 2004. *Françoise de Graffigny. Her life and works.* Oxford: Voltaire Foundation: Voltaire Foundation.

Siess, J. 2008. Image de la philosophe et égalité des sexes dans la correspondance de Mme Du Châtelet. In *Émilie Du Châtelet: Éclairages & documents nouveaux,* ed. Ulla Kölving and Olivier Courcelle, 37–52. Ferney-Voltaire: Centre International d'Étude du XVIIIe Siècle.

Silver, M.-F., and M.-L. Giroud Swiderski, eds. 2000. *Femmes en toutes lettres: les épistolaires du XVIIIe siècle.* Oxford: Voltaire Foundation.

Simond, A.-M. 2000. Madame du Châtelet et la quête du bonheur. *La graphologie* 240, no. 4:43–53.

Simonin, C. 2008. Pompon Newton versus Marie Chiffon? Émilie Du Châtelet et Françoise de Graffigny en miroir, et au miroir de l'histoire littéraire. In *Émilie Du Châtelet: Éclairages & documents nouveaux,* ed. Ulla Kölving and Olivier Courcelle, 61–83. Ferney-Voltaire: Centre International d'Étude du XVIIIe Siècle.

Simonin, C., and W. D. Smith. 2005. Du nouveau sur Mme Denis. Les apports de la correspondance de Mme de Graffigny. *Cahiers Voltaire* 4:25–56.

Singal, A. R. 1986. Women mathematicians of the past: some observations. *Math. Ed.* 3, no. 1:9–18.

Sirois, A. 1997. *Les femmes dans l'histoire de la traduction: de la renaissance au XIXe siècle français: domaine français.* Ottawa: UMI Dissertation services.

Slachman, V. 1990. *Women mathematicians.* Palo Alto/Calif. Dale Seymour Pub.

Smelding, A. von. 1932. *Die göttliche Emilie. Voltaire, ihr Freund – Fridericus, ihr Feind. Dichtung und Wahrheit aus dem Leben der drei Feuergeister.* Berlin: Schlieffen-Verlag.
Smith, H., ed. 1998. *Women writers and the early modern British political tradition.* Cambridge/ New York: Cambridge University Press.
Smith, D. E. 1921. Among my autographs. 12. The marquise du Châtelet. *American mathematical monthly* 28, no. 28:368–69.
Smith, D. E. 1923. Historical-mathematical Paris. Ile de la Cité and the Voltaire-Châtelet Paris. *American mathematical monthly* 30:107–13.
Smith, D. E. 1923–1925. *History of mathematics.* Boston/New York: Ginn and Company.
Smith, D. W. 2004. Nouveaux regards sur la brève rencontre entre Mme du Châtelet et Saint-Lambert. In *The Enterprise of Enlightenment. A Tribute to David Williams from his friends,* ed. T. Pratt and D. McCallam. Oxford: Peter Lang.
Société typographique de Bouillon, ed. 1779. *Huitième recueil philosophique et littéraire.* Bouillon: Société typographique de Bouillon.
Socolow, E. 1988. *Laughing at Gravity Conversations with Isaac Newton.* Boston: Beacon.
Sotelo, E. d., ed. 1998. *Feministische Hochschuldidaktik. Materialien zu den Koblenzer Frauenstudien.* Münster: Lit-Verlag.
Spencer, S. I., ed. 1984. *French Women an the Age of Enligthenment.* Bloomington/ IN: Indiana University Press.
Spencer, S. I., ed. 2005. *Writers of the French Enlightenment.* Farmington Hills/ Mich.: Thomson Gale.
Spiess, O. 1948. Voltaire und Basel. *Basler Zeitschrift für Geschichte und Altertumskunde* 47: 105–35.
Stedman, G., and M. Zimmermann, eds. 2006. *Höfe - Salons - Akademien. Kulturtransfer und Gender im Europa der Frühen Neuzeit.* Hildesheim: Olms.
Suard, J.-B.-A. 1858. *Mémoires et correspondances historiques et littéraires inédits 1726 à 1816: Publiés par Charles Nisard.* Paris: Michel Lévy frères.
Suard et Bourlet de Vauxcelles, ed. 1796. *Opuscules philosophiques et littéraires, la plupart posthumes ou inédits.* Paris: De l'Imprimerie de Chenet.
Summer, M. 1900. Le salon de Mme du Châtelet à Paris et à Cirey. In *Quelques salons de Paris au XVIIIe siècle,* ed. Mary Summer, 29–59. Paris: L.-Henry May.
Sutton, G. V. 1995. *Science for a Polite Society: Gender, Culture and the Demonstration of Enlightenment.* Boulder/Colorado: Westview Press.
Szabo Compare, I., ed. 1997. *Geschichte der mechanischen Prinzipien und ihrer wichtigsten Anwendungen.* Basel: Birkhäuser.
Szabo Compare, I. 1997. Der philosophische Streit um das wahre Kraftmaß im 17. und 18. Jahrhundert. In *Geschichte der mechanischen Prinzipien und ihrer wichtigsten Anwendungen,* ed. István Szabo Compare. Basel: Birkhäuser.
Tarte, M. 1955. Mme du Châtelet à la cour du roi Stanislas. *Revue médicale de Nancy* 76:658–75.
Taton, R. 1969. Madame du Châtelet, traductrice de Newton. *Archives internationales d'histoire des sciences* 22:185–210.
Taton, R. 1970. Isaac Newton. Principes mathématiques de la philosophie naturelle, trad. de la marquise du Chastellet, augmentée des commentaires de Clairaut. *Revue d'histoire des sciences et de leurs applications* 23:175–80.
Taton, R. 1971. Châtelet, Gabrielle-Émilie le Tonnier de Breteuil, Marquise du. In *Dictionary of scientific biography,* ed. Charles Coulston Gillispie, 215–17. New York: Charles Scribner's Sons.
Tee, G. J. 1983. The pioneering women mathematicians. *The Mathematical Intelligencer* 5:27–36.
Tee, G. J. 1987. Gabrielle-Emilie Le Tonnelier De Breteuil, Marquise du Châtelet (1706–1749). In *Women of Mathematics. A Bibliographic Sourcebook,* ed. Louise S. Grinstein and Paul J. Campbell, 21–25. New York: Greenwood Press.
Ternisien d'Haudricourt, F. 1788. *Femmes célèbres de toutes les nations avec leurs portraits...II, Madame la Marquise du Châtelet.* Paris: Gattey.

Terrall, M. 1990. The culture of science in Frederick the Great's Berlin. *History of Science* 28:333–64.
Terrall, M. 1992. Representing the earth's shape: The polemics surrounding Maupertuis's expedition to Lapland. *Isis* 83:218–37.
Terrall, M. 1994. Gendered Spaces, Gendered Audiences: Inside and Outside the Paris Academy of Sciences. *Configuration* 2:207–32.
Terrall, M. 1995. Emilie du Châtelet and the gendering of science. *History of Science* 33, no. 3:283–310.
Terrall, M. 1999. Metaphysics, mathematics, & the gendering of science in eighteenth-century France. In *The Sciences in Enlightened Europe*, ed. William Clark, Jan Gollinski, and Simon Schaffer, 246–71. Chicago: University of Chicago Press.
Terrall, M. 2002. *The Man who flattened the earth: Maupertuis and the sciences in the Enlightenment.* Chicago: University of Chicago Press.
Terrall, M. 2003. The Uses of Anonymity in the Age of Reason. In *Scientific Authorship: Credit and Intellectual Property in Science*, ed. Mario &. Galison, Peter Biagioli, 91–112. New York: Routledge.
Terrall, M. 2004. Vis viva revisited. *History of Science* 42, no. 2:189–209.
Thomann, M. 1979. Voltaire und Christian Wolff. In *Voltaire und Deutschland. Quellen und Untersuchungen zur Rezeption der französischen Aufklärung: Internat. Kolloquium der Universität Mannheim zum 200. Todestag Voltaires*, ed. P. Brockmeier, R. Desné, and E. Voss, 124–36. Stuttgart: Metzler.
Thor, W. 1914. Die Marquise du Châtelet und Voltaire. *Velhagen und Klasing's Monatshefte*:572–76.
Todhunter, I. 1962. *A History of the Mathematical Theories of Attraction and The Figure of the Earth*, Vol. I. New York: Dover Publications.
Tonelli, G. 1987. *La pensée philosophique de Maupertuis*. Hildesheim/New York: Olms.
Toulmonde, M. 2008. Le Commentaire des Principes de la philosophie naturelle. In *Émilie Du Châtelet: Éclairages & documents nouveaux.* ed. Ulla Kölving and Olivier Courcelle, 309–15. Ferney-Voltaire: Centre International d'Étude du XVIIIe Siècle.
Trapnell, W. 1997. Le manuscrit 'Voltaire 80 221' de Saint-Pétersbourg. In *La philosophie clandestine à l'âge classique: actes du colloque de l'Université Jean Monnet Saint-Etienne du 29 septembre au 2 octobre 1993*, ed. Anthony McKenna and Alain Mothu, 233–44. Oxford: Voltaire Foundation.
Trousson, R., and J. Vercruysse, eds. 2003. *Dictionnaire général de Voltaire*. Paris: H. Champion.
Tunstall, K. E. 2007. Emilie Du Châtelet: rewriting Enlightenment philosophy and science. *Modern Language Review* 102, no. 1:235–36.
Turgot, A.-R.-J., and J.-A.-N. C. de Condorcet. 1883. *Correspondance inédite de Condorcet et de Turgot, 1770–1779, publiée avec des notes et une introductiond'après les autographes de la collection Minoret et les manuscrits de l'Institut, par M. Charles Henry*. Paris: Charavay frères.
Utermöhlen, G. 1987. Die gelehrte Frau im Spiegel der Leibniz-Korrespondenz. In *Res publica litteraria. Die Institutionen der Gelehrsamkeit in der frühen Neuzeit*, ed. Sebastian Neumeister and Conrad Wiedemann, 603–18. Wiesbaden: Harrassowitz.
Vaillot, R. 1978. *Madame du Châtelet: Préface de René Pomeau*. Paris: Albin Michel.
Vaillot, R. 1988. *Avec Madame Du Châtelet, 1734–1749*. Oxford: Voltaire Foundation.
Vaillot, R. 1988. *Voltaire en son temps 2: Avec Madame du Châtelet (1734–1749)*. Oxford: Voltaire Foundation.
Valentin, M. 1998. *Maupertuis: un savant oublié*. Rennes: La Découvrance.
Vallée, G. J. 1747. *Lettre sur la nature de la matière et du mouvement à l'auteur des Institutions de physique*. Paris: Thiboust.
Valogne, C. 1954. Soixante-treize lettres de la marquise du Châtelet éclairent la maturité de Voltaire. *Les lettres françaises*:1–5.
Vamboulis, E. 2001. La discussion de l'attraction chez Voltaire. In *Cirey dans la vie intellectuelle: la réception de Newton en France*, ed. François de Gandt, 159–70. Oxford: Voltaire Foundation.

Van Crugten-André, V. 2003. Du Châtelet, Gabrielle Émilie Le Tonnelier de Breteuil, marquise (1706–1749). In *Dictionnaire général de Voltaire*, ed. R. Trousson and Jeroom Vercruysse, 373–84. Paris: H. Champion.
Van den Heuvel, J. 1967. *Voltaire dans ses contes. De Micromégas à L'Ingénu.* Paris: Armand Colin.
Van den Heuvel, J. 1980. In der Schule Émilies – Geometrie und Vergnügen. Platons Traum, die erste philosophische Erzählung. In *Voltaire*, ed. Horst Baader, 321–47. Darmstadt: Wissenschaftliche Buchgesellschaft.
Van Gansen, J. 1871. Coup d'oeil historique sur Beeringhen, à propos d'une lettre de Voltaire. *Société chorale et littéraire des Mélophiles de Hasselt, Bulletin de la section littéraire*, no. 7:73–91.
Veluz, L. 1969. *Maupertuis.* Paris: Hachette.
Vercruysse, J. 1961. La marquise Du Châtelet, prévôte d'une confrérie bruxelloise. *SVEC* 18:169–71.
Vercruysse, J. 1966. *Voltaire et la Hollande.* Genève, Suisse: Institut et Musée Voltaire.
Vercruysse, J. 1978. Cinq actes notariés bruxellois relatifs à Voltaire, à Helvétius et à la marquise du Châtelet (1741–1742). *LIAS. Sources and documents relating to the early history of ideas* 5:167–76.
Vercruysse, J. 1979. Quinze lettres inédites, oubliées ou rectifiées de la correspondance de Voltaire. *SVEC* 182:203–18.
Voisenon, C.-H. F. de. 1781. Anecdotes littéraires. In *Œuvres complètes de M. l'abbé de Voisenon.* Paris: Moutard.
Voisenon, C.-H. F. de. 1781. *Œuvres complètes de M. l'abbé de Voisenon.* Paris: Moutard.
Völger, G., ed. 1998. *Sie und Er. Frauenmacht und Männerherrschaft im Kulturvergleich.* Köln: Rautenstrauch-Joest-Museum.
Voltaire. 1736. Épître à madame la marquise Du Châtelet. In *Alzire ou les Américains. Tragédie de M. de Voltaire. Représentée pour la première foi le 27 janvier 1736*, 5–12. Paris: Jean-Baptiste-Claude Bauche.
Voltaire. 1736. Épître (à Madame la marquise Du Châtelet) sur la calomnie. In *La Mort de César, tragédie de M. de Voltaire: Seconde édition revue, corrigée et augmentée par l'auteur*, 3–70. Amsterdam: Etienne Ledet & Compagnie (ou) Jacques Desbordes.
Voltaire. 1737. Le Philosophe, à madame la marquise du Ch… In *Reflexions sur les ouvrages de literature*, ed. François Granet, 74–78. Paris: Briasson.
Voltaire. 1738. À Madame la Marquise du Ch**. In *Éléments de la philosophie de Newton*, 9–13. Amsterdam: Etienne Ledet & Compagnie (ou) Jacques Desbordes.
Voltaire. 1738. *Éléments de la philosophie de Newton.* Amsterdam: Etienne Ledet & Compagnie (ou) Jacques Desbordes.
Voltaire. 1739. Extrait de la dissertation de Mad. L.M.D.C. sur la nature du feu. *Mercure de France*, no. 2:1320–28.
Voltaire. 1741. Exposition du Livre des Institutions de physique dans laquelle on examine les idées de Leibnits. *Mercure de France*, no. 2:1274–310.
Voltaire. 1741. À madame la Marquise Du Chastelet. In *Éléments de la philosophie de Newton, contenant la métaphysique de Newton, la théorie de la lumière, et celle du monde.* Paris: Prault.
Voltaire. 1742. Doutes sur la nature des forces motrices, et sur leur nature, présentés à l'Académie des sciences de Paris, avec le jugement que l'académie en a porté. In *Nouvelle bibliothèque ou histoire littéraire des principaux écrits qui se publient*, 219–34. Vol. IX. La Haye: Pierre Poupie.
Voltaire. 1742. Mémoire sur un ouvrage de physique de Madame la marquise du Châtelet, lequel a concouru pour les prix de l'Académie des sciences en 1738. In *Nouvelle bibliothèque ou histoire littéraire des principaux écrits qui se publient*, 414–22. Vol. IX. La Haye: Pierre Poupie.
Voltaire. 1748. Épître dédicatoire à Madame la Marquise Du Chastellet. In *Œuvre de M. de Voltaire*, 15–16. Dresden: George Conrad Walther.

Voltaire. 1752. Eloge historique de madame Du Châtelet, pour mettre à la tête de la traduction de Newton, par M. de Voltaire. In *Oeuvres*, 136–46. Leiden: Bibliothèque Impartiale.

Voltaire. 1752. Essai sur la nature du feu et sur sa propagation. In *Recueil des pièces qui ont remporté les prix de l'Académie royale des Sciences, depuis leur fondation jusqu'à présent, avec les pièces qui y ont concouru: Tome quatrième, contenant les pièces depuis 1738 jusqu'en 1740*, ed. Académie royale des Sciences, 171–219. Paris: Martin, Coignard, Guérin, Jombert.

Voltaire. 1754. Éloge historique de Madame du Châtelet pour mettre à la tête de sa traduction de Newton. *Mercure de France*, no. 1:6–18.

Voltaire. 1756. Préface historique. In *Principes mathématiques de la philosophie naturelle: par feue Madame la marquise Du Chastellet*. 2 vols. Paris: Desaint & Saillant.

Voltaire. 1764, ²1765. *Lettres secrèttes de Mr. De Voltaire: Publiées par M.L.B.* Genève.

Voltaire. 1765. *Geheime Briefe des Herrn von Voltaire: Herausgegeben von dem Herrn L*** B***, aus dem Franz. übersetzt von Johann Friedrich Seyfart*. Frankfurt, Leipzig.

Voltaire. 1766. *Monsieur de Voltaire peint par lui-même, ou Lettres de cet écrivain, dans lesquelles on verra l'histoire de sa vie, de ses ouvrages, de ses querelles, de ses correspondances, & les principaux traits de son caractère: avec un grand nombre d'anecdotes, de remarques & de jugements littéraires*. Lausanne: Compagnie des libraires.

Voltaire. 1775. *Des Herrn von Voltaire Geheime freundschaftliche Briefe*. Hamburg: Buchenröder und Ritter.

Voltaire. 1781. *Lettres de M. de Voltaire à M. l'abbé Moussinot son trésorier, écrites depuis 1736 jusqu'en 1742, pendant sa retraite à Cirey, chez madame la marquise Du Châtelet, et dans lesquelles on voit quelques détails de sa fortune, ses bienfaits; quelles furent alors ses études, ses querelles avec Desfontaines etc. Publiées par l'abbé D*** (Duvernet)*. Paris: Moutard.

Voltaire. 1782. *Lettres de M. de Voltaire et de sa célèbre amie. Suivie d'un petit poëme, d'une lettre de J.J. Rousseau & d'un paralèlle entre Voltaire et J.J. Rousseau*. Genève, Paris: Cailleau.

Voltaire. 1784. *Mémoires de M. de Voltaire, écrits par lui-même*. Genève.

Voltaire. 1959. *A Madame du Chatelet: three english versions*. Oxford: Tudor Rose Press.

Voltaire. 1968–1977. *Complete works of Voltaire. Les oeuvres complètes de Voltaire*, ed. Theodore Besterman. Vols. 135. Genève: Institut et Musée Voltaire.

Voltaire. 1997. *Elemente der Philosophie Newtons: Übers. d. Texte "Elemente der Philosophie Newtons" u. "Verteidigung des Newtonianismus" aus dem Franz. von Christa Poser*. Berlin: de Gruyter.

Voltaire. 2003. Doutes sur la mesure des forces motrices et sur leur nature. In *Les œuvres complètes de Voltaire*, ed. Robert L. Walters and David Beeson, 359–447. Vol. XXA. Oxford: Voltaire Foundation.

Voltaire. 2003. Exposition du livre des Institutions de physique. In *Les œuvres complètes de Voltaire*, ed. Robert L. Walters and William H. Barber, 213–62. Vol. XXA. Oxford: Voltaire Foundation.

Voltaire. 2003. Mémoire sur un ouvrage de physique de Mme la marquise du Châtelet. In *Les œuvres complètes de Voltaire*, ed. Robert L. Walters,189-212. Vol. XXA. Oxford: Voltaire Foundation.

Voltaire. 2006. *Écrits autobiographiques*, ed. Jean Goldzink. Paris: Flammarion.

Voltaire. 2006. Eloge historique de Mme la Marquise du Châtelet. In *Les Œuvres complètes de Voltaire*, ed. William H. Walters, 365–91. Vol. XXXIIA. Oxford. Oxford: Voltaire Foundation.

Voltaire. 2006. Mémoires pour servir à la vie de Monsieur de Voltaire. In *Écrits autobiographiques*, ed. Jean Goldzink. Paris.

W. 1922. Die göttliche Émilie. *Neue Freie Presse, Wien*, November 5.

Wade, I. O. 1938. *The Clandestine organization and diffusion of philosophic ideas in France from 1700 to 1750*. London: Oxford University Press.

Wade, I. O. 1941. The Intellectual Atmosphere at Cirey. In *Voltaire and Madame du Châtelet: An essay on the intellectual activity at Cirey*, ed. Ira O. Wade, 13–47. Princeton Publication in romance languages. Princeton: Princeton University Press. Repr.: 1967. New York: Octagon Books.

Wade, I. O., ed. 1941. *Voltaire and Madame du Châtelet: An essay on the intellectual activity at Cirey.* Princeton: Princeton University Press. Repr.: 1967. New York: Octagon Books.

Wade, I. O., ed. 1947. *Studies on Voltaire. With some unpublished papers of Mme. du Châtelet.* Princeton: Princeton University Press. Repr.: 1967. New York: Russell & Russell.

Wade, I. O. 1958. The Search for a New Voltaire: Studies in Voltaire Based upon Material Deposited at the American Philosophical Society. *Transactions of the American Philosophical Society* 48, no. 4:1–206.

Wade, I. O. 1959. *Voltaire and Candide: A study in the fusion of history, art, and philosophie; with the text of the La Vallière manuscript of Candide.* Princeton: Princeton University Press.

Wade, I. O. 1967. *The Clandestine Organization and diffusion of philosophic ideas in France from 1700 to 1750.* New York: Octagon Books.

Wade, I. O. 1968. Voltaire and Madame du Châtelet. In *Voltaire: a collection of critical essays*, ed. William F. Bottiglia, 64–68. Englewood Cliffs/N.J.: Prentice-Hall.

Wade, I. O. 1969. *The intellectual development of Voltaire.* Princeton: Princeton University Press.

Wade, I. O. 1994. *The intellectual development of Voltaire.* Repr. Princeton, 1969. Ann Arbor, Mich: Bell & Howell.

Wade, I. O. 1994. *The intellectual origins of the French enlightenment.* Repr. Princeton, 1971. Ann Arbor, Mich: Bell & Howell.

Wahsner, R. 2001. Die Kant'sche Synthese von Leibniz und Newton und deren Konsequenzen für den Mechanik-Begriff des deutschen Idealismus. In *Kant und die Berliner Aufklärung: Akten des IX. Internationalen Kant-Kongresses*, ed. Volker Gerhard, 381–91. Berlin: de Gruyter.

Wahsner, R., and H.-H. v. Borzeszkowski. 1980. *Newton and Voltaire: Zur Begründung und Interpretation der klassischen Mechanik.* Berlin: Akademie Verlag.

Waithe, M. E., ed. 1989, ²1991. *A History of Women Philosophers (Vol. III). Modern Women Philosophers, 1600–1900.* Dordrecht/Boston/London: Kluwer Academic Publishers.

Waithe, M. E., ed. 1989, ²1991. Gabrielle Émilie Le Tonnelier de Breteuil Du Châtelet-Lomont. In *A History of Women Philosophers (Vol. III). Modern Women Philosophers, 1600–1900*, ed. Mary E. Waithe, 127–152. Dordrecht/Boston/London: Kluwer Academic Publishers.

Waithe, M. E., ed. 1989. On Not Teaching the History of Philosophy. *Hypatia* 4, no. 1:132–38.

Walpole, H. 1769. *To Madame du Châtelet.* Strawberry Hill.

Walsø, M., and Å. Birkenheier. 2009. *Geliebter Voltaire.* München: Dt. Taschenbuch-Verl.

Walters, R. L. 1954. *Voltaire and the Newtonian universe: a study of the Eléments de la philosophie de Newton.* PhD. Diss.

Walters, R. L. 1967. Chemistry at Cirey. *Studies on Voltaire and the 18th Century* 58:1807–27.

Walters, R. L. 1997. Voltaire and Mme du Châtelet's continuing scientific quarrel. In *Voltaire et ses combats.* Actes du congrès international Oxford-Paris 1994, ed. Ulla Kölving and Christiane Mervaud, 889–99. 2 vols. Oxford: Voltaire Foundation.

Walters, R. L. 2001. La querelle des forces vives et le rôle de Mme du Châtelet. In *Cirey dans la vie intellectuelle: la réception de Newton en France*, ed. François de Gandt, 198–211. Oxford: Voltaire Foundation.

Warnock, M., ed. 1996. *Women Philosophers.* London: Dent.

Warren, v. de. 1888–1889. Les descendants des Du Châtelet. *Bulletin de la Société philomatique vosgienne* 14:269–74.

Wehinger, B. 2003. Gemeinsame Experimente auf der Suche nach dem Glück – Voltaire und Emilie du Châtelet in Cirey (1734–1749). In *Das literarische Paar – Le couple litteraire. Intertextualität der Geschlechterdiskurse – Intertextualité et discourses des sexes*, ed. Gislinde Seybert, 89–110. Bielefeld: Aisthesis Verlag.

Wehinger, B. 2005. *Geist und Macht.* Berlin: Akademie Verlag.

Wehinger, B., and H. Brown, eds. 2007. *Übersetzerinnen im 18. Jahrhundert.* Hannover-Laatzen: Wehrhahn Verlag.

Wertheim, M. 1995. *Pythagoras's trousers: God, physics, and the gender wars.* New York: Times Books.

Whitehead, B. 2006. The singularity of Mme Du Châtelet: An analysis of the *Discours sur le bonheur*. In *Emilie Du Châtelet: rewriting Enlightenment philosophy and science*, ed. Judith P. Zinsser and Julie C. Hayes, 255–76. Oxford: Voltaire Foundation.
Whitfield, A. 2002. Emilie du Châtelet, traductrice de Newton, ou la 'traduction-confirmation'. In *Portraits de traductrices*, ed. Jean Deslisle, 87–116. Ottawa: OttawaPU d'Ottawa.
Wilbur, R. 1995. *Three Poems*. Tuscaloosa/Alabama: Dugdemona Press.
Winter, U. 1972. *Der Materialismus bei Diderot*. Genf/ Paris: Droz et Minard.
Winter, U. 2001. Naturphilosophie und Naturwissenschaften. In *Die Wende von der Aufklärung zur Romantik*, ed. Horst A. Glaser and György M. Vajda. Amsterdam/Philadelphia: John Benjamins Publishing.
Winter, U. 2004. Diderot und Leibniz. Die Leibniz-Rezeption in der Naturphilosophie der französischen Aufklärung. *Studia Leibnitiana* 36:57–69.
Winter, U. 2006. Vom Salon zur Akademie. Émilie Du Châtelet und der Transfer naturwissenschaftlicher und philosophischer Paradigmen innerhalb der europäischen Gelehrtenrepublik des 18. Jahrhunderts. In *Höfe - Salons - Akademien. Kulturtransfer und Gender im Europa der Frühen Neuzeit*, ed. Gesa Stedman and Margarete Zimmermann, 285–306. Hildesheim: Olms.
Winter, U. 2007. Übersetzungsdiskurse der französischen Aufklärung: Die Newton-Übersetzung Emilie du Châtelets. In *Übersetzerinnen im 18. Jahrhundert*, ed. Brunhilde Wehinger and Hilary Brown. Hannover-Laatzen: Wehrhahn Verlag.
Winter, U. 2008. *Das vernetzte Universum. Die Leibniz-Rezeption in der Naturphilosophie der französischen Aufklärung*. Stuttgart: Franz Steiner Verlag.
Wolff, C. von. 1988. *Gesammelte Werke* 28, Abt. 3: Materialien und Dokumente. Hildesheim/ Zürich/New York: Olms.
Woolston, T. 2001. *Six discours sur les miracles de notre Sauveur*. Paris: Champion.
Yount, L. 1999. *A to Z of women in science and math*. New York: Facts on File.
Zacharias, K. L. 2006. Marquise Du Châtelet. In *Great Lives from History. The 18th Century 1701–1800*, ed. John Powell, 240–43. Pasadena/CA: Salem Press.
Zan, M. de. 1987. Voltaire e Mme du Châtelet, membri e corrispondenti dell'Accademia delle scienze di Bologna. *Studi e memorie per la storia dell'Universitá di Bologna* 6:141–57.
Zinsser, J. P. 1998. Emilie du Châtelet : genius, gender and intellectual authority. In *Women writers and the early modern British political tradition*, ed. Hilda Smith, 168–90. Cambridge/New York: Cambridge University Press.
Zinsser, J. P. 2001. Translating Newton's 'Principia': The Marquise du Chatelet's Revisions and Additions for a French Audience. *Notes and Records of the Royal Society of London* 55, no. 2:227–45.
Zinsser, J. P. 2002. Entrepreneur of the "Republic of Letters": Emilie de Breteuil, Marquise Du Châtelet, and Bernard Mandeville's *Fable of the Bees*. *French Historical Studies* 25, no. 4:595–624.
Zinsser, J. P. 2003. A Prologue for La Dame d'Esprit: the biography of the marquise du Châtelet. *Rethinking History, the journal of theory and practice* 7, no. 1:13–22.
Zinsser, J. P. 2003. The ultimate comentary: a consideration of I. Bernard Cohen's Guide to Newton's *Principia*. *Notes and records of the Royal Society* 57:231–38.
Zinsser, J. P. 2004. A Prologue for La Dame d'Esprit: the biography of the marquise du Châtelet. In *Experiments in rethinking history*, ed. A. Munslow and R. A. Rosenstone, 195–208. London/ New York: Routledge.
Zinsser, J. P. 2004. The Marquise Du Châtelet. In *Enlightenment and Revolution, 1690–1815 : An Interdisciplinary Biographical Dictionary*, ed. Kevin E. Dodson. Santa Barbara/CA: Greenwood Pub Group Inc.
Zinsser, J. P. 2005. *Men, women and the birthing of modern science*. DeKalb (IL): Northern Illinois University Press.
Zinsser, J. P. 2005. The many representations of the marquise Du Châtelet. In *Men, women and the birthing of modern science*, ed. Judith P. Zinsser, 48–67. DeKalb (IL): Northern Illinois University Press.

Zinsser, J. P. 2005. Why are there no 18th century women scientists and philosophers? The Marquise du Châtelet and the categorization of knowledge. In *Geschlecht und Wissen = Genre et savoir = Gender and knowledge. Beiträge der 10. Schweizerischen Historikerinnentagung 2002*, ed. C. Brosshart-Pfluger, D. Grisard, and C. Späthi. Zürich: Chronos.

Zinsser, J. P. 2006. *La Dame d'Esprit: A Biography of the Marquise Du Châtelet*. New York: Viking.

Zinsser, J. P. 2006. Madame Du Châtelet et les historiens. In *Madame du Châtelet. La femme des Lumières*, ed. Elisabeth Badinter and Danielle Muzerelle, 118–21. Paris: Bibliothèque Nationale de France.

Zinsser, J. P. 2007. *Emilie du Châtelet: Daring Genius of the Enlightenment*. New York: Penguin.

Zinsser, J. P. 2007. Mentors, the marquise Du Châtelet and Historical Memory. *Notes and records of the Royal Society* 61, no. 2:89–108.

Zinsser, J. P. 2008. Mme Du Châtelet: sa morale et sa métaphysique. In *Émilie Du Châtelet: Éclairages & documents nouveaux*, ed. Ulla Kölving and Olivier Courcelle, 219–29. Ferney-Voltaire: Centre International d'Étude du XVIIIe Siècle.

Zinsser, J., and B. S. Anderson. 1988, ²2000. A history of their own: women in Europe from prehistory to the present. 2 vols. New York: Harper & Row.

Zinsser, J. P., and O. Courcelle. 2003. A remarkable collaboration: the marquise du Châtelet and Alexis Clairaut. *SVEC* 12:107–20.

Zinsser, J. P. and J. C. Hayes, ed. 2006. *Emilie Du Châtelet: rewriting Enlightenment philosophy and science*. Oxford: Voltaire Foundation.

Zinsser, J. P. and J. C. Hayes, ed. 2006. Rereading W.H. Barber. In *Emilie Du Châtelet: rewriting Enlightenment philosophy and science*, ed. Judith P. Zinsser and Julie C. Hayes, 3–4. Oxford: Voltaire Foundation.

Zinsser, J. P. and J. C. Hayes, ed. 2006. The Marquise as Philosophe. In *Emilie Du Châtelet: rewriting Enlightenment philosophy and science*, ed. Judith P. Zinsser and Julie C. Hayes, 24–31. Oxford: Voltaire Foundation.

Zinsser, J. P. and I. Bour, eds. 2009. *Emilie Du Châtelet. Selected Philosophical and Scientific Writings*. Chicago: Chicago University Press.

About the Authors

Prof. Dr. Ruth Hagengruber holds a chair in Philosophy at the University Paderborn, where she is head of the research areas History of Women Philosophers and Scientists and Philosophy and Computation Science. She studied Philosophy and History of Natural Sciences, with a concentration in the History of Geometry at the Ludwig-Maximilian-University in Munich. She received her Master's degree with a thesis on Plato's Symposium, her Ph.D. deals with Renaissance Metaphysics and Mathematics in the philosophy of Tommaso Campanella, Academia 1994. Ruth Hagengruber earned a certificate in Economics and wrote on *Nutzen und Allgemeinheit. Zu einigen grundlegenden Problemen der Praktischen Philosophie*, Academia 2000. She was awarded a grant from the Deutscher Akademischer Austausch Dienst to support her studies on Renaissance philosophy and mathematics in Naples, Italy, and held a scholarship from the Studienstiftung des Deutschen Volkes as well as of the Istituto per gli Studi Filosofici, Naples, Italy a.o. Since 2001 she has been engaged in research in the area of Philosophy and Computation Science, documented also in *Philosophy's Relevance in Information and Computing Science*, 2011. She is editor and author of several books and essays contributing to the History of Women in Philosophy, as *Von Diana zu Minerva. Philosophierende Aristokratinnen des 17. Und 18. Jahrhunderts*, 2010 Akademie; *Klassische Philosophische Texte von Frauen. Texte vom 14. Bis zum 20. Jahrhundert*, Deutscher Taschenbuch Verlag 1998 et al.; Major contributions on Emilie du Châtelet, Edith Stein, on the idea of a history of women philosophers from ancient times up to now. Currently she is preparing a textbook on women philosophers focusing on Economic Aspects in Political Philosophy.

Dr. Hartmut Hecht studied Physics and Philosophy and is Associate Professor for Philosophy at Humboldt University in Berlin. As director of the Berlin Leibniz edition of the Berlin-Brandenburg academy, he assumes the position of main editor for the academy's publication of Gottfried Wilhelm Leibniz' work on natural sciences, medicine and technical issues, namely Leibniz' *Sämtliche Schriften und Briefe*. His main areas of research are natural philosophy and science history with a focus on

Leibniz and his reception during the times of the European Enlightenment. His publications in the field of Leibniz studies include a biography with an emphasis on mathematics and the natural sciences; Teubner 1991, a translation of Leibniz *Monadology* and several edited volumes on Maupertuis, La Mettrie and on the French Enlightenment.

Dr. Fritz Nagel studied Mathematics, Physics, Philosophy and Philosophy of Science at the universities of Heidelberg and Basel. In 1981 he received his Ph.D. from the University of Basel for a thesis on "Nicolaus Cusanus und die Entstehung der exakten Wissenschaften". He subsequently worked as a teacher of mathematics and physics. Since 1987 he has been a senior researcher at Basel where he has been acting as editor for an edition of Bernoulli's correspondence. He is also the main editor of the *Basler Inventar der Bernoulli-Briefwechsel*, which contains more than 3,000 letters written by or addressed to Bernoulli (http://www.ub.unibas.ch/spez/bernoull.htm#Inventar) and he is an associate editor of a special edition of the *Studia Leibnitiana* and of a work on the Basel mathematicians for the former St. Petersburg Academy. Fritz Nagel is currently preparing an edition of Johann I Bernoulli's correspondence with Christian Wolff.

Dr. Nagel has contributed several articles to the volumes he edited on Bernoulli. In addition, he has also had his research published in numerous encyclopedias and journals. He is a former director of the Swiss "Gesellschaft für die Erforschung des 18. Jahrhunderts" and is a member of the Executive Committee of the International Society for eighteenth Century Studies. He is a member of the "Euler-Kommission der Schweizerischen Akademie der Naturwissenschaften", the "Kommission für die Erforschung des 18. Jahrhunderts und der Aufklärung in der Schweiz" of the Swiss Academy for the Humanities and a correspondent member of the "Académie internationale d'histoire des sciences" in Paris.

Mag. Andrea Reichenberger, cand. Ph.D. studied Philosophy, Art History and Philosophy at the University of Salzburg from 1996 to 2001. After graduating in 2001, she continued her studies in Philosophy with a focus on theoretical philosophy and logic at the universities of Konstanz and Hagen. From 2006 to 2007 she has been working as a research assistant in Prof. Dr. R. Hagengruber research area of Philosophy and Computation science at the University of Paderborn. Andrea Reichenberger prepares her Ph.D. at the University of Bochum and publishes on a wide range of subjects including the history of mathematics, logic and aesthetics.

Ana Rodrigues, M.A., cand. Ph.D. studied Philosophy, French and English Literature at the University of Paderborn. 2006 she started collaborating in Prof. Dr. R. Hagengruber's research project on the History of Women Philosophers and Scientists. After her graduation in 2008 she continued her work within the project as a research assistant. Since then she is preparing her Ph.D. on Emilie du Châtelet's moral philosophy.

Dr. Dr. Dieter Suisky studied physics at the Martin-Luther University in Halle-Wittenberg and he received his Ph.Ds in 1971 (Dr. phil.) and in 1981 (Dr. rer.nat.) from Humboldt-University in Berlin. His research areas lie within the fields of solid state physics, history and the philosophy of natural sciences. Since 1972 he has been working as a researcher and teacher at Humboldt University. He has been invited for research visits by the Lomonossow University in Moscow in 1980–1981 and 1983–1984 and by the University of São Paulo in 1989. Since 1999 Dieter Suisky has been focusing his research on the genesis of Leonhard Euler's contributions to mathematics and mechanics and the influence of his work on the eighteenth and nineteenth centuries.

Dr. Ursula Winter is currently preparing a publication on Leibniz as part of a DFG research project led by Professor Poser on Gottfried Wilhelm Leibniz' influence on the natural philosophy of the French Enlightenment.

She studied Philosophy, Comparative Philology, Romance Studies and History. at the Freie Universität Berlin and at the Université Libre de Bruxelles. Her thesis *Der Materialismus bei Diderot* was published by Droz and Minard in 1972 and achieved wide recognition in numerous international reviews.

After working as a research assistant at the Freie Universität in Berlin, Ursula Winter acquired a secondary school teaching qualification for German before taking up a position as lecturer at Humboldt University and the TU in Berlin. She has published several essays on the philosophy of the French Enlightenment in both German and international publications. She has given guest lectures at the universities of Heidelberg and Essen, and she has also contributed to various international conferences on the Enlightenment and on Leibniz in particular.

Index

A
Absorption, 110, 130
Academy of Bologna, 6
Academy of Dijon, 4
Academy of sciences and arts, 5, 27, 160, 166, 168, 169
Alétophiles, 25
Algarotti, Francesco, 29, 31, 79, 105
Algorithm, 115, 124, 128, 129, 146
Amsterdam, 6, 79, 82, 84, 91, 105, 161, 187, 191
Analytical, 121, 125, 127–129, 134, 145, 148, 149, 160
A priori, 8, 12, 13, 23, 24, 53, 195
Architecture, 22, 161
Aristotle, 22, 106, 148
Axiom, 4, 10, 28, 115, 122–128, 162, 163, 167, 180

B
Basel, 30–32, 97–111
Berkeley, George, 132, 139
Berlin, 28, 32–34, 51–53, 65, 69, 113, 173, 177
Berlin Academy, 28, 52, 65, 177
Bernoulli, Daniel, 31, 101, 127, 159
Bernoulli, Johann I, 30–33, 35, 36, 97–99, 101, 103, 160, 161
Bernoulli, Johann II, 32–34, 36, 97–104, 110, 111, 165
Bernoulli, Nicolaus I, 97, 103
Bible, 5, 6
Bodies, 19, 20, 36, 37, 41–44, 49, 70, 73, 89, 108, 109, 115, 116, 120, 121, 123–126, 128, 131–142, 146–151, 158–160, 163, 166, 179, 180, 191, 197
Burnett, Thomas, 25

C
Cartesianism, 86, 87
Cassini, Jacques, 30
Christine of Sweden, 2
Clairaut, Alexis-Claude, 30, 99, 165
Clarke, Samuel, 30, 77–93, 115, 153, 160, 165
Condorcet, Marie Jean Antoine Nicolas Caritat, Marquis de, 56, 101, 117
Contradiction, 8–10, 13, 19, 23, 36, 46, 67, 108, 127, 129, 130, 135, 161, 162, 169
Coste, Pierre, 8, 84, 186

D
D'Alembert, Jean Baptiste Le Rond, 158
Descartes, René, 158, 159
Deschamps, Jean, 27
Determinism, 157
D'Holbach, 54
Diderot, Denis, 185
Dynamics, 99, 121, 122, 126, 127, 137, 143, 163, 174, 178, 196, 203

E
Einstein, Albert, 195, 201
Elisabeth of Bohemia, 2
Empiricism, 8, 23, 47
Empirist, 3, 12, 54, 78

Encyclopédie, 7, 25, 86, 88, 180, 184–186, 196, 200
English, 8, 14, 25, 30, 31, 78, 80–82, 84, 85, 87, 92, 160, 184, 191
Enlightenment, 2–5, 8, 29, 31, 46, 53, 62, 64, 67, 77, 78, 84, 91, 99, 160, 169, 174, 175, 178–180, 184–187, 196, 198, 201, 203
Equilibrium, 118, 121, 122
Euler, Leonhard, 21, 29, 113–153, 159, 167

F

Fontenelle, Bernard le Bovier de, 31, 51, 178
Force, 2, 63, 83, 97, 103, 114, 119–122, 135–146, 157–169, 174
Formey, Jean Henry, 29
Frederick, King of Prussia, 2, 7, 53, 79, 85, 88

G

Galilei, Galileo, 21, 25, 70–72, 87, 134, 151, 158
Gender, 206, 217, 218, 228–230, 234, 238, 240, 245, 246
Geometry, 2, 18, 35, 69, 99, 129, 137
Gottsched, Luise Adelgunde Victoria, 25
Gravensande, Willem Jacob s', 187
Gravitation, 16, 72, 73, 83, 89, 92, 107–109, 122, 140, 173, 181, 182, 191
Gravité absolue, 70
Gravité respective, 70

H

Haller, Albrecht von, 34, 47, 52
Helvétius, Claude-Adrien, 2, 188
Hobbes, Thomas, 92
Homer, 4
Horaz, 4
Hume, David, 54
Huygens, Christiaan, 17, 19, 21, 72, 110, 158, 183, 184
Hypotheses, 11, 15–21, 23–26, 38, 39, 43, 46, 54, 64, 73, 74, 86, 87, 107, 109, 110, 116, 122, 127, 130–134, 153, 161, 168, 183, 191, 198, 199

I

Inertia, 14, 42, 73, 114, 115, 118–123, 125, 126, 130, 132, 136, 143, 144, 148, 161, 163, 165, 180, 182, 197

Integrals, 116, 134, 145
Invariance, 128, 151, 152
Invention, 69, 114, 118–120, 195

J

Jurin, James, 165

K

Kant, Immanuel, 28, 179, 184
König, Samuel, 31, 32, 34, 46, 52, 53, 99, 100, 102, 168, 188–190, 192, 193

L

Lagrange, Joseph-Louis, 115, 129, 137
La Mettrie, Julien Offray de, 47–53
Lapland, 30, 31, 34, 35, 99, 179
La Rochefoucauld, duc de, 101
Leibniz-Clarke correspondence, 28, 36, 78, 79, 82–84, 87–89, 91–93, 120, 138, 148, 198
Leibniz, Gottfried Wilhelm, 80, 158, 169
Leibnizianism, 3, 29, 53, 78, 86, 169, 189, 190, 192, 194
Living force, 2, 6, 14, 27, 28, 30, 32, 35–37, 40, 44, 97, 114–116, 120, 121, 126, 133–135, 144–147, 158, 164, 167, 168
Locke, John, 3, 5, 7–13, 23–25, 53, 81, 84, 87, 186, 203

M

Magnitude, 120, 128, 134–137, 145, 147, 149
Mairan, Jean Jacques d'Ortous de, 160, 165
Malebranche, Nicolas, 111
Materialism, 26, 47, 50
Materialist, 8, 11, 51
Mathematics, 3, 5, 15, 18, 24, 30, 31, 46, 68, 72, 84, 85, 91, 92, 98, 100, 113, 117, 119, 129, 131, 152, 153, 158, 161, 174, 182, 185, 193–196
Maupertuis, Pierre Louis Moreau de, 29, 30, 47, 66, 79, 92, 98, 159, 161, 167
Mechanics, 17, 28, 32, 33, 36, 37, 44, 91, 107, 113–153, 157–161, 163, 166, 167, 169
Metaphysical calculus, 99, 113–116, 118, 126, 129–132, 134–136, 146, 148, 152, 164, 191, 200
Metaphysical principle, 10, 69, 166, 167, 195

Index 253

Metaphysics, 1–54, 61, 63, 72, 74, 77–80, 82, 83, 87, 91, 158, 159, 161, 163, 166, 169, 181, 182, 190, 193–195
Methodology, 12, 29, 37, 46, 62, 64, 66, 69, 74, 85, 91, 114, 129, 132, 147, 200
Microbiology, 196
Motion, 15, 28, 30, 35–37, 40, 41, 43, 45, 49–51, 53, 63, 64, 72–74, 90, 108, 113, 115, 116, 120–129, 131, 132, 134–143, 145–152, 157–160, 162, 163, 165–167, 169, 180, 182, 197
Musschenbroeck, Pieter van, 110, 187

N

Natural philosophy, 1, 5, 6, 12, 16, 24, 63, 67–70, 74, 81, 84–87, 93, 117, 118, 124–126, 136, 140, 157, 174, 175, 180, 181, 185, 196, 200, 203
Natural science, 17, 26, 29, 43, 61, 62, 66, 74, 110, 114, 117, 129, 131, 144, 173, 178, 179, 181, 187, 188, 194, 195, 198, 200
Newtonianism, 3, 5, 26, 27, 29, 30, 35, 53, 78–83, 86, 87, 89–93, 119, 122, 185, 186
Nollet, l'abbé, 105

O

Orthodoxy, 80, 157

P

Pemberton, Henry, 79, 160, 178
Perpetuum mobile, 37
Philosophical Transactions, 79, 188, 191
Physics, 1, 3, 10, 16, 17, 25, 29, 34, 46, 61, 63, 68, 70–74, 81, 82, 85, 87, 89–92, 98, 104, 105, 110, 113, 116, 117, 119, 152, 153, 157, 159, 161, 163, 167, 169, 173, 174, 178, 179, 182, 185, 191–196, 200, 201, 203
Poleni, Giovanni, 160, 165
Prault, Laurent-François, 79, 83, 84, 177, 187, 188, 190
Principia, 6, 16, 28, 29, 77, 79, 82, 84, 87, 91, 92, 107, 113, 115, 118, 120, 123–125, 127–129, 131, 137–139, 143, 146, 152, 153, 162, 173, 174, 176–178, 180–186, 188, 191, 195, 197, 198, 200, 203
Principle of least action, 44, 45, 52, 61, 67–69

Principle of sufficient reason, 18, 28, 67, 73, 74, 78, 83, 88, 91, 92, 124, 125, 129, 161, 162, 164, 166, 169, 198
Principle of the identity of indiscernibles, 83, 162, 197

R

Relativism, 9, 121, 140, 141
Relativity, 128, 148, 157, 201–203

S

Scholastic, 3, 13–15, 17, 23, 27, 53, 54, 83
Scientific system, 128
sensualist, 7, 50
Space, 42, 44, 49, 69, 82, 83, 87–89, 92, 111, 114, 119, 120, 131–133, 136–141, 143, 145, 148, 157, 169, 174, 179–181, 183, 190–192, 195, 200–203
Steinwehr, Wolf Balthasar Adolph von, 85, 184
Synthesis, 143

T

Tencin, cardinal de, 98

U

Universe, 14, 21, 26, 40, 63, 83, 88, 90, 147, 148, 159, 163, 164, 173–175, 178–181, 194–200, 203

V

Validity, 15, 16, 29, 37, 40, 43–45, 73, 126, 134, 151
Varignon, Pierre de, 70, 72, 115
Velocity, 71–73, 90, 106, 109, 110, 116, 127, 128, 134, 135, 138, 145, 149, 158–160, 166, 167
Verification, 13, 19, 53, 67, 131, 188
Vis viva controversy, 30, 157–169
Voltaire, François Marie Arouet, 2–16, 21, 23, 25–31, 33, 34, 37, 47, 51–54, 77–84, 88, 98, 101–103, 105, 118, 119, 165, 169, 176, 177, 179, 185, 186, 188–190, 196, 203

W

Wolff, Christian, 1, 13, 26, 27, 31, 80, 81, 102, 169